Cambridge Studies in the History of Psychology
GENERAL EDITORS: WILLIAM R. WOODWARD AND MITCHELL G. ASH

The professionalization of psychology
in Nazi Germany

Cambridge Studies in the History of Psychology

This new series provides a publishing forum for outstanding scholarly work in the history of psychology. The creation of the series reflects a growing concentration in this area by historians and philosophers of science, intellectual and cultural historians, and psychologists interested in historical and theoretical issues.

The series is open both to manuscripts dealing with the history of psychological theory and research and to work focusing on the varied social, cultural, and institutional contexts and impacts of psychology. Writing about psychological thinking and research of any period will be considered. In addition to innovative treatments of traditional topics in the field, the editors particularly welcome work that breaks new ground by offering historical considerations of issues such as the linkages of academic and applied psychology with other fields, for example, psychiatry, anthropology, sociology, and psychoanalysis; international, intercultural, or gender-specific differences in psychological theory and research; or the history of psychological research practices. The series will include both single-authored monographs and occasional coherently defined, rigorously edited essay collections.

Also in this series

Constructing the subject: historical origins of psychological research
KURT DANZIGER
Metaphors in the history of psychology
edited by DAVID E. LEARY
Crowds, psychology, and politics, 1871–1899
JAAP VAN GINNEKEN

It has been widely believed that the discipline of psychology in Germany was attacked, or even ceased to exist, under National Socialism. Yet in *The Professionalization of Psychology in Nazi Germany*, Ulfried Geuter shows that, rather than disappearing, German psychology rapidly grew into a fully developed profession during the Third Reich. Presenting his argument in the larger context of German military and economic history, Geuter makes it clear that the rising demands of a modern industrial nation gearing up for war afforded psychology a unique opportunity in Nazi Germany: to transform itself from a marginal academic discipline into a profession recognized and sanctioned by the state.

Ultimately, Geuter shows how the history of the professionalization of German psychology – its emphasis on characterology and its fraternization with the military establishment – displayed diacritical flashes of German history itself. Yet the relevance of this book goes far beyond the history of German psychology. Its conclusion – that psychology in Germany grew through its alliance with the interests of the army, industry, and the ruling regime – points toward the larger picture behind the particulars: the tangled relations between science, professional expertise, and state power in modern society.

The professionalization
of psychology
in Nazi Germany

Ulfried Geuter

Translated by Richard J. Holmes

CAMBRIDGE
UNIVERSITY PRESS

CAMBRIDGE UNIVERSITY PRESS
Cambridge, New York, Melbourne, Madrid, Cape Town, Singapore, São Paulo, Delhi

Cambridge University Press
The Edinburgh Building, Cambridge CB2 8RU, UK

Published in the United States of America by Cambridge University Press, New York

www.cambridge.org
Information on this title: www.cambridge.org/9780521332972

Originally published in German as *Die Professionalisierung der deutschen Psychologie im Nationalsozialismus* by Suhrkamp Verlag, Frankfurt, 1984

First published in English by Cambridge University Press 1992 as
The Professionalization of Psychology in Nazi Germany
© English translation Cambridge University Press 1992

This digitally printed version 2008

A catalogue record for this publication is available from the British Library

Library of Congress Cataloguing in Publication data
Geuter, Ulfried, 1950–
[Die Professionalisierung der deutschen Psychologie im
Nationalsozialismus. English]
The professionalization of psychology in Nazi Germany / Ulfried
Geuter; translated by Richard J. Holmes.
p.   cm. – (Cambridge studies in the history of psychology)
Translation of: Die Professionalisierung der deutschen Psychologie
im Nationalsozialismus.
Includes bibliographical references and index.
ISBN 0-521-33297-4 (hardcover)
1. Psychology – Germany – History – 20th century.   2. National
Socialism.   3. Germany – Politics and government – 1933–1945.
4. Psychology – history – Germany.   I. Title.   II. Series.
[DNLM: 1. Political Systems – history – Germany.   BP   108.G3   G395p]
BF 108.G3G4813   1992
150′.943′09043 – dc20
DNLM/DLC   91-26576
for Library of Congress   CIP

ISBN 978-0-521-33297-2 hardback
ISBN 978-0-521-10213-1 paperback

Contents

Tables

Note to readers of the English edition

This translation follows the revised German edition published by Suhrkamp Taschenbuch Verlag in 1988. The text has been abridged without affecting the argumentation or structure of the book. The reader may find more detail on some points by referring to the German text.

I would like to thank Richard J. Holmes, who not only translated the book, but also helped to condense the text. I am also very grateful to Mitchell G. Ash, who revised the entire translation and checked it with regard to the use of scholarly English terminology. At Cambridge University Press, the manuscript was read by Mary Racine, whom I also thank.

Note to readers of the English edition

Preface

In undertaking to produce a theoretical and historical critique of test diagnostics, I began by looking at the history of psychological tests and the reasons for their development and practical use. In the course of the work it became clear that the Nazi period was the least well researched, but at the same time the one that posed the most questions: What did psychologists do in the Third Reich? How was the field of psychology able to develop? How was it obstructed or encouraged, supported or abandoned? These questions led me into uncharted areas on which few reports existed and about which the German postwar generation knew little, except through hearsay.

A first look at the material, especially scientific publications from the Nazi period, confronted me with a multitude of facts that defied organization. There had been racist typology, but it did not seem possible to understand the history of German psychology in this period solely in terms of ideological Nazification. There had been practical, diagnostic psychology, particularly in the armed forces; there had been professional psychologists; and – in the middle of the war – examination regulations had been introduced for a certificate recognized by the state. But the history of German psychology could also not be described simply as an instrumentalization for the goals of expansion and oppression. There had been dismissals, and some scientists had been persecuted; but the common opinion that psychology had been politically subjugated did not seem to explain all the facts. What, then, were the forces acting on psychology in the Third Reich? An examination of the available sources – particularly interviews, archives, and files of universities, ministries, the armed forces, and the psychological societies – showed that a number of points could be clarified if the problem was approached from the aspect of professionalization. This seemed to have been a driving force for the development of the discipline in the Nazi period.

Such an approach was not usual. Indeed, traditional historiography of psychology was long content with an immanent treatment of general psychological theories. Few links were drawn between theories and application, between political circumstances and developments in the discipline, or between external demands on the discipline and the attitudes and opinions of its representatives. This book can be seen to some extent as a case study showing the value of such questions.

Earlier studies on the development of science in the Third Reich frequently concentrated on leading representatives and spectacular events, such as political dismissals and the abuse of science. But just as general history has discovered the history of ordinary people, so with professionalization the history of science has begun to consider everyday research, applied science, and professional behavior. For the Third Reich this can be less disturbing than having to come to terms with the use of terror and the suffering of the victims, or with race-psychological investigations of the Polish population. It can be easier than being continually confronted with the language of the "Master Race" masquerading as science, which would be the consequence of concentrating on the relationship between Nazi ideology and psychology. But at the same time, with this approach one is confronted with the problem of the political instrumentalization of everyday science and with the frightening normality of the production and use of science under totalitarian rule.

The choice of professionalization also has a topical aspect. In recent years psychologists have been concerned to advance the professionalization of the field, but the unconsidered activities of some representatives gave rise to critical studies of this process. However, neither the critics nor the activists seemed aware of the importance of the Third Reich in this aspect of the history of German psychology.

Though many aspects of the development of psychology during the Nazi period can be understood in the light of professionalization, not all of them can. After the war not one piece of research was published on psychology in the Third Reich, so that this book represents only a first attempt to trace aspects of the history and to provide a theoretical evaluation. It does not claim to be a comprehensive analysis of the development of psychology in this period. Such an analysis would have to be based on other work. At the time of completion of the German manuscript, in November 1983, Professor Carl-Friedrich Graumann organized a meeting entitled "Psychology in National Socialism," the first academic forum at which this topic was discussed. There I discovered that Angela Benz had already written an unpublished diploma thesis in 1980 at the University of Trier ("Psychologie und Nationalsozialismus: Versuch eines Paradigmas von Wissenschaftsgeschichte"), which considered the relationship between the scientific development of the subject and Nazi ideology and politics as shown by the change in content of congress papers and the development of psychological schools. In some respects, her results overlap mine, but in only one respect (the account of conflicts within the psychological associations) has Benz drawn on documents not used here.

Psychology in the Nazi period is a piece of the history of German psychology. History cannot be undone; at the most it can be forgotten, repressed, argued away or it can be accepted and understood as one's own history. This book is therefore aimed at psychologists. But it also seeks a place in the debates of sociologists and historians of science, which will be considered in Chapter 1.

There are still all sorts of problems peculiar to work on the Third Reich. The Berlin Senate refused me permission to consult the files of the Berlin Document Center, which include those of National Socialist German Workers' Party (NSDAP) members. Some psychologists refused to be interviewed or failed to reply when I wrote to them. I was compared to the devil for working on this topic, and more besides. But others provided much-needed encouragement. I would like to thank everyone who helped, as well as all those who were able to put up with me through the ups and downs of the project.

It was important for this work that the majority of those asked were willing to talk or write about their experiences and the development of psychology. It added detail to my picture of the period and helped me to see these people not just as objects of research, but as individuals. Various documents were kindly provided from private collections. I received generous assistance from libraries and institutions in the course of my search for documents, particularly the psychological and central libraries of the Free University of Berlin, the Berlin State Library, and the Library of Congress in Washington, D.C. I would like to thank the archivists who helped and advised me, in particular Rudolf Absolon, Aachen; Professor Dr. Laetitia Boehm, Munich; Mr. Glaser, Potsdam; Ms. B. Klaiber and Mr. Loos, Freiburg; Professor Dr. E. Meuthen, Cologne; Rohtraut Müller-König, Münster; Uwe Plog, Hamburg; Anne Ryfn, Poznan; Dr. Volker Schäfer, Tübingen; Dr. H. Schwabe, Halle; Professor Dr. G. Schwendler, Leipzig; Mr. Teschauer, Frankfurt/M; Günther Thomann, Erlangen; Dr. Volker Wahl, Jena; and Dr. Werner, Koblenz.

Numerous discussions with friends, colleagues, and students helped me to clarify my ideas. I thank especially Erich Lennertz, who recommended focusing my original research project on the Third Reich; he followed my research with patience and useful suggestions. Mitchell G. Ash, Siegfried Jaeger, and Rudolf G. Wagner frequently discussed my ideas and methods. After coming to Berlin in 1981, Irmingard Staeuble provided untiring help in finalizing my concept.

Siegfried Jaeger, Ursula Reinhart, and Irmingard Staeuble read the original manuscript and suggested a number of improvements. Alexandre Métraux made some important comments in the final stages.

Ulfried Geuter
Berlin, November 1983

Preface to the 1988 revised German pocketbook edition

Since the publication of the first German edition of this book, a number of studies on the history of science in the Third Reich have been published. It has become clearer and clearer that in the Nazi period science was neither the victim of systematic persecution nor abused against its own intentions and perceptions. Many scientists were persecuted for their origins or their convictions, while their discipline – carrying on business as usual – compromised itself and their colleagues willingly placed themselves in the service of the new regime.[1]

The persecution of scientists was often mistaken for the persecution of science. For example, because, as Jews, many psychoanalysts suffered under Nazi terror, the opportunities that remained for psychotherapy were ignored by many. Recent publications have made it necessary to reconsider this, giving rise to much controversy.[2] It seems that psychoanalysis, which was willing to accommodate itself to Nazi power, received official support.

The history of psychology in the Third Reich is confused not only with that of psychoanalysis, but also with that of psychiatry. In discussions, someone always asks whether psychologists were, like psychiatrists, involved in selecting victims for sterilization or euthanasia. In a comprehensive study of forced sterilization in the Third Reich, Gisela Bock (1986) showed how psychiatrists and other doctors made such decisions about unwilling victims at health offices and courts for hereditary health. Much material has also been collected that shows that medicine helped to rationalize the machinery of killing (Aly, Masuhr et al. 1985; Kudlien 1985), and that civil servants,

---

1 See Lundgreen (1985). On the ideological continuity in applied psychology see the contribution by Alexantre Métraux in Graumann (1985).
2 See Lohmann (1984), Lockot (1985), Cocks (1985), and Brecht et al. (1985). See also *Psychoanalyse...*, 1984.

judges, doctors, and medical staff all played a part (Aly, Ebbinghaus et al. 1985). However, none of these recent works has produced reliable evidence that psychologists were involved in these deeds.[3] Perhaps psychologists were just lucky that psychiatry was still completely in the hands of physicians, so that they had no professional contact with the business of killing, at least as far as we know. It remains an open question whether individual psychologists already worked in psychiatric clinics and were involved in the selection of people for euthanasia.[4]

With the killing of undesired members of the population, eugenics and racial hygiene came into their own. As Weingart (1985) notes, these were not *mis*used by the Nazis, but their ideas were applied with murderous effectiveness (see also Müller-Hill 1984). Geneticists, anthropologists, judges, and psychiatrists all participated in the process. These professional groups therefore now approach this period in the history of their disciplines with different questions than psychologists. And yet, even though psychologists did not place their professional competence in the service of crime and murder, there is a comparable question to be asked: How does one reflect, as an expert, on the conditions and aims of one's work? To what extent is one willing to make one's expertise available to increase the effectiveness of a political or military apparatus? As Peter Lundgren remarks in his introduction to a collection of essays on science in the Third Reich (1985, p. 28), no science proved able to offer resistance to the Nazi regime solely as a consequence of its own content, despite the fond belief to this effect nurtured by some psychologists, psychoanalysts, and sociologists.

Carsten Klingmann (1985, 1985a, 1986) and Otthein Rammstedt (1985, 1986) have shown that those sociologists who stayed not merely were involved with providing the right ideology for social theory, but also were in demand as experts, primarily for gathering empirical information and for opinion research. Sociologists were active in numerous institutions of the Nazi state, right up to the Reich Security Service. Despite the high proportion of emigrants and a decline of theory, sociology was able to continue to establish

---

3 Aly has claimed that at the institute at Brandenburg-Görden under Professor Hans Heinze (a senior expert for the Reich Committee on Hereditary Defects) a psychologist worked on one team: a fifteen-year-old was tested by the institute's psychologist (Aly, Masuhr 1985, p. 39). Aly wrote to me that this would follow from the files of the state prosecutor in Ansbach; he had no name. The psychiatrist Carl Schneider had also requested that a post for a psychologist be established (letter 30 January 1986). Hence, the term "psychologist" was still used loosely at that time; this deserves closer inspection. The work of Gisela Bock, the most carefully researched of the recent studies, does not indicate any cooperation by psychologists in forced sterilization.

4 In the *Psychological Register* published in 1932 by Carl Murchison, 187 psychologists are listed for Germany (Vol. 3, Worcester, Mass.: Clark University Press), of which only one non-physician was active clinically, Kurt Gottschaldt, who headed the psychological laboratory of the Rhenish Provincial Children's Hospital for the Mentally Ill in Bonn from 1929 to 1933. For prisons, the only entry is a woman psychologist as head of a women's prison.

itself academically and professionally during the Third Reich, findings that are along the same lines as those presented in this book.[5]

Some works on psychology in the Third Reich have appeared, in particular the essays edited by Carl Friedrich Graumann (1985), only some of which were made available to me as manuscripts while I was writing this book. They concentrate on the history of theory, particularly of holistic (*Ganzheit*) and Gestalt psychology.[6] In terms of *theoretical* development in the Third Reich, some psychological theories allied themselves with Nazi racial doctrines (see Geuter 1987, pp. 87ff.). This is most obvious in the case of Erich Jaensch, who made a systematic attempt to include anti-Semitism in his psychological theory and about whom two essays have appeared (Geuter 1985; Pinn 1987). However, taking the texts by the social psychologist Willy Hellpach as an example, Horst Gundlach (1985, 1987) has shown the extent to which nuances must be registered in the vocabulary of that period. What seem to modern readers to be clearly racist expressions turn out on closer inspection to be artful criticisms of Nazi views.

The psychological service of the West German army has commissioned a history of military psychology, mainly from former Wehrmacht psychologists, which seems to confirm my comments in Chapter 9 on the mentality of army psychologists (*Deutsche Wehrmachtpsychologie . . . 1985*). This book completely overlooks the fact that psychology in the German army was able to thrive as a result of the war started by the Nazis. As of old, the authors seem to see the policies of militarization and war more as a gift of the gods than as a cause for reflection.[7]

It remains for future research to establish whether psychologists worked in psychiatric establishments and were involved in inhumane crimes there. Also, little, if anything, is known about the lives of those psychologists who stayed in Germany and offered political resistance, such as Kurt Huber and Heinrich Düker; nor is much known of someone like Theodor Lessing, who published psychological articles in the twenties and was murdered in Czechoslovakia after he fled there in 1933 because of his Jewish origins and political beliefs. There were also psychologists such as Otto Bobertag, Otto Lipmann, and Martha Muchow who broke under the burden of circum-

5 Warsewa and Neumann (1987), in a study of racial questions, arrived at the same conclusions as mine, namely, that ordinary scientific methods were used to increase productivity in the firms, but that racist ideology served largely as a cloak, though not, of course, where it led to people being worked to death. On labor psychology see also Geuter (1987a) and Métraux (note 1).
6 This book also includes a study by Mitchell G. Ash and myself ("NSDAP-Mitgliedschaft und Universitätskarriere in der Psychologie," written after this book was finished). We showed that party membership did not generally make a university career more probable (see Chapter 2). There were cases, however, that of Kurt Wilde, for instance, where political activity provided an enormous boost for a career.
7 See my review in *Psychologie heute*, H 4, (1987), pp. 78–9; a book by Riedesser and Verderber (1985) is only superficially researched in the parts concerning Germany.

stances and put an end to their lives in 1933–4. Were there other such examples?

Joachim Wohlwill (1987) has shown that individuals could make a variety of responses; whether a psychological journal followed a path of enthusiastic conformity or carefully sought to maintain its independence depended on the attitude of the editors.

Research on the professionalization of psychology should pay more attention to the importance of psychology in teacher training,[8] which I barely touched on since it is an academic or semiacademic teaching activity. However, it was one of the few professional opportunities in the twenties outside the universities, although it was nowhere near as influential as military psychology was later to become.[9]

One last idea for further research arises from a note by Trudy Dehue (1988): in the Netherlands the occupying Germans introduced the first full university training in psychology, the "Academisch Statuut," and Austria was forced to accept the Diploma Examination Regulations of 1941, though these were abandoned after the war. What effect did Nazism have in these and other countries on the development of psychology?

The opportunity has been taken to introduce a number of minor corrections, particularly relating to the history of the institutionalization of psychology in the universities.[10]

Ulfried Geuter
La Playa, Christmas 1987

8 See Otto Ewert, "Erich Stern und die Pädagogische Psychologie im Nationalsozialismus," in Graumann (1985, pp. 197–219).
9 Nine people in the *Psychological Register* (note 4) were active in teacher training; a further six were in youth and school psychology, two at the Central Institute for Education and Teaching in Berlin, one as a local school psychologist, and three in local youth counseling. Psychology in teacher training also affected the institutionalization of psychology in the universities. In Thuringia, which had the first chair in psychology at Jena in 1923, a law was passed in 1922 transferring teacher training to the universities, as later happened in Saxonia, Hesse and Hamburg.
10 For more detailed data see Geuter (1986).

Abbreviations

(Additional abbreviations of sources are listed in Comments on Sources.)

Amt BuB	Amt für Berufserziehung und Betriebsführung (Office for Vocational Training and Works Management)
BDP	Berufsverband deutscher Psychologen (Professional Association of German Psychologists)
DGfPs	Deutsche Gesellschaft für Psychologie (German Society for Psychology)
DINTA	German Institute for (Nazi) Technical Training
DPO	Diplom-Prüfungsordnung (Diploma Examination Regulations)
DWEV	*Deutsche Wissenschaft Erziehung und Volksbildung*
Ind. Pst.	*Industrielle Psychotechnik*
JHBS	*Journal of the History of the Behavioral Sciences*
NSD	National Socialist Lecturers' League
NSDAP	National Socialist German Workers' Party
NSV	National Socialist People's Welfare Organization
ObdH	*Oberbefehlshaber des Heeres* (commander in chief of the army)
ObdL	*Oberbefehlshaber der Luftwaffe* (commander in chief of the air force)
ObdW	*Oberbefehlshaber der Wehrmacht* (commander in chief of the armed forces)
OKW	Wehrmacht Supreme Command (Armed Forces Office)
Prakt. Psych.	*Praktische Psychologie*
PsRd	*Psychologische Rundschau*
RABL	Reichsarbeitsblatt
RGBL	Reichsgesetzblatt

RMWEV	Reich Ministry for Science and Education
SS	Schutzstaffel
UA	University Archive
VdpP	Verband deutscher praktischer Psychologen (Association of German Practical Psychologists)
WPsM	*Wehrpsychologische Mitteilungen*
Z. ang. Ps	*Zeitschrift für angewandte Psychologie*
Z. päd. Ps	*Zeitschrift für pädagogische Psychologie*
Z. Ps	*Zeitschrift für Psychologie*

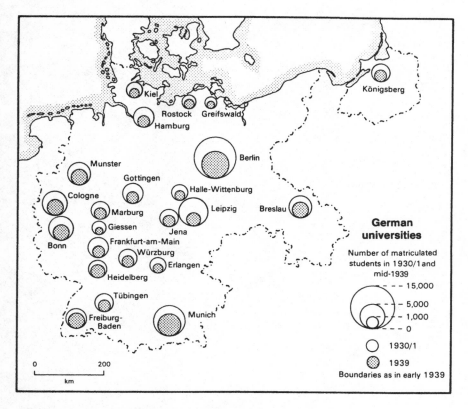

Drawn by Michael Freeman. Reproduced from *The Atlas of Nazi Germany* (London: Croon Helm, 1987), p. 85, by permission of the publisher.

xxi

1

Introduction

Only when it is responsible for providing psychological diagnoses for state purposes does psychology really become important.

Max Simoneit, scientific director of
Wehrmacht Psychology, 1938

It is becoming... plain that psychology has ceased to be a science for connoisseurs. With activities such as selection, evaluation, control, guidance, and care for the mental hygiene of the healthy members of our people, with aid and advice for the susceptible, the endangered and the inefficiently functioning, it is becoming deeply involved in the necessary tasks of regulating, maintaining, and strengthening the *Volkskraft* as a whole.

Oswald Kroh, chairman of the German
Society for Psychology, 1941

It is widely believed that the Nazis were opposed to science in general and to psychology in particular, with the result that they obstructed the development of psychology in every way or indeed threatened its very existence. In fact, the history of the professionalization of psychology during the National Socialist period was not one of setbacks and defeats, but one of gains and successes. This is certainly not easy for psychologists to admit, which is perhaps one of the reasons this aspect of the history of German psychology has often been passed over. After the Second World War German psychologists were more concerned with reestablishing their profession than with raising the question of the relationship between psychology and Nazism.[1] Within the discipline there was some controversy, but public discussion was

---

1 Baumgarten (1948) posed this question. It has not been taken up further; von Allesch (1950) merely rejected it out of hand (see Geuter 1980). One point in my 1980 article is in need of correction: the reticence of emigrants is not the result of a professional consciousness maintained across the Atlantic, as I then supposed. Cf. the archive records of my conversations with emigrants.

prevented by professional considerations and the politics of scholarly rivalries, not to mention the general difficulties of "reappraising the past" in Germany (see Adorno 1968).[2] In the opinion of Schunter-Kleeman (1980), the absence of any discussion in the German Democratic Republic was due to the fact that, as in the Federal Republic, some university staff from the previous era had continued lecturing there after 1945.

When, as a reaction to the vehement student criticism of the sixties, some universities organized lecture series on the sciences during the Nazi period, psychology was not represented. Nor did the development of psychology under National Socialism receive much attention from the student movement, although the question of the social responsibility of science and the relationship between science and power was discussed.

By the end of the sixties it had become clear that science had ceased to be the preoccupation of a small elite. As early as 1963 in the United States, Derek de Solla Price had diagnosed the transition from "Little Science" to "Big Science." The sciences had developed into large organizations and had aligned themselves with the state and – as became so apparent during the Vietnam War – with the military establishment. During the student movement students were no longer willing to leave the sciences to those in power. They criticized both the theoretical content of science and the uses to which it was put. Often this criticism was historically oriented, not least because in their search for a political theory many students rediscovered Marx and the Marxist tradition.

In Germany the students' criticism of psychology concentrated on two questions. First, how was it possible to explain why social conditions had not been included as a theoretical category in psychology? Second, what was the social role of practical psychology; how was it used and abused for industrial, military, and state purposes? These questions were directed toward the U.S.-style psychology dominant at that time (Geuter and Mattes 1984). Because there were only a few cases of confrontation with professors who had already taught during the Nazi period, for example in Mainz, the development of psychology in the Third Reich was scarcely investigated. It was only in 1979–80 that the journal *Psychologie- und Gesellschaftskritik* took up the topic.[3] It was first dealt with at a congress of the German Society for Psychology in 1982 (Geuter 1983a).

After the waning of the student movement, the history of psychology

2 Mild criticism by the young psychologist Ferdinand Merz (1960) of the role played by Leipzig holistic psychology (*Ganzheitspsychologie*), and a more outspoken criticism of that role in the East Berlin journal *Forum* ("Zur Situation..." 1960) was rejected by Wellek (1960) as though one essay by a professor who had lived through that period was enough to put an end to the topic of "psychology and National Socialism" once and for all (see Geuter 1980, 1983).
3 Cf. the contributions by Chroust, Geuter, Hantel, and Kienreich (all 1979), and Geuter, Mattes, and Schunter-Kleemann (all 1980).

became mainly the domain of psychology's fundamental critics. It was logical that they should turn to the beginnings of psychology in their historical studies, focusing on the general relationship between the history of society and the history of forms of thought in psychology.[4] Critical examination of the more recent history of psychology, however, deals with an institutionalized and professionalized discipline. It has to be borne in mind that we find a whole series of additional links between the general history of society and the history of science, some of which are of the most concrete nature. Sciences are no longer just theoretical systems; they are also social institutions. They intervene through their knowledge and their experts in social reality. They are called on to solve problems whose relevance has not been defined by science but by those outside it. At the universities they not only are research units, but are also responsible for teaching; their development depends not only on progress in the realm of theory, but on the state budget, research funds, education plans, the demand for scientific experts, and other factors.

This aspect has been receiving more and more attention from critical historians of psychology.[5] A number of works have been published, especially in the United States, that have opened up new approaches and whose use of archival material has provided new sources for historians of psychology (see Buss 1979). This development of the critical historiography of psychology coincided with a general trend in science studies in which – at least since the work of Thomas Kuhn – sociological methods were increasingly used in the study of scientific communities. The relationship between social developments and interdisciplinary developments then received more attention.

A part of this relationship is the professionalization of academic disciplines. The influence of this process on the development of the sciences is being increasingly regarded as important, but it is seldom treated historically (MacLeod 1977, p. 166; Rüschemeyer 1980, p. 311). When authors do write about the history of professionalization, at least in the U.S. literature, they concentrate on the academic institutionalization of the discipline and do not consider the total process of professionalization as I use the term. By *professionalization* of academic disciplines I mean the process whereby specific areas of application of knowledge in a given discipline, including the associated professional roles and the education related to this activity, become institutionalized. A professionalized discipline must have a corresponding scientific occupation, a *profession*. I take this to mean a field of occupational activity requiring the application of systematic scientific knowledge, that has previously been acquired in an educational institution for this purpose and that has been documented by a certificate. This then entitles the individual to carry out certain professional activities, or is at least viewed

4 Bruder (1973), Grünwald (1980), Jaeger and Staeuble (1978), Schmid (1977), and Staeuble (1972).
5 For an overview of the contributions see Geuter (1981) and Geuter and Mattes (1984).

on the employment market as a prerequisite for certain employment. The professionalization of academic disciplines is influenced by the history of a society and its demand for expertise. It also depends on the theoretical development of a discipline and in turn will itself influence that development. Especially in a country like Germany, where the university system is state-organized and where many professional roles are in the public sector, this process is subjected to a tension between science and power.

During the Third Reich, when the institutions of power were endeavoring to utilize the sciences for the consolidation of their ideological, political, and military dominance, German psychology progressed on its way from being a scientific discipline to becoming a profession. Today it is a professionalized discipline, a science with its own theoretical and methodological contours, with a secure position at the universities, its own professional role, and its own system of training and qualification. But in the Weimar Republic it was divided into quarreling schools, hardly established as a profession, occupying a weak position in the universities, and without a commonly accepted educational concept. It was only with reluctance that the academic psychology of that period began to tackle practical psychological problems. Not much more than models and instruments for selection diagnostics had been developed for practical psychological use. At the Vienna Congress of the German Society of Psychology (Deutsche Gesellschaft für Psychologie – DGfPs) in 1929, the committee of the society drafted a declaration entitled "On the Fostering of Psychology at German Institutions of Higher Education," in which it lamented the academic situation of psychology. It also demanded that, in view of the scientific advances, methodological progress, and practical importance of psychology, the number of chairs of psychology be increased ("Kundgebung..." 1930). At the Hamburg Congress in 1931, the society's president, Karl Bühler, again drew attention to the fact that psychology was under pressure. Industrial psychotechnics, the major field of applied psychology in the Weimar period, was less in demand following the economic crisis. The army, which had begun to use psychology during the First World War, primarily to select specialists, was now restricted to 100,000 men by the Treaty of Versailles. Heavy weapons and military planes were forbidden, as was all cooperation with universities. In the field of pedagogical psychology the only professional openings were for lecturers at teacher training institutions. In the universities psychology was restricted by the fact that most of its proponents occupied chairs of philosophy specified for psychology or chairs of psychology together with philosophy or pedagogy. There was only one university that had a full professorship defined exclusively for psychology, in Jena. The only possible qualification was a doctorate, although at a number of universities even this was not possible for psychology as a subject in its own right. Teaching at the various institutes was not unified and was oriented largely toward the special interests and views of each professor. At the technical colleges psychologists taught in various faculties, but only as part of the special training for engineers. They were not able to offer their own courses in psychology. At some technical colleges and universities there were no representatives of psychology whatsoever.

This situation was not in the interests of the discipline or its members. A glance at developments from 1933 to 1945 shows that the position of psychology changed very much during this period. It was able to consolidate its position as an academic discipline, to establish its own courses and examinations, and to institutionalize the first career for psychologists.

The most obvious feature was the enormous expansion of military psychology in the Wehrmacht where a large number of psychologists found employment. It was the first time that a socially important institution had created a career that was open only to qualified psychologists. At employment offices the selection procedures of the twenties were continued; the use of methods of industrial and managerial psychology in industry continued, and the first psychologists received full-time positions. In the German Labor Front industrial psychology was even given its own institute. As far as pedagogical psychology was concerned a new field of activity was created, advising on education in the National Socialist People's Welfare Organization (NSV).

In the universities the Law for the Reconstruction of the Civil Service of 7 April 1933 certainly forced many psychologists out of their jobs, but during the Nazi period as a whole the number of professors of psychology increased. The opportunities for obtaining a doctorate in psychology also improved. Finally, during the war – at a time of greatest coercion, subjugation, and complete rationalization of public life – examination regulations for a professional certificate were introduced that were based on a general consensus on the internal structuring of the subject expressed in the examination topics.

In the long struggle to become an independent university subject and a recognized profession, the Diploma Examination Regulations (Diplom-Prüfungsordnung – DPO) were a real breakthrough for psychology. In the midst of war psychology became an independent teaching subject. Every university was supposed to have its own chair of psychology. It could become a subject for doctorates and "habilitation" examinations in its own right. It provided a professional qualification to show on the job market. Yet only a year after the DPO had been decreed, the largest employers, the army and air force psychological departments, whose needs had helped to bring about its development, were disbanded. The political developments of the Third Reich nonetheless continued to leave their marks on the history of the subject.

For an investigation of the professionalization of psychology, the Third Reich is a doubly interesting period. On the one hand, there were, of course, rapid institutional developments in this period, but at the same time extreme pressure was being exerted on science. This makes the external influences and political constraints on the development of the discipline and profession more perceptible, leaving little room for a self-image (found not only in psychology) in which professional activity per se is reformist and humane.

Did the special relationship between science and power have anything to do with the professionalization of psychology at that time, or was it merely a temporal coincidence? Would psychology have developed at the same speed, in the same direction, and in the same way without the political changes? Why was psychology – a young, weak, and relatively unknown discipline – able to achieve professionalization in the Nazi state? Who had an interest in it?

This study will try to answer these questions. It will show how far the professionalization of psychology progressed in the Third Reich and whether it was successful in every respect. It will also consider the positive and negative factors affecting the professional development of psychology. A theoretical goal is to clarify these factors as a step toward establishing the general relationship between the history of a society, a profession, and a science.

Approaching the history of psychology as the history of a profession has two implications. First, the concept of psychology is limited. Since I treat the development of the academic discipline and at the same time the social enterprise of psychology, as it defined itself at that time, I do not consider all theoretical or practical initiatives that could be termed psychological. This also means that the study does not touch on the development of psychoanalysis in the Third Reich, which went its own scientific and professional way. In psychology there were both open and cautious theoretical borrowings from psychoanalysis, but there was hardly any critical examination of it.[6] The psychotherapists pursued their own professionalization in the Third Reich – with some success, to judge from the finding of Cocks (1985).[7] Psychoanalysis was not represented at the universities either in psychology or medicine; then, as now, teaching and practical training were given at their own institutions, which were closed by the Nazis. The three major groups (Freudians, Jungians, and Adlerians) were forcibly united in the German Institute for Psychological Research and Psychotherapy, where psychotherapists were trained. I will return to this in Chapter 4.

A second implication of investigating the professionalization of psychology is that it excludes other interesting questions, such as the ideological influence of Nazism on psychology as a science and the susceptibility or resistance of

---

6 Brodthage and Hoffmann (1981) consider the critical treatment of psychoanalysis by psychology for the period until 1940. There was, however, implicit adoption of psychoanalytical ideas. Rothacker (1938), e.g., used Freud's theory of the unconscious without mentioning his name. The main work of personality theory, or characterology, by Lersch (1938) is also inconceivable without Freud (see Chapter 3, this volume).

7 On psychoanalysis and psychotherapy see also Dräger (1971), Lockot (1985), and Lohmann and Rosenkötter (1982).

psychological theories to Nazi ideology. Nor will I deal with or classify the political behavior of individual university teachers. When I deal with race psychology, typology, characterology, or the political behavior of individuals, it is only in terms of questions relevant to my subject – for example, which theoretical elements made psychology practically useful or how the discipline attempted to legitimate itself in its quest for professionalization.

The historiography of psychology has placed no great emphasis on theory in the past (Graumann 1983). The reader will perhaps forgive me if, before presenting my findings, I include some general theoretical and methodological considerations. In the next section I consider earlier views of the development of psychology and neighboring disciplines in the Third Reich, as well as a more general reflection on the demands of the Nazi period on science. Then I discuss the extent to which a theoretical view of the character of the Nazi system influences the view of the development of a science, and explain my own frame of reference. The theoretical and methodological approach of this study is then developed in a section on the theory of professionalization, which also includes a model for the historical investigation of professionalization. A final section discusses the available sources.

Decline or continuity

The literature up to 1984 includes only one attempt to provide a comprehensive picture of the history of psychology in Nazi Germany. This study, written by two emigrants, Frederick Wyatt and Hans Lukas Teuber, and published in the United States in 1944, remains the best investigation of the subject. However, it is restricted to the period up to 1940 and to an evaluation of the psychological literature – necessarily so at the height of the war. It was hardly possible for Wyatt and Teuber to examine the practical activities of psychology in the war or the discipline's professionalization and institutionalization. They concentrated instead on the theoretical development of psychology and reached the conclusion that existing irrational tendencies in German psychology were strengthened by the rapidly expanding dictatorship. The greater the philosophical character of theories, in their view, the greater was the influence of Nazi ideology. For industrial, Wehrmacht, and physiological psychology they found that more exact methods were used. Apart from this paper only a few shorter essays have dealt explicitly with aspects of psychology in the Third Reich, such as the political behavior of psychologists or their adaptation to Nazism.[8] Anglo-American studies during

---

8 Baumgarten (1948) studied the political behavior of university teachers of psychology, Wellek (1960) the political constraints on psychological schools, and Chroust (1979) the ideological-theoretical conformity of Pfahler. Geuter (1979) examined the political and theoretical behavior of psychologists faced with Nazi "seizure of power" in 1933.

8 *The professionalization of psychology*

and even after the war evaluated Wehrmacht Psychology[9] and industrial psychology[10] methodologically only for possible use in Britain and the United States.[11]

This period does not fit well into the view of history held by psychologists. Where the history of psychological theory is seen as the continual accumulation of knowledge, and the history of the application of psychology is understood as the progressive humanizing of social life, periods such as the Third Reich must be disavowed. Without explicit intention, a linear view of development thus emerges. The history of psychology in the Third Reich has been very largely nonhistory, or at least not part of the history of German psychology.[12] In the only book dedicated to the history of applied psychology in Germany, the Nazi period is discounted as an "unhappy chapter in the history of psychology" and as a period of "exaggerations forced more or less successfully by political events" (Dorsch 1963, pp. 76, 90). Even in the critical literature of the seventies we find the conviction that psychology had not existed as a profession before the Second World War, and is only now in the process of establishing itself as a profession.[13]

There seems to be a commonly held opinion among psychologists that the history of psychology under the Nazi dictatorship was only a history of suffering. We find sweeping statements about discrimination and decline. There is talk of a "serious setback," of a "*disciplina ingrata*," of demands to do away with psychology as a "Jewish science," or there being no decline worse than that during fascism.[14] Prejudice and a lack of knowledge about

9 Ansbacher (1941, 1941a, 1949), Davis (1946), Dunlap (1946), Farago (1942), Feder, Gulliksen, and Ansbacher (1948), Fitts (1946), Gerathewohl (1950), and Kreipe (1950).
10 Ansbacher (1944, 1950) and Dunlap and Rieffert (1946). These studies were part of a U.S. and British attempt during and after the war to evaluate German science and technology for their own use. Most comprehensive was the 84-volume *FIAT Review of German Science*, which did not include psychology. (The agency responsible for this study, the Field Information Agency Technical, was attached to the Office of the Military Government for Germany–U.S.) The "FIAT Final Report 930" was a study by Viteles and Anderson (1947) on psychological training and selection of supervisory personnel at IG Farben.
11 On the German side the scientific head of the Wehrmacht Psychology, Max Simoneit, presented a rather apologetic account of its development (Simoneit 1972). Hinrichs (1981) claims to cover industrial psychology in the period from 1871 to 1945, but only touches briefly on the Nazi period, and considers only its program not its practical application.
12 "To judge from psychology's own history books, the involvement of German psychologists with Nazism has been almost totally forgotten" (Billig 1978, p. 15). A list of such books readily comes to mind. The statement does apply not only to the treatment in textbooks. In a history of psychology at Kiel University, 1898–1965, the years from 1933 to 1945 are simply nonexistent (Volkamer 1965).
13 Cf. Kardorff and Koenen (1981, pp. 26, 92ff.), Ottersbach (1980, p. 8), and Zillmer (1980, p. 86).
14 See Schultz (1969, p. 323), Pechhold (n.d., p. 310), Wellek (1960, p. 181), and Hiebsch (1961, p. 16).

the development of psychology in the Third Reich also seem common among historians.[15] Hans Peter Bleuel (1968, p. 220), for example, assumes that psychology was decimated as a branch of science, and Hans Mommsen recently claimed that it was practically liquidated.[16]

The idea that fascism brought about the total decline of German psychology may stem from the fact that Nazi policies led to the loss of important German-speaking scientists such as Karl Duncker, Adhemar Gelb, David Katz, Kurt Lewin, Wilhelm Peters, Otto Selz, William Stern, Heinz Werner, Max Wertheimer, and later Karl Bühler, as well as Wolfgang Köhler, who left Germany in protest. With their expulsion and emigration, certain theoretical developments in Germany came to an abrupt end. The emigrants included most of the better-known Gestalt psychologists and the leading developmental psychologists (see Ash 1984; Mandler and Mandler 1969; Metzger 1976).

Less common are criteria or detailed statements concerning this hypothesis of decline. Misiak and Sexton (1966, p. 113) talk of a reduction in the number of professorships; Adler and Rosemeier (1970) assess the falling off of experimental methods in the thirties as a sign of a politically induced decline. How easily superficial conclusions are drawn is shown by Nussbaum and Feger (1978), who simply blame the general decrease in the number of publications in psychological journals on the "animosity towards science in the Third Reich" (p. 108), neglecting all other factors such as the military enlistment of psychologists, the ending of the publication requirement for dissertations in June 1941 (which had often been published in journals), and paper rationing. There was also the growing importance of publications outside the traditional spectrum, particularly *Wehrpsychologische Mitteilungen* and *Soldatentum*. Such quantitative results must seem dubious anyway in the light of other available findings. Treuheit (1973) reports that during the Nazi period there was a higher yearly average of psychological dissertations than during the Weimar Republic, or in West Germany until 1973.

Occasionally it is explicitly claimed that psychology and Nazism are incompatible. Metzger writes that it is characteristic of autocratic regimes "that they are, without exception, full of distrust and aversion against this science" (1965, p. 112). Damage to psychology under such a regime must then have something to do with the nature of the discipline itself. Metzger (1979) goes on to specify the incompatibility of Gestalt psychology and Nazism, although

---

15 Gerhard Grimm (1969), in a survey of German universities in the Third Reich, mentions Oswald Kroh only for psychology under the heading "NS-Ideology and Science." Richard Grunberger (1971) cites the example of Erich Jaensch only to illustrate the relationship between academic interests and political attitudes.

16 In a talk held at the Dies academicus of the University of Hamburg on the fiftieth anniversary of the book burning, 16 May 1983.

forty years earlier he had tried to demonstrate their compatibility.[17] Arnold (1970, p. 29) also sees a fundamental contradiction between democracy/ psychology and dictatorship/animosity to psychology, and draws parallels between conditions in the Soviet Union and Nazi Germany. Elsewhere, he cites the dissolution of psychological services in the army and Luftwaffe in 1942 as proof that the practical influence of German scientific psychology after 1933 had also been suspended (1959, p. 298).

We find similar views in neighboring disciplines. Lepsius (1979) claims that the history of professorships and scientific societies shows that sociology was nonexistent in the Third Reich.[18] Schelsky calls this view the creation of a historical and disciplinary legend (1980, p. 417); Klingemann (1981, 1986) contradicts the hypothesis, claiming that sociology also served the practical functions of securing order and providing ideological support. Psychoanalysts are also beginning to separate themselves from the traditional idea of their discipline's nonexistence in the Third Reich (Lockot 1985; Lohmann and Rosenkötter 1982). Cocks's investigation (1985) shows that psychotherapy became stronger in the Third Reich because of its practical usefulness and, in his view, was able to pursue its three professional tasks of research, practice, and training.

The idea of a discipline's incompatibility with Nazism guards against unpleasant questions. When a political dictatorship cannot admit psychology as a scientific discipline (as Arnold indicates), then we do not even need to ask whether there was any demand for ordinary psychology, whether its use furthered the development of the profession, or whether there were perhaps mechanisms governing the relationship between the discipline and society that were not limited to that period.

It is strange that the question of normality (and thus of continuity) is also disregarded in the historiography of psychology by those who see psychology as having been completely subservient to the Nazi state. Fritsche speaks of the "*Gleichschaltung* of psychology in the Third Reich, i.e. its total ideological and political subjugation as a science" (1981, p. 9). The Nazis used psy-

---

17 Metzger (1938, 1942) then wanted to use the Gestalt laws of perception to back up the *völkisch* view of the state (see Chapter 5, this volume). These essays and one other I found (Metzger 1938a) are not mentioned in the Metzger bibliography (*Psychologische Beiträge*, 5 (1960), pp. 283ff.). Asked about the 1942 essay in 1979 (I did not yet know about the others) Metzger replied that this had been his "school-boy" position; he had not written the essay to get the professorship in Münster, 1942 (Z, f. 189). Metzger wanted to send me a comment when he had seen the essay again – he no longer had a copy. Unfortunately, he died on 20 December 1979 before he could do so. The two essays in 1938 appeared in the journal of the Nazi Teachers' League in Halle-Merseburg. In 1937–8 Metzger was stand-in for a chair in Halle. A first interpretation of these essays appeared in Geuter (1983a, p. 102); cf. in more detail Prinz (1985).

18 See also Schoenbaum (1968, p. 307, n. 64), who claims that though other scientists fell from grace as individuals, sociology was the only discipline to meet with displeasure as a whole. I think that academic theology and philosophy deserve to be examined in this respect.

chology "as a political tool" (p. 95). Schunter-Kleemann sees a "comprehensive, propagandist and practical utilization by the National Socialists";
there was "not one tendency in German psychology that – after the expulsion
of numerous well-known Jewish representatives ... did not place itself in the
service of the Fascists" (1980a, p. 49). The point of agreement with the
incompatibility hypothesis is the claim that there was no normal psychology.
These authers also find themselves in agreement in exaggerating the
importance of psychology to the Nazis; if they used it as a political tool, or
if it suffered particular persecution, then it must have been very important.
The declaration that psychology was a tool or a victim involves a reconstruction of the field's significance.

The idea of complete utilization is also alive in· the talk of "fascist
psychology," rather than of psychology in the Nazi period (Fritsche 1981,
p. 140). Here again, we find similar views in related subjects. Wuttke-
Groneberg (1980) always speaks of "National Socialist" medicine when
referring to medicine in the Third Reich. Güse and Schmacke diagnose a
"fascist phase of psychiatry," which they see as the "peak of a continuous
development" (1980, p. 86). This talk of "fascist science" or a "fascist phase,"
however, makes one blind to the spectrum of scientific theories and practices
that precisely in their normality, were most certainly functional for the
system, perhaps inevitably so. Talk of National Socialist science would mean
accepting the fantasies of those Nazis who wanted to shape the whole of
science after their fashion. The problem of the responsibility of science can be
discussed adequately only when the continuity of science in the Third Reich
is also a topic, alongside its apologetic adaptation or its use for terror and
murder.

The theme of continuity has been discussed mainly in terms of those earlier
theoretical views that could merge with Nazi ideology. For medicine, according to Baader (1980), this was the ideology of social Darwinism; for psychiatry, Güse and Schmacke (1976) identify concepts of order and security.
For the ethological theory of Konrad Lorenz, Kalikow (1980) shows that
Lorenz's biologistic ideas before, during, and after the Third Reich were
similar to those of Nazi ideology. As already mentioned, Wyatt and Teuber
(1944) pursued this question for psychology. The political and ideological
situation was certainly also reflected in psychological theories; answering the
question whether psychological concepts of the pre-Nazi period tended to
coincide with basic concepts of Nazi social theory will require further
research.[19]

This study examines the continuity problem at a different level: To what
extent did psychology continue to be a discipline oriented toward practical
application during the Nazi dictatorship, perhaps even for its own interests
in the process? In the postwar Anglo-American literature previously

19 Cf. Geuter (1985) and Prinz (1985).

mentioned, this continuity was treated as a fact, not as a problem, since the aim of that literature was to examine whether the methods developed in German applied psychology could enrich U.S. psychology.[20] The U.S. psychologist Ansbacher, a German-Jewish immigrant who made the major contribution to these studies, investigated the use of industrial and organizational psychology for the selection of deportees and concluded that "when and as psychology is employed, even within a totalitarian state and dealing with forced labor, it is still essentially the same psychology one would find in a voluntary situation within a free society, thus indicating that the same set of psychological principles holds universally" (1950, p. 48).

For Ansbacher psychology was a universal enterprise that remained unchanged in a totalitarian system – for Metzger and Arnold, the two were mutually exclusive. Did psychology measure up to the demands on it, or was it necessarily in contradiction with them? Nazism required two things of the sciences. They were to form a weltanschauung and to deliver purposive, practically useful knowledge. Both Metzger and Ansbacher may therefore have been right to some extent, since their generalizations correspond to these two sides: Metzger for the general theories of psychology that had been embedded in philosophical-ideological systems even before the Nazi period, Ansbacher for the methods of applied psychology. If psychology wished to continue to develop theoretical formulations with ideological content, then it had to ally itself with the Nazi ideology and either adopt this for its own theories or reformulate its results so that it could function as a support for the Nazi weltanschauung. But it could continue to be practically useful without particular adaptation to Nazi ideology.

Wyatt and Teuber (1944) had already established a contradiction between the irrationality of psychological theory in the Third Reich and the rationality of practical scientific procedures. They explained this with a contradiction in Nazi ideology itself – one between an expanding industrial imperialism and a romantic *ressentiment* at the type of world this had created. They also saw in this an expression of the concrete requirements of war. Lerner (1945) opposed their hypothesis that the influence of Nazi ideology was greater on more philosophical systems. He did not wish to limit Nazi influence to ideology and pointed out that researchers, in "objective" fields such as physiological psychology, chemistry, and physics were also victims of the emotional frenzy of the time.

Unfortunately, this problem was not pursued. On the contrary, some psychologists even suggested that psychology, on the strength of so-called objective methods, was already resistant to cooptation by the Nazis, as could

20 See notes 9 and 10. The studies were done in part for the U.S. government and military authorities. (cf. Z, ff. 170–1; APA I, G4: Dept. of the Navy, Office of Naval Research, Research Proposals; and APA I, H 2: Dept. of Commerce). German psychologists like Gerathewohl, Kreipe, and Rieffert worked on such projects.

be seen from the low level of ideological influence on psychotechnics (see Geuter 1983, p. 208f.). Such a reduction of the relationship between psychology and National Socialism to the ideological level, however, is not only unable to explain the professionalization of psychology. It is also unable to explain why a part of psychology remained unchanged, and yet was seen as useful and neither persecuted nor banned. The normality and continuity of psychology in the Third Reich should be discussed, and not ignored by the history of the subject as "the other." The history of psychology in the Third Reich is a part of the history of German psychology.

Science in the Third Reich and historical theory

Views on the development of psychology in the Third Reich have also depended on views about the character of the Nazi system. The literature on science in National Socialism in general shows the influence of various interpretations of German fascism.[21] We can identify three main views of the Nazi system: totalitarian, monopoly capitalist, and polycratic. The theory of totalitarianism arose before German fascism, as a theory that Social Democrats employed against the aims of the Communists (Paschukanis 1928, pp. 311–12; cf. Kühnl 1979, p. 122). It became widespread only in the wake of the Nazi concentration camps and Stalin's labor camps. Italian Fascism, German National Socialism, and Stalinism were all interpreted as a new state form. Carl J. Friedrich, who developed this theory of totalitarianism the furthest, was of the opinion that totalitarian dictatorship is historically unique and sui generis. Totalitarian systems resembled one another more then they resembled any other autocratic system and shared the following six features: an ideology, a party, a terrorist secret police, an information monopoly, a monopoly of arms, and a centrally controlled economy (1957, p. 19).

The totalitarianism theory was generally accepted by scholars in the fifties. Research on science under totalitarian conditions concerned itself for the most part with the Soviet Union, although a similar approach can also be found in the work on science in the Third Reich.[22] If, as Hannah Arendt

21 Since I am considering the professionalization of an academic discipline, I will discuss the influence of theories of fascism only on history of science investigations, not on studies of the attitudes of intellectuals or the behavior of individual professional groups. The German debate on fascism has been criticized for restricting itself to National Socialism (Schieder 1976, p. 12). Since I am not concerned with general statements on psychology under fascist conditions, I restrict myself to Nazism. When the term "fascism" is used here, German fascism is meant.

22 For example, at a conference of the American Academy of Arts and Sciences in 1953 in Boston, science under totalitarian conditions was also discussed. The three contributions (Part V: Muller, de Santillana, Wolfe) dealt with science in the Soviet Union, Galileo's conflict with the Catholic church, and historiography in the Soviet Union. Science in the Third Reich was only touched on in the discussion (Friedrich 1954).

writes, terror, propaganda, one-party rule, and an ideology are the most important features of totalitarianism, and if the reality of the Third Reich can be described in these categories, then the analysis of science must concentrate above all on the way science was subjugated by the system and its ideology. The openly Nazi elements of science were sought out, and this was Nazi ideology. The analysis of Nazi ideology may well be the strength of this approach compared with other theories, since ideology was correctly seen as a tool of domination.[23] For the work on the Third Reich, however, this approach was restrictive, because it reduced the relationship between Nazism and science to ideological dominance, so that overcoming Nazism was restricted to overcoming its ideological influence. Having reached the conclusion that, in comparison with the Soviet Union, the Nazis provided natural scientists with quite reasonable research conditions after a severe initial blow, Hirsch (though writing as late as 1975) failed to ask whether this might not have corresponded to the very aims of the Nazi system.[24]

Once the theoretical peculiarities of Nazi ideology (Haug 1977, p. 126) had been rejected after the war, the old understanding of science was restored, without questioning whether this might not have been precisely what made it functional for the fascist system. If National Socialism, as a variant of totalitarianism, was a social form sui generis, as Friedrich believes, then there was no problem of continuity, and science shriven of ideology could play its part in the "struggle between the free world and totalitarianism" (Hofer 1964, p. 7). Science, which had remained uninvolved with Nazi race ideology, was now magically transformed into antifascist science, even if it had met demands placed on it by the system. In psychology we can find this approach adopted to some extent by Herrmann (1966), who sees the adherence to the "solid craft tradition" of psychotechnics as a sign that psychotechnicians had avoided enforced ideological alignment. We have already seen Arnold's parallels between National Socialism and the Soviet Union. This pattern is also displayed in the claim that the incompatibility between knowledge gained from natural laws and Nazi ideology proves that a contradiction existed between individual scientists or schools and the Nazi state, as made by Metzger (1979) for Gestalt psychology or Scherzer (1965) for physics.

It is characteristic of the situation in West Germany after World War II that the treatment of science in the Third Reich did not go beyond ideological de-

23 Bauer (1955), who took part in the conference mentioned in the preceding note, produced one of the best demonstrations in the older works on Soviet psychology of the connection between political development and psychological theory. For an assessment of Bauer and a more exact discussion see Thielen (1984).
24 Timasheff (1966) sees no free development of the professions in either communist or fascist societies. He cites Italy, Germany, and the Soviet Union as examples, although the only example for Germany is the Reich Chamber of Culture. Unfortunately, he provides only illustrations of his theory but no empirical basis for it.

Nazification, apart from the question of the political behavior of individual scholars.[25] Thus, the blatantly Nazified psychology could be forgotten without any bother, but no one concerned himself with the normality of its functioning in the Nazi system. Perhaps the dominance of the totalitarianism theory helped to prevent debate on this question in psychology.

The totalitarianism theory also fits well with the ruling scientific view of epistemological progress unaffected by external factors, in which society and science confront each other as disparate entities, whose only conceivable link is the influence that science has on society. Where the influence of society on science was as obvious as in the Nazification of psychological theories, then this could only be the result of compulsion or the political prostitution of individual scientists. Thus, the totalitarianism theory favored the exclusion of the Nazi period from the continuity in the history of psychology.

A contrary view of a whole science completely functional for Nazism without any degree of independence from social conditions results from the rigid application of a theory of fascism as the total dictatorship of monopoly capitalism developed by Dimitroff in 1935. He saw fascism primarily as a political instrument of finance capital. Large-scale industry did indeed back Hitler, and it drew considerable profit from Nazi rule. Research has shown that, on the whole, Nazism met the needs of capital by conquering new sources of raw materials and goods, by creating new markets, and by controlling the workers (see Kühnl 1979). It also seems to be established that territorial expansion was a way of solving the crisis of German industry (see Sohn-Rethel 1973). This does not mean, however, that Nazi politics can be seen solely as the product of the ambitions of large-scale capital (see Hennig 1974). Before 1933, August Thalheimer (1926), theoretician of the KPD opposition, had defined fascism as a system in which the executive was independent of both major classes, but society was dominated by the bourgeoisie and private property. After the war Marxists were faced with the problem of explaining why the Nazi state intervened in the economy against the profit interests of industry. If the state was a superstructure over an economic basis, was fascism no longer a form of capitalist dominance? How was the relationship between industry and politics to be understood?

In 1941, Alfred Sohn-Rethel had already suggested that this relationship was a dialectical one: "The two parts are chained to each other in a sort of mutual dependence. They do not stick together because they like each other, but despite the fact that they hate each other. Each side is dominated by the wish that their dominance might be possible without the other" (1973, p. 198). This makes it possible to dispense with the idea that the science that was free from ideological influence was protected by industry from Nazi pressure. The Nazis adopted the interests of industry as their own, while

25 See the typologies of university teachers' behavior in Erdmann (1967), Grimm (1969), Nolte (1965), and Miehe (1968).

capital was interested in an ideology for its "following," the workers. Thus, the two partners had differing but also common interests in science.

Just as authors who assume the forced decline of psychology are obviously following totalitarianism theory, so we find that those who claim it was pressed into service show an orientation toward Dimitroff's theory of fascism. For Schunter-Kleemann (1980) National Socialism is omnipresent in psychology. Fritsche (1981) refers to the total subordination of psychology and state control. There are no interstices where contradictions could lodge, no room for differentiation. In the history of psychiatry by Güse and Schmacke (1976) psychiatry and its reactionary ideology are simply counterposed to the resistance to extermination, but contradictions at the theoretical level and at the level of practical day-to-day treatment are barely touched on.[26] Although in her essay Schunter-Kleemann criticizes the fact that East German psychologists have not dealt with psychology in the Third Reich, she shares the official East German concept of fascism as a monolithic power center in the service of finance capital, confronted by a monolithic resistance movement.[27] If science was completely in the service of this omnipresent system, then its nature can be determined only by its function for Nazi dictatorship. This point of view also provides the basis for Epstein's (1950) criticism of Ansbacher in the leftist *Benjamin Rush Bulletin*, or for the hypotheses of Bergmann et al. (1981, p. 22) on sociology in fascism.[28]

All the works oriented toward the two theories discussed have a relatively monolithic view of the Nazi system. They recognize only pro-Nazi science and non-Nazi or anti-Nazi science. As a critical reaction to these ideas, especially to totalitarianism theory, which in contrast to Marxist theory dominated historical scholarship for a long period before becomig outdated, recent historical research has considered the various bearers of Nazi rule (see Hüttenberger 1976). It is frequently claimed that the sources do not show the Third Reich as a uniform power structure, but as a confusion of responsibilities and a chaos of squabbles. Specialized research has displaced the concept of the monolithic superstate and recognized "the lack of system,

26 There are exceptions, e.g., on pp. 401 and 427.
27 Jenak's attachment to the Soviet theory of fascism is even plainer. For him fascist politics are "only the most ruthless pursuit of the greatest profits of the largest German monopolies" (1964, p. 70f.). His only comment on psychology in his work on the Technical College of Dresden is the daring generalization that in philosophy and psychology "the interests of the capitalist order ... [were] at the center of teaching and the research there" (p. 126).
28 The latter also claim the class nature of science. There are "two sciences," one for the workers and the other dedicated to producing profits. Although science is doubtless influenced by specific interests, this formulation smacks of the Stalinist "proletarian science" of the forties and fifties, which legitimated the persecution of undesirable scientists, as well as heralding the spectacular rise of Lysenko. It reemerged in the Chinese Cultural Revolution, when among other things sociology and psychology were abolished. Theoretically it reduces the analysis of science to the question "For whom?" The answer is usually a pragmatic one in accordance with the distribution of power.

the improvisation and the unevenness of the Nazis' use of power" (Broszat 1975, p. 9).[29] Recent historical writings have done much to provide a more differentiated picture of the chaotic or heterogeneous nature of Nazi rule.[30] At the same time, however, less and less attention has been paid to theoretical explanations of Nazi dictatorship. Hans Mommsen, for example, identifies the essence of Nazi dictatorship only as an unclear "tension between the inherited state constitution and the overlying zones of dominance of the Party elites" (1976, p. 32). Hüttenberger feels that Nazi rule can at most be understood in terms of a "formally descriptive and typifying" concept (1976, p. 422). He suggests the concept of a "polycracy" in which a number of largely autonomous powers compete with one another; he sees Nazi polycracy essentially as "continually varying relationships and constellations" (p. 427). If the essence is seen only as variance in time, and theory merely as the description of these variations, then all attempts at theoretical explanation have been relinquished.

Political arguments were also advanced against this position. Bracher (1976), protagonist of a modern totalitarianism theory, accuses the polycracy theoreticians of underestimating Nazism. The Nazis followed certain aims consistently to the end; their use of various means did nothing to alter the matter. Mommsen (1976) replies that they hardly had any aims, and their whole system of domination was temporary and incomplete. Stollberg's (1976) criticism of Bollmus (1970) is perhaps also valid here: the social function of fascist rule is not considered, nor who was in fact ruled over. One could also ask at whom this supposedly directionless politics was aimed, and in whose interests.

---

29 It would be inadequate to explain the retreat of the doctrine of totalitarianism solely in terms of the research findings on Nazism. It was primarily a retreat from the use of such theories to describe the Soviet Union and China at the end of the fifties. Three reasons seem to be important. First, there was a genuine change in the Soviet Union after Stalin's death away from totalitarianism. Second, there was the schism between China and the USSR after the Twentieth Communist Party Conference of the CPSU in 1956; both required an explanation. Third, the focus of U.S. foreign policy shifted from the confrontation with Russia toward increasing infiltration and domination of the Third World, which required more refined theoretical concepts for social analysis.

At times of transition, the pressures and the functions of theories are sometimes expressed more clearly. For example, Almond, a leading political scientist of the period, wrote that since U.S. interests were global, political systems must be explained comparatively, for which existing concepts were inadequate. His categorization of "totalitarian" systems draws interestingly on Neumann, who had not adopted a monocratic view of Nazism (Almond 1956). The use of totalitarianism theories in studies of both the Soviet Union and China became increasingly inappropriate as cooperation began between the United States and these communist states, in place of the earlier confrontation. This in turn affected the analysis of fascist rule (Stollberg 1976, p. 92).

30 Among others Aleff (1976), Bollmus (1970), Broszat (1975); see also P. Diehl-Thiele, *Partei und Staat im Dritten Reich* (Munich 1969); H. Höhne, *Der Orden unter dem Totenkopf* (Frankfurt/Hamburg 1969); and P. Hüttenberger, *Die Gauleiter* (Stuttgart 1969).

As Bollmus had already remarked in 1970, despite the chaotic leadership the effect on those ruled was no less totalitarian (pp. 9, 236, 250). More recently Bollmus has tried to combine the concept of totalitarianism with polycracy, by "defining it from the point of view of the ruled and from the threat" perceived by them (1980, p. 127). The social Darwinist competition among offices further increased the anxiety. Bollmus sees the two theoretical approaches as complementary, one relating to the programming of Nazi goals and the other to their realization. However, it seems doubtful whether scholarly considerations are sufficient to bridge differences in position that are not solely scholarly in nature.

The hypothesis that a uniform fascist rule dissolved into a chaos of rival responsibilities is the basis of some studies on science in the Third Reich, such as that of Bollmus (1973) on the dissolution of the Mannheim Trade College. According to Bollmus, administrative conflicts hindered "the performance in the field of the Natural Sciences required by the regime" (p. 135); this would contradict postwar U.S. studies that attest to the very high level of German science.[31] In Losemann's work on archeology the organs of Nazi science policy form a maze of differing approaches and rivalries, with only one common characteristic: adherence to race ideology. In view of the plethora of departments involved in appointments, Losemann feels unable to establish general assessment criteria and finds it more appropriate to judge each case individually (1977, p. 61). Studies such as Losemann's or Cocks's on psychotherapy are empirically fruitful, but one misses the theoretical generalizations that could be drawn from the accounts of fractions and rivalries. It is as if they had capitulated in the face of such a confusing profusion of data.

According to the polycracy hypothesis each individual interest group beat its own path through the dense undergrowth of the dictatorship. All we would then need to do for our study is to track the path cleared by psychology. To be sure, the empirical studies based on this theory have pointed out details and contradictions of the Nazi system, and these would lead us to expect that psychologists in the Third Reich also enjoyed a degree of relative independence. Nevertheless, the question remains how the interests of scientific disciplines and groups could be fulfilled. The answer is that they had to conform to the interests of decisive power centers that could set up courses, provide jobs, or throw people out. Therefore, a theory must be able to explain who had power in Nazism and what the social function of their power was.

It is surprising that since 1942 a work has existed that offers just such a differentiated approach – *Behemoth* by Franz Neumann. It was no doubt ignored for a long time because it did not fit the postwar climate of anti-

---

31 E.g., Simon (1947, p. 12): "The research establishments of the German Air Force were the most magnificent, carefully planned, and fully equipped that the world has ever seen."

Marxism. Neumann sees monopoly capitalism as the economic basis of the Nazi dictatorship but does not interpret the Nazi state as an instrument of capital. He sees the codified system, the "Leviathan," as having undergone a transformation in Nazi Germany to become the lawless nonstate, or "Behemoth." With these concepts Neumann refers first to Hobbes's *Leviathan*, which describes domination in a state through law, and then to his *Behemouth*, in which he describes the lawlessness that obtained during the English civil war. For Neumann the Third Reich was a similarly lawless nonstate, a social form in which the ruling groups directly controlled the rest of the population without the intervention of any sort of rational apparatus of coercion previously identified with the state. Legality was eliminated by the will of the Führer. This is close to the idea of the totalitarian rule of terror, but Neumann's analysis of the economic basis of the dictatorship and of its various pillars is far removed from the theory of totalitarianism.

The ruling class, according to Neumann, is not homogeneous. It is made up of dominant groups from the party, army, bureaucracy, and industry that share no common loyalties but that need one another for their own interests. They are held together by the "regime of terror" and the fear that the collapse of the system would be their mutual end. The army needed the party to organize society for total war. At the same time, the party depended on the army to win wars and to establish and expand its own power. Both needed a monopolized industry that could ensure continual expansion. All three needed the bureaucracy to attain the technological rationality without which the system could not function. Each of the four groups was sovereign and authoritarian; each had its own legislative, administrative, and judiciary power and was thus in a position to reach necessary compromises rapidly and without scruples. The weakest of the four was the ministerial bureaucracy – the more so the longer the war lasted.

For my purposes I would like to conclude from the theories discussed that National Socialism had a monopoly economy as its basis and that, by and large, it furthered the interests of monopoly capitalism.[32] Thus, it is not something different, cut off from previous and subsequent history. Solving economic problems with the twin strategy of external expansion and internal pacification of the classes was in the interests of various social power blocs. They agreed on fundamental goals, but each followed its own specific ends and could exert its own leverage. It was the existence of one center of power – the Nazi Party, with its totalitarian claims and its own apparatus – that above all sets this structure of domination apart from those before and after it.

Neumann's analysis of the structure of dominaton under Nazism makes it possible to determine more precisely the relationship of science to the various power blocs exercising pressure on it, and with which it entered

32 See also Bettelheim (1974) on the economy in the Third Reich, in which he refutes the theory that the interventionist policies of the Nazis put an end to the capitalist form of ownership.

either into coalitions or into disputes. Mehrtens (1980, p. 45) has already drawn attention to this. It seems important to investigate the relationships between psychology and all four power blocs.[33] In general, the development of a profession results from the interaction of the prospective professional group, its clients, and the state as guarantor of training standards. To gain social recognition the discipline has to win the approval of socially relevant forces. For the Third Reich it thus seems appropriate to bear in mind the four power blocs when considering the coalition that led to the professional advancement of psychology.

From discipline to profession: a historical model of professionalization

If we wish to judge the extent of the professional advancement of psychology in the Third Reich, then in addition to a theoretical understanding of the social framework for this advancement we also require criteria with which to assess it. If we also wish to trace the reasons for advancement, then we must have some idea of the factors determining the professionalization of a discipline. On the basis of previous work on this topic, I will try to present a model for investigating the historical process of the professionalization of psychology.

The emergence and increasing importance of professionals can be seen as the most important social transformation in the contemporary vocational system (Parsons, 1968). This change caught the attention of scholars some forty years ago (Rüschemeyer 1980, pp. 213–14), but it was only as new professions began establishing themselves that the literary floodgates opened. In pedagogy, sociology, and psychology such literature was a response to the crisis of professionalization, such as the problems of finding jobs for graduates.[34] Some of the recent literature also draws on the critique of experts.[35]

33 Ansbacher, in his study of German industrial psychology in the war, opposed a bloc approach: "... even a totalitarian state is far more complex than its stereotype" (1950, p. 48). He points out the discontinuities and contradictions, but seems to turn a blind eye when considering the use of psychology in the selection of foreign workers. He sees nothing oppressive or fascist and concludes that psychology played a basically democratic role. This earned him accusations of pro-Nazism. For his critics, psychology used in the Third Reich was "Nazi psychology," and its proponents were "Nazi psychologists" (Epstein 1950).
34 See the many new studies relating to the professionalization of clinical psychology: Fichter and Wittchen (1980), Kardorff and Koenen (1981), Keupp and Zaumseil (1978), and Wittchen and Fichter (1980). For sociology, see Kluth (1966), Matthes (1973), Siefer (1972–3), and Schlottmann (1968); for pedagogy, which became a diploma subject (and thus a new profession) in 1969, see Koch (1977), Müller (1979), and Weiss (1976). Reviewing the literature, Hesse (1972) notes that most of the U.S. investigations only try to establish whether a certain vocation is a "profession" (in terms of certain criteria). An example for psychology is Peterson (1976a).
35 Freidson (1975), Kardorff and Koenen (1981), Ottersbach (1980), and Zillmer (1980). The criticism of experts also takes the form of antiprofessionalization literature (e.g., Nowotny 1979).

Whereas the history of science hardly investigated the professionalization of academic disciplines before 1980, English-speaking sociologists have examined in detail the characteristics of professions or the sequence of professionalization with widely varying concepts and models.[36] In the literature "occupations" are generally vocations in which the necessary practical knowledge can be taught in the course of work. The term "profession" was originally applied to medicine, law, and the priesthood. There are different views about their special characteristics. Parsons (1968) names the requirement of formal training, as well as the ability and institutional means of controlling the socially responsible application of competence, and adds that professions are based on humanistic disciplines. Wilensky speaks of professions wherever a substantial proportion of a scientific discipline is applied (1972, p. 199). In contrast, Hesse makes no reference to the scientific disciplines; he concentrates on the monopoly of certain social opportunities and the expectations placed on qualification (1972, p. 69). Ottersbach, a critic of professionalization in German psychotherapy, also defines professions solely in terms of their "special social and economic status" (1980, p. 77). Where the emphasis is on knowledge gained in academic institutions, the concept of "academic professions" is prominent, while emphasis on the responsible application of expertise focuses on the concept of "free professions" (cf. Hartmann 1972).

The term "professionalization" is also used in different ways (Hesse 1972, pp. 34–35). It covers (1) the process of transformation from an occupation to a profession; (2) the replacement of laypersons by experts and a relatively rapid increase in the proportion of professionals in the labor force; (3) the strategy of the professional group on the labor market; and (4) the transformation of a previously theoretical science into an applied one for which a profession is created. This last definition, which applies to cases like psychology, pedagogy, sociology, and political science, is the sense in which I use the term.[37] Schmitz and Weingart (n.d.) have developed four criteria for a successful or completed professionalization of this type based on Parsons and Wilensky:

1. the existence of a clearly definable subject matter in the system of systematic knowledge (for the classic professions identical with academic disciplines, e.g., law, medicine);
2. a system of formal training by which students are socialized into that system of knowledge (academic studies, curricula);
3. an institutionalized pattern of granting access to the practice of the

---

36 G. Millerson found twenty-three characteristics of "true" professions in the works of twenty-one authors. Not a single one of them was felt to be essential by all the authors (Johnson 1972, p. 23).
37 In the United States the psychologists were the first occupational group to professionalize directly from the universities (Napoli 1975, p. 10).

profession by awarding credentials and determining the conditions
for obtaining them (examinations and titles);
4. an institutionalized pattern of applying the knowledge (the
regulation of practice). (pp. 5–6)

This list does not include other criteria named in the literature, such as the
organization of a self-administered professional body, altruistic motives, or
the development of professional ethics.[38] These, however, are more relevant
to free professions or to the description of the results of professionalization.
When I speak of "development" or "progress," then, I am referring to these
four criteria, and it is not my intention to make a value judgment.

For a historical investigation it is important not only to trace the indicators
of professionalization in psychology, but also to ascertain the factors responsible for this development. In the literature we come across two controversies: is professionalization a strategy for the optimization of the social
opportunities of an academic group or the final stage of the institutionalization of a discipline, and does it depend more on the activities of the
professional group itself or on state regulation?

Concerning the first controversy, in accordance with their definition of
profession Hesse (1972) and Ottersbach (1980) see professionalization solely
in terms of increasing social prestige and income.[39] For Ottersbach professionalization is then independent of the nature of professional activity, and
the establishment of psychology as a profession is simply the reservation of
professional abilities against competing offers from others. But if an offer
is to find acceptance, it must be useful. In the case of academic professions
there must be a certain body of knowledge and a resultant competence. If
a need is to be met, the knowledge required for the work must be available.
A historical investigation of professionalization must therefore deal with this
knowledge, its specifics, and its relevance. This is a perspective opened up
by Weingart in his research on scientific disciplines. For him professionalization, the "establishment of special fields of application for systematic knowledge" (1976, p. 61), is the result of a process leading from the development
of a new cognitive structure in science through its social institutionalization
to professionalization.

The position adopted in the second controversy, whether professionalization is furthered by self-regulation or by the state, appears to depend on the
type of professionalization and the country addressed by the theory. Both

38 See Elliott (1972, p. 96), Jackson (1970, pp. 5–6), and Johnson (1972, pp. 11–12).
39 "We understand professionalization as the conscious, planned effort... to secure or increase
   income, prestige and influence (control of social relationships), which has a relatively
   autonomous association as a precondition, and which by means of the normative structuring
   of work performance, especially by means of the specialization and monopolization of
   performance, as well as the expansion, theoretization and specialization of training, attempts
   to secure or increase reimbursement" (Hesse 1972, p. 73).

Ottersbach and Wilensky emphasize the independent activity of the professional group as a decisive factor. However, both refer to free professionals, businessmen on the psychotherapeutic market, and professionals dealing with individual clients. Hesse explicitly refers to self-structuring as the criterion for distinguishing "professionalization" from the "construction of a vocation." When Burchardt (1980) used the latter term for chemists in Germany (because of external influences on the process), Meyer-Thurow (1980) commented in a critique that such a conceptual differentiation was inappropriate for the German Kaiserreich, since questions of academic qualification were regulated by the state and controlled by the bureaucracy. In Germany any professionalization takes place against a backdrop of a state training system and a uniform state system of certification (cf. Rüschemeyer 1980, p. 324). In contrast, the model of self-control of professionalization is based largely on the U.S. example, where neither university training nor certification is regulated from above, but is to a large extent dependent on the activities of the profession itself.[40]

The example of graduation certificates demonstrates how difficult it is to make comparisons among different countries using a single model. In the United States, universities draw up their own curricula, so that the qualification acquired does not automatically lead to a professional certificate, as is the case for many subjects in Germany. Nor is psychology taught in professional schools like law, business, or medicine. The American Psychological Association therefore had to set up its own certificate before federal and state certification laws could be drawn up, which in turn influenced the university curricula (Napoli 1975). Much the same is true in Britain (Hearnshaw 1964, p. 279; Oldfield 1956). In Germany, however, a university diploma or certificate of graduation governed by state examination regulations is the formal entitlement to exercise a certain profession. In some cases there is

---

40 Napoli's study of U.S. psychology from 1920 to 1945 is the only historical investigation known to me of the professionalization of psychology in another country. It presents extensive material on the history of applied psychology, mostly chronologically, in terms of the expansion of practical activity, creation of a professional organization, development of an ethical code, banning of charlatans, and, briefly, the development of standard training criteria. The study concentrates on the first two points. Napoli is of the opinion that in the period under investigation there were no decisive methodological developments in psychology, so that it is superfluous to analyze the links between available knowledge, social demands, and the application of psychology. He can thus restrict himself to describing social needs, especially the effects of the First World War and the Depression and the demands arising from the Second World War. His central hypothesis is that psychologists made use of each situation for their professionalization. "Psychologists saw themselves travelling with the tides of history, and they realized that they were not merely fabricating requests for their services out of a desire to attain professional status" (Napoli 1975, p. 333). It is a pity that the author is content to depict the ebb and flow of requests for psychological services without providing some approach to a theoretical analysis.

an official state examination with a state examination commission, for example, for physicians, jurists, and teachers.

The academic professions in Germany are dependent on the state as guarantor of training and certification. The state does not, however, always turn to these professions for the optimal solution to social problems, at least not always to psychology; the positions of lawyers and physicians in their fields are certainly stronger. This makes it necessary to investigate the specific relationship between the initiatives of professional groups and the state.

Schmitz and Weingart have named four factors of professionalization that take into account both the knowledge of a discipline and state training policies:

(1) the development of systematic knowledge;
(2) the application of systematic knowledge as expressed in the demand for trained manpower;
(3) infrastructural policies by government; and
(4) status policies by occupational groups and governments. (n.d., p. 11)

An additional factor, in my opinion, would be the level of social institutionalization of a discipline at the training institutions, the universities, which is not dependent merely on the development of systematic knowledge.

The model I will use for the historical investigation of the professionalization of psychology takes these factors as a starting point. It should not be regarded as an invariant sequence of events.[41] Such an approach only leads one to view the history of a discipline in the course of professionalization as an immanent process of advancement of an entire scientific or occupational group. The points that I will outline here are to be understood, rather, as a heuristic model of those dimensions that should be considered in a historical investigation. They are not independent variables responsible for certain variances of professionalization; in reality the processes that are separated in the model are in fact linked.

I would like to add two further points to those already mentioned. First, it seems important to include one factor that hinders the process, namely, the rivalry of competing occupations or professions. Second, the professionalization of a discipline places subjective requirements on the members of the group and has subjective effects. These seven points will be considered in the following sections:

1. the institutionalization of the academic discipline as a social activity of research and training;
2. the development of independent, applicable knowledge within the scientific system;

41 See the criticism by Johnson (1972, p. 28) of Wilensky's model of five typical stages of professionalization (1972, p. 202). See also the general criticism by Kuklick (1980) of professionalization models in the U.S. social sciences.

3. the level of institutionalization of the application of knowledge and of the demand for its professional manpower;
4. the policies of the occupational group for establishing the recognition of the discipline;
5. the regulation of the qualifications and the state training policies;
6. the hindrance of professionalization by competitors; and
7. the subjective factors of professionalization.

The institutionalization of the discipline

Since the universities became the main institutions for the production and transmission of scientific knowledge in our society, this knowledge has become organized in academic disciplines. Such disciplines should not only provide the knowledge needed by a profession, but also function as loci of professional socialization, by teaching that knowledge to those who will later use it professionally. A professionalized discipline must therefore have a certain level of university institutionalization. Professionalization requires this as a precondition, but at the same time also influences it.

But when can we speak of a discipline? Does the founding of the first psychological institute mark the beginning, or are other characteristics also necessary, such as the establishment of a training system?[42] With Whitley (1974) we can distinguish the "cognitive" from the "social" institutionalization of a discipline. Cognitive institutionalization is the creation of an intellectual consensus among scientists regarding the problem area, the concepts and methods of the field, and the extent to which an individual scientist sees his or her research in the framework of this consensus. Social institutionalization refers to the degree of internal organization of a research field and the definition of belonging to it, as well as the degree of its integration in the governing system of legitimation and funding, that is, in the university. Whereas Whitley relates social institutionalization above all to the internal communication structure (e.g., societies, publications) and the external integration of fields of research, for our purpose the institutionalization of courses and examination opportunities are also important.

In the literature on professionalization the key importance of the universities is rightly emphasized (e.g., Jackson 1970, pp. 4–5; Parsons 1968). At times the concept of professionalization is equated with academic institutionalization, for example, in the study by Hardin (1977) on sociology in Germany and the United States, and in an essay by Camfield (1973) on psychology in the United States. The term "academic professionalization" found in the U.S. literature expresses more aptly the idea that academic institutionalization is only a part of professionalization, though an important one, if only because

---

42 See Clark (1972) and van den Daele and Weingart (1975).

the university provides the first labor market for the graduates of the new subject.

In opposition to Ben David and Collins (1966), Danziger (1979) argues that the fact that the professional-academic group had to legitimate its field of activity to those social groups that had a fairly direct influence on it – that is, groups with influence at the universities – was essential for the establishment of psychology as an academic discipline. In the United States, psychology had to convince university boards of trustees made up of businessmen and politicians. In Germany, where it was a latecomer to an established university system, it had to persuade leading members of the other disciplines in the philosophical (arts and sciences) faculties. Ash (1980, 1981) has analyzed the academic context of the emergence of experimental psychology in Germany, drawing attention particularly to the relationship between psychology and philosophy.

Within the existing "network" of disciplines (Spiegel-Rösing 1974, p. 18) the relationship to philosophy was decisive for psychology. It not only institutionalized itself largely via philosophy, but also became its major academic rival. In many countries, not only in Germany, the separation of psychology from philosophy meant its disciplinary emancipation.[43] The conflict between psychology and philosophy will therefore be considered in the investigation of the state of psychology's academic institutionalization. That institutionalization can be investigated in terms of criteria such as the number and definition of chairs, as well as the opportunities to graduate and obtain higher qualifications. For this investigation, social institutionalization is more important than cognitive institutionalization. The cognitive precondition for professionalization is the existence of a definite type of disciplinary knowledge, which will be considered in the next section.

The development of independent, applicable knowledge within the scientific system

To meet external demands a science must have a body of knowledge that is distinguishable from that of other sciences and that can lead to methods

43 In Australia, for example, there were initially "schools of psychology" within the philosophical institutes, and independent psychological institutes only from 1950 onward. Since 1923 there had been a joint *Australian Journal of Psychology and Philosophy*. Separate journals appeared in 1948–9. In Finland psychology broke away from philosophy in 1946 at the University of Helsinki. Siguán begins his review of the history of psychology in Spain thus: "In Spain, as in the rest of the Western world, psychology has traditionally been linked with philosophy, and a formal separation of the two has only occurred recently." In France Huteau and Roubertoux refer to the consolidation of the separation of psychology and philosophy as one of the most prominent features of the history of psychology over the previous twenty-five years. The same applies to Ireland, where the separation took the form of the establishment of a chair for applied psychology at the University of Cork, and to Norway. See the essays in Sexton and Misiak (1976) and also "Psychology in Europe" (1956).

of solving the problems with which it is confronted. Van den Daele and Weingart (1975) call this the "cognitive receptivity" of a science. Often, the social sciences can offer little established, practical knowledge, the technological success of which could be predicted from experimental testing: instead, they offer promises. The science on which these promises are based is not gained from a uniform research program, but represents a conglomeration of differing proposals, which may in turn help a nonparadigmatic science like psychology to adapt more readily to changing demands. Not all psychological theories and methods can be applied to practical problems.

The knowledge offered by psychology is not to be understood as the result of scientific progress alone, as implied in many histories of the subject. There is also a social component. This applies not only to the scientists as a social group who are the object of externalistic historiography, but also to the theories themselves, which are often discussed by internalistic historiography without consideration of this aspect (cf. Kuhn 1978). In a field stretching from psychophysiology of the sense organs through the formation of the character to totemism and mass psychology, the types and the extent of such social influences are certainly varied. Krohn (1976) sees no reliable general theoretical approach capable of resolving this internal–external debate and providing an account of the social conditions of scientific knowledge. For psychology, Jaeger and Staeuble have made a useful suggestion. They want to interpret the genesis of scientific concepts as themselves historically dependent on real societal problems. In their investigation of German psychology before Wundt they proceed from the assumption that the science of psychology took up those problems that arose from the discrepancy between objective social demands for action, and their inadequate realization "by empirical subjects" (1978, p. 50). The analysis of the content of psychology can therefore proceed from these problems.

For the history of science in general, Burrichter has suggested considering problems as social action situations with knowledge deficits "at the level of orientation or at the level of means" (1979, p. 8). A study of the professionalization of psychology is concerned with the aspects of the science relevant to the practical solution of socially defined problems – the question of means. The relationship between external requirements and internal methodological or theoretical developments will therefore be considered by examining the problem-solving knowledge provided by psychology and the problems that this could be used to solve. Was the problem-solving knowledge that furthered professionalization specific and new, or did professionalization advance by the extension of expert activities that were based on already existing knowledge? Since professionalized knowledge must be applicable in the same way by each expert, we must also ask to what extent it can be operationalized. This investigation will concentrate on the psychological knowledge available at the stage of its development when professionalization started. The focus will be on the general link between the alteration of solutions proposed in psychology and changes in the corresponding social problems.

Institutionalized professional roles and societal demand

A social science undergoing professionalization not only is linked to social problems through its theory, but also intervenes in these problems through its experts and their methods. There is a selection of the problems to be considered in the science. However, that selection is carried out not by the science itself, but by those bodies responsible for bringing in a science to solve social problems. The social problems selected then influence the development and structure of the science, which preserves the new areas of investigation as subdisciplines. As a result, problems are then perceived and classified in the categories of these subdisciplines, so that the science is slow to adapt to new problem areas.

One criterion advanced for the state of professionalization is the extent to which corresponding expert roles are institutionalized. The demand for the practical implementation of scientific knowledge, and thus the demand for experts, is expressed by the institutions of social life that deal with problems, such as schools, the army, and employment agencies. In their history of postwar West German psychology, Maikowski, Mattes, and Rott (1976) have classified such institutions as the intermediate levels for the processing of scientific knowledge in practical social contexts.[44] In their view, the way in which psychological problems are approached institutionally constitutes the discipline's conception of its subject matter. Psychology's problems do not arise from the relationship between a theoretician and the "person" in the abstract; they arise from the institutionally created relationship in which individuals confront the psychologist as worker, pupil, or patient to be selected, guided, or treated. Social problems are thus refracted by the demands of institutions, and become the problems of practical psychology. In addition to the social problems that a science is expected to deal with, it is therefore necessary to define the institutions that will support professional activity, their interests, their demands on the science, and their resources for making their tasks into science.

These institutions create a demand for psychologists on the labor market. Alterations in their needs and the reasons for their demands are thus important dimensions of professionalization. In public institutions in Germany, the sign of a legal establishment is the setting up of a career profession (*Laufbahnberuf*). Setting up a public career path requires the prior establishment of training programs; conversely, those who are trained aim to have their occupation recognized as a career profession. This point will also have to be considered in the investigation of the demand for professionals in psychology.

---

44 The authors do not extend their analysis of these institutions to include the university.

The policies of the occupational group to achieve professional recognition

The literature on professionalization places great emphasis on the role of the occuptional group and its organizations, which is hardly surprising since many consider the activities of these groups to be the driving force of the professionalization process. The character of the free professions has stirred memories of the medieval guilds and led some sociologists to refer to the guild character of the professions (Goode 1964). In the professionalization of academic disciplines, the academic group precedes the occupational group. As research became a matter for groups or large organizations, literature on the history of science paid more attention to the significance of such academic groups for the development of science. Kuhn (1962) recategorized the history of science in terms of paradigms, which function as normative regulators of these groups. According to Kuhn, scientific communities operate within a network of conceptual, theoretical, methodological, and instrumental commitments. They then form a group with common intellectual interests through which they can develop a link between social problems and scientific concepts, as was mentioned in the second section of this model (see Danziger 1979). Since they have their own disciplinary interests, the scientists also make a selection from among these problems. [45]

The interests and politics of psychologists and their organizations had a profound influence on the professionalization of psychology. In the Third Reich the politics of psychology were largely organized by the Executive Committee of the DGfPs. The analysis will therefore concentrate on this body. But a scientific community is never a monolith. Groups and factions are always competing with each other, and not only at a scientific level (cf. Thomae 1977). What internal contradictions existed in relation to the institutionalization and professionalization of psychology, and what were the common interests?

Scientific communities and groups meet with success only when they can present the appropriate scientific and practical accomplishments to those bodies that control the resources for the discipline. This seems to me to be particularly true when a discipline tries to establish itself as a profession, with corresponding demands for improving its academic status. The first addressee is the university. In Germany by the end of the preceding century, the universities had become large-scale, state-regulated "factories" (Busch 1959). The scholar, who could work at the university or elsewhere, was replaced by the scientist who had taken up an academic "career." Professors became civil servants (*Staatsbeamte*), whose appointment was the responsibility of the education minister, although he generally followed the faculty's

---

45 According to Böhme et al. (1974, 1978) the choice of problems occurs according to different criteria in the various phases of scientific development – the explorative, the paradigmatic, and the postparadigmatic. The disciplinary aspect dominates in the paradigmatic phase.

recommendation. The relevance of a science had to be demonstrated to the bureaucracy if a new institute or chair was to be set up, new staff employed, or a library or laboratory replenished. The faculty, which according to Danziger (1979) played the main role in granting legitimation, remained the negotiating partner for such applications, but these had to be passed to the ministry by way of the dean and the rector.

In general the problem of legitimation is posed at a new level when the science is used for the solution of practical social problems (see Weingart 1976, pp. 216–17). In addition to the criterion of scientific relevance there is the criterion of social relevance. Since nonscientific methods can often replace social sciences for practical applications, and the latter are often confronted with their lack of success, which they continually cover up with promises, they are under pressure to demonstrate their usefulness. If a social science discipline tries to gain state recognition as a training discipline for an extramural profession, then it has to demonstrate its relevance and competence. Scientific administrators are then no longer the only ones who have to be convinced. And if the discipline enters the market, it also has to prove itself in the face of competing professions.

This investigation will therefore consider whether there was a transition from the scientific legitimation of psychology in the universities to a practical legitimation directed more toward bodies outside the universities; it will also consider which bodies these were. These aspects can be studied from the arguments advanced in favor of professorial appointments or research projects or in project reports. Published literature will also be examined in an attempt to determine whether it expresses legitimation strategies. Such strategies will be interpreted as elements of the politics of the entire professional group, or of certain group members.

Certification and state training policies

Scientific products are assessed and used primarily by scientists themselves. Quality control can therefore be carried out informally by members of the scientific community. Professional activities, however, are purchased by clients or institutions, who should be able to trust the competence of a professional without reservation. This makes it necessary to have formal control over competence. State systems of certification or licensing protect clients against charlatans and protect professionals against doubts concerning their competence (Hughes 1952).

If education and the professional application of the knowledge acquired are separated from each other, there is a general problem of proving qualifications. In the professions, examinations are supposed to bridge the gap. They certify qualification in a scientific subject, but not the qualifications for a certain professional activity. At the same time they assign a certain social status and define the membership of a professional group. In the career system of the German civil service, they are the formal requirement for certain posts. The examination syllabus can also give some idea of the minimum qualifications for the profession. The introduction of the diploma examination for psychology in 1941 provides material documenting views on professional qualification at that time.

Since in Germany the educational system is subject to state control and since graduates of certain courses were employed by the state as civil servants, for whose training the state had a monopoly, the courses linked to such occupations were regulated by state examinations or by university examination regulations requiring state approval. Such state-regulated courses were then regarded as guaranteeing a certain level of competence. The state came in when training was no longer solely in the science as such, but also for a profession. The classical university examinations, the doctorate (*Promotion*) and the master of arts, whose requirements were determined internally, were then replaced by the state examination or diploma. In contrast to university examinations, state examinations entitle their holders to certain categories of office. The principle of the qualifying examination was introduced in Prussia in 1794. The first examinations were for theologians and lawyers, followed by physicians, schoolteachers, and technicians (Jastrow 1930). The first diploma courses were in the field of higher technical training (see, e.g., I B Ebert 1978). Chemistry was the first university subject for which the replacement of the doctorate as a qualification was discussed (see Burchardt 1980; Meyer-Thurow 1980).

Establishing that state-regulated examinations were instruments for guaranteeing a minimum level of qualification does not help to explain why a certain examination was established at a certain time. For psychology we must ask why the science bureaucracy set up the DPO in 1941, why the need for a unification of training arose under these specific conditions, how it expressed itself, and who supported it. A further question is how it affected the standing of the subject relative to other disciplines once it had been recognized as preparation for a profession.

These last two points of our model raise questions that are largely absent from the literature on the history of psychology. The only similar approach I have found is that of Samelson (1979), who has investigated the relationship between psychology and the armed forces in the United States and the extent to which they have benefited each other.[46] The investigation of professionalization thus introduces another dimension to the historiography of psychology; its two poles are marked, on the one hand, by the psychologists as a professional group and, on the other, by those state bodies chiefly responsible for the recognition of a profession.

46 Samelson (1979) distinguishes between the technical efficacy of psychology and its political promises, coming to the conclusion that it was above all the latter which advanced the development of psychology as a profession in the United States. This brought it recognition, although its effectiveness could scarcely be proved.

Professional rivalries

In addition to alterations in the network of academic disciplines, profession-alization leads to alterations in the network of occupational groups. At the same time, the relationship among these groups and their efforts to establish boundaries between their areas of activity are themselves an important factor in deciding when and to what extent a discipline can professionalize. Goode (1960), for example, has described how psychologists, sociologists, and physicians quarrel about areas of competence and shares in the market. If problem areas that had previously been claimed by other disciplines are now transferred to a new profession, then the representatives of these disciplines criticize the competence of that profession. When problems that had previously been solved nonscientifically are dealt with by scientifically trained profes-sionals, this will lead to the so-called lay practitioners being edged out and to a campaign by the profession against quackery.

As Wilensky (1972, p. 204) has described, the social sciences that deal with the problems of everyday life face the dilemma that lay people find it hard to see the need for any special competence to solve questions in which everyone is an "expert." This is a criticism that psychology has continually faced when trying to establish itself as a profession, both from lay people and from other professionals. With its special scientific methods it encroaches on social problems that have hitherto been treated by others. Professional psychology replaces the personal experience or the unsystematic knowledge of priests, doctors, officers, managers, parents, mothers-in-law, old people, neighbors, and friends with the expertise of a professional body. It is therefore to be expected that psychology would defend itself against amateurs and quacks, and that the skepticism and rivalry of established professional groups such as engineers, physicians, and officers would be a hindrance to the professionalization of psychology. The conflict between these professional groups and psychology is therefore a further point for investigation.

The subjective factors of professionalization

The final point is one that receives little attention in sociological consider-ations of professionalization, but that seems to me to be important, not least for self-reflection in the social sciences, namely, the psychological prerequi-sites and effects of professionalization. The use of the terms "professionals," "members of a discipline," and "academic group" suggests uniform interests or actions. Goode, who speaks of the development of a "guild" in clinical psy-chology, cites as the first feature of such a development a common occupational identity and common values (1960, 1972, p. 157). Does this not presuppose a developed profession along the lines of medicine or law?

Perhaps in a profession such as psychology a comparable homogenous self-image has not yet properly developed. In Germany the professional group remained a small one until the sixties, with only a few hundred members, and between the world wars its first members were just beginning to undertake extramural occupational activities. But even at such a stage, the development of a profession may be dependent on the subjective state of its members. The treatment of this point is necessarily preliminary, thus less complete and more hypothetical than that of the others discussed.

The possibility of putting one's science into practice may be perceived subjectively as an intellectual stimulus, thus masking the institutional and political context of its use. This would ease the opening of the science to anyone who expressed an interest in its practical services. Perhaps new opportunities for practically applying one's science create new forms of loyalty toward those institutions that provide the science with space for its professionalization. Or alternatively, perhaps the scientific interests of the members of a discipline actually restrict the discipline from addressing practical matters.

An academic discipline produces graduates who are then forced to sell their specific scientific qualifications. They therefore have a personal interest in the institutionalized application of their science. Furthermore, progress in professionalization increases the prestige of the discipline, and thus also of each of its members. It is therefore to be supposed that professionalization is also subjectively perceived as an important event, which would then contribute to the development or establishment of a guild mentality. Perhaps the possibility of applying psychology practically and its establishment at the universities led to a change of attitudes that needs to be considered very critically in the light of the period concerned. For this final aspect of the model a psychological approach may prove valuable.

The problem of sources

Works on the history of scientific disciplines are often profoundly influenced in both their intention and their methodology by the background of the author. Members of a discipline usually write with the intention of providing legitimation or an identity for the discipline or subdisciplines within it; the history of the historiography of psychology shows this.[47] Psychologists are only too keen to turn to the history of psychological theories. When they go beyond this, they tend to apply their own methods, rather than those of history. This can be clearly seen in the emphasis on the biographical approach in the historiography of psychology; a recent social history of psychology is written in terms of social psychology, as a history of groups and group processes (Thomae 1977). However, when confronted with problems such as the relationship between the politics of an academic group and the politics

47 Lepenies (1978). Graham et al. (1983), Ash (1983), and Geuter (1983).

of the state, the use of historical knowledge and methods becomes essential.[48] This is not to say that the investigation should be guided solely by the disciplinary considerations of historians. Whereas the members of the discipline are interested primarily in the history of their theories, so we find that when historians tackle the history of science in the Third Reich they tend to concentrate on the external political history of subjects or on the political behavior of leading representatives.[49] Historians of science also have a typical approach, influenced by the problems and controversies of their discipline, concentrating, for example, on the question of the relationship between internal and external factors in the development of a given science.

These various approaches are reflected in the different handling of source material. Histories of theory are usually based solely on analyses of the scientific literature. A history of professionalization cannot restrict itself to this material, because some of the relevant motives and facts leave at best only indirect traces in the scientific literature. It is therefore necessary to draw on unpublished sources, archival material, and the memories of contemporary psychologists. The task of a critique of these sources is to recognize the intentions of the author of a document in the light of the knowledge of overall developments and to evaluate the significance of a source for a certain question. Details on the sources used in the present study are given in the comments on sources.

It should be borne in mind that, in the period considered, articles aimed at impressing the state and the party were placed strategically; they are not always to be found in the psychological journals. Furthermore, the alteration in the structure of scientific work in turn altered the structure of publication. The institutional utilization of research in the Wehrmacht or the Labor Front meant that results were often published internally, or in house journals such as *Wehrpsychologische Mitteilungen*. Moreover in studying the Third

---

48 See here primarily the previous research on science in the Third Reich. On the whole the level of this research is relatively low – compared, for example, with the field of foreign policy (see Kent 1979, pp. 49, 66). There is no study of the policies of the Ministry of Science (RMWEV); as far as university politics is concerned there are some descriptions from a personal point of view and systematic treatments in the discussion on the university and National Socialism from the sixties; see Flitner (1965), (*Die*) *Deutsche Universität*...(1966), and *Nationalsozialismus und deutsche Universität* (1966). Empirical studies on the politics of science are rare. Hartshorne (1937) is still an important source; Seier (1964) investigates the Nazi restructuring of the universities, Kelly (1973) Nazi attempts to bring professors into line politically. For a number of recent smaller studies, see Heinemann (1980). On specific universities Adam's study of Tübingen (1977) deserves special mention. The most extensive bibliography, though concentrating on natural sciences, can be found in Mehrtens (1980); that of Hüttenberger (1980) is most imcomplete with regard to universities and is strangely selective. Studies in other disciplines have already been mentioned; see especially Beyerchen (1977), Cocks (1985), Klingemann (1984), Losemann (1977), and the essays in Mehrtens and Richter (1980), including those on mathematics, physics, and chemistry.

49 See, e.g., the studies by Losemann (1977) on archaeology and by Beyerchen (1977) on physics.

Reich we come across the common problem of trying to trace controversies that are either concealed or only alluded to in the scientific literature – except that here it is particularly acute. Under different circumstances a word or argument has different meanings. In the Third Reich, there was not only the usual sort of concealment, but also forced hypocrisy, or a subtle criticism cloaked in general approval, at least among those who did not kowtow to the new age out of political conviction. A text can be interpreted incorrectly unless detailed knowledge of the circumstances is drawn on to deduce the unnamed adversaries or the undisclosed aims. Finally, when investigating the professionalization of psychology the general contradiction between the issues considered in scientific publications and the problems of professional psychology must be taken into account. The representation of social problems in the literature does not correspond to their representation in the relevant professional activity. Social problems may give rise to scientific investigation as well as to institutional treatment, but by no means simultaneously or in the same sense. It is possible to come closer to the psychology actually practiced if internal records from military and industrial psychology institutions, such as instructions in the use of aptitude tests, are also consulted. Such documents are often available only from private sources.

I shall look in more detail at the value of archival records and interviews for the history of psychology, since the views on this issue clearly demonstrate the link between disciplinary research interests and the treatment of sources. The evaluation of archival records, a natural source for the general historian, is not always an obvious option for historians of academic disciplines. In articles by representatives of individual sciences on their history in the Third Reich, which are concerned more with the theoretical history of the subjects and their relationship to Nazi ideology, we find hardly any archival material.[50] However, empirical studies by historians on science in the Third Reich make use of records without exception.[51]

In the United States, historians of psychology have recently made more use of archival material and have also expressed views on the value of such material. Psychologists have hitherto concentrated on factors that were important for the development of the scientific works of certain authors, on the interaction between scientists as disclosed in their correspondence, on the reconstruction of unpublished scientific material, and on the augmentation of biographies of psychologists, with their links to scientific work.[52] Historians are more interested in other questions, such as using correspondence to study how scientists sought and found a forum for their ideas (Ash

50 This is true for the studies of Bechstedt (1980) on chemistry, Bergmann et al. (1981) on sociology, Güse and Schmacke (1976) on psychiatry, Lindner (1980) on mathematics, and Richter (1980) on physics.
51 See Beyerchen (1977) and Bollmus (1980) on the project of a party college, as well as Cocks (1985), Losemann (1977), Ludwig (1979), among others.
52 Cadwallader (1975), Bringmann (1975), Bringmann and Ungerer (1980), and Brožek (1975).

1980a), or what processes within a profession are to be found in the course of professionalization (Napoli 1975).

For my investigation it was above all necessary to draw on the files of the ministerial bureaucracy, Wehrmacht Psychology, and the German Society for Psychology in order to understand problems of academic institutionalization (especially new professorships), the background to the new diploma examination, the arguments presented to the Wehrmacht in support of practical psychological work, the internal politics of the society, and so on. Ministries and armies do not present their motives in the scientific literature, but this literature can sometimes appear in a different light when read against a background of unpublished strategies. Finally, the records, especially the files of the Nazi Party's ideological office, the Amt Rosenberg, provide an insight into the attitudes of the Nazis toward psychology.

However, the value of files is also limited. It often gives decisions, but no motives (Bollmus 1973, foreword). A faculty or a rector may give motives for an appointment to a chair, but these are not necessarily the real motives of the writer. An argument may be advanced in a letter solely to win over the recipient. Reports of Wehrmacht Psychology are not only reports on the reality of the work there, but also documents aimed at establishing support for that work. Intentions intervene as a filter between reality and document. Records can be a more important source for studying legitimation strategies than for studying the actual development of psychological work. Therefore, the interpretation is just as dependent on context in the case of files as in the case of scientific literature (cf. Bringmann 1975, p. 24).

A further source is the memory of psychologists active at that time. Therefore, autobiographies and interviews were used. The method of interviewing is finding increasing use in U.S. research on the history of psychology, but as yet we find little systematic consideration of the type of historical information the interviews represent. Usually autobiographies and interviews are seen as providing additional information about the work of individual scientists. Oral history mostly means that reminiscences of prominent psychologists are taped and stored in some collection (see Berman 1967; McPherson 1975; Myers 1975). In Germany, Pongratz, Traxel, and Wehner (1972, 1979) asked for written autobiographical reports.

If psychologists tend to treat subjective reports as reality, historians tend to underestimate their value and to believe only the records. Some historians do use interviews in their research on science in the Third Reich (e.g., Beyerchen 1977; Cocks 1985; Heiber 1966), but Adams (1977) makes scant use of them, and Losemann does not use them at all. He argues that the selection of interviewees occurs against the will of the authors and infringes on the rights of those "who do not want to face an interview, or are no longer able to" (1977, p. 17). But the fact that they cannot be comprehensive is surely at most a reason to view the information gathered critically, but not a reason for doing without this source altogether.

Interviews offer a certain type of subjective information that cannot be obtained from so-called objective sources. They give an impression of how those interviewed experienced the period and their professional activity, including a few old squabbles. They often include facts that cannot be found in the files. They also train the eye of an observer approaching the Third Reich forty years later with a head full of hypotheses. When one views this period retrospectively, it seems almost as if the tendency is to overlook numerous little details, whereas those who lived through it all seem regularly to lose sight of the overall context in a welter of petty incidents.

In the course of my investigations, forty-two people were interviewed and a number of written communications were gathered. The interviews were of differing relevance to this investigation; for instance, those with emigrants were used very little. The attempt was made to talk with psychologists from universities, the army, and industry. The interviewees included the last surviving member of the commission to draw up the regulations for the diploma examination, Carl Alexander Roos, all the survivors from the headquarters of Wehrmacht Psychology, called the "Inspectorate for Aptitude Testing,"[53] Heinz Dirks, Gotthilf Flik, Leonhard von Renthe-Fink, and Josef Schänzle, and also two of the four survivors of the Institute for Work Psychology and Work Pedagogy at the Office for Vocational Training and Works Management of the German Labor Front – Elisabeth Lucker and C. A. Roos – as well as a former head of this office, Albert Bremhorst. Two of the surviving five occupants of chairs in 1945 were also interviewed – Wolfgang Metzger and Johannes Rudert. Thirty-nine transcripts of these structured interviews, mostly signed by those interviewed if they were still alive, are deposited in the archives of the Institut für Zeitgeschichte in Munich. A list of these appears in the sources at the end of the book. In some cases, it was not possible to agree on a transcript. The older colleagues were most willing to speak of their memories. Two psychologists refused, one because he had been interviewed frequently, the other without giving a reason; two did not respond at all. A more extensive description of the material is included in the collection of transcripts (see Comments on Sources).

Perhaps due to my background in psychology, I began the interviews before studying the historical records. It is interesting to observe how in the

53 In the present book, this is often simply referred to as the "Inspectorate for Aptitude Testing" when used generally. In fact it had this name only from 1939 to 1941, at the height of Wehrmacht Psychology. At first it was called the "Psychotechnisches Laboratorium" of the Army Ministry, then the "Psychologisches Laboratorium" of the War Ministry, then from June 1938 after the dissolution of the ministry, the "Hauptstelle der Wehrmacht für Psychologie und Rassenkunde." It was the Inspectorate from 23 August 1939 under the head of Army Personnel. At this time the Luftwaffe Psychology became independent. On 1 January 1941 the army psychologists became "personnel evaluators" and on 27 August the Inspectorate became the "Inspectorate for Personnel Evaluation," until it was dissolved on 30 June 1942.

course of two years my questions grew more precise. It is also interesting to see how the interviewees responded by taking up sensitive issues to a greater or lesser degree. A personal aspect is also of methodological relevance. Sitting opposite me were living people, usually very friendly, whose occupational activities under the Nazis represented a large part of the reality of the profession of psychology. It would have been easier to criticize them without knowing them, just as Wilhelm Wundt did not go to congresses in order to avoid personal contact with his adversaries, which might blunt his polemics against them. Once I got to know them, I found myself wondering why someone behaved in a certain way, for example, going to the Wehrmacht Psychology to work as a psychologist, and I felt obliged to show understanding. It is possible to use this situation to create a more realistic description and more appropriate, credible criticism, without excluding the self-perception of the interviewees. But it can also lead to deception. Much sounded plausible, and I cannot exclude the possibility that one thing seemed right rather than another simply because it was put across in a better way (see Heiber 1966, p. 11). Just as with the scientific literature and the archival records, the information taken from the interviews must be critically evaluated, and the subjective reports must be interpreted in their historical context, but also accepted as experienced history. Oral history, as it is, is not an adequate background for a history of psychology, but it represents an important source for this study, not least for investigating the subjective factors of professionalization.

One might consider including a further source by consulting those who were tested by the psychologists – a history "from below." It would be important to know how these people saw the psychological examinations and how they talked about them, in order to get an idea of the general climate in which the professionalization of psychology took place. However, methodologically, this is a virtually insoluble problem. It is hard to come to terms with empirically, and since it is not decisive for the issue of professionalization I have not included it. Anyway, my conclusions are only the result of an initial interpretation of the material examined, which may need to be corrected in the light of additional material.

The next five chapters correspond to the first five points of the model described, that is, the institutionalization of psychology at the universities (Chapter 2), the state of practically applicable psychological knowledge in the 1930s (Chapter 3), the development of professional applications (Chapter 4), the politics of the psychologists to gain recognition (Chapter 5), and the establishment of uniform state qualifications (Chapter 6). The Diploma Examination Regulations of 1941 and the disbanding of the army and air force psychology units are so important that the following two chapters are devoted to them. The penultimate chapter deals with the subjective aspects of professionalization, the last point of the model, and in the final chapter I discuss some consequences of the Nazi period for psychology.

2

On the way to becoming an independent discipline: the institutionalization of psychology in the universities to 1941

In the history of the academic institutionalization of psychology in Germany the National Socialist period is part of the phase of its establishment as an independent discipline. In Germany this was largely a matter of setting up new professorships in the universities. These usually bring institutes, personnel, budgets, libraries, and so forth in their wake. This chapter therefore examines the extent to which psychology gained or lost chairs during the Third Reich. This quantitative question is accompanied by a qualitative one: What reasons were there for the loss or creation of chairs in psychology, and what teaching and research contents were to be institutionalized? This question of the content of institutionalization will also be pursued by studying the appointments to chairs; this will act as a gauge of the development of demands placed on the subject and the orientation of the subject itself. While the investigation concentrates on universities, the technical colleges (*Technische Hochschulen*) are also considered. The representation of psychology in pedagogical academies and other teacher training institutes will be considered in Chapter 4 on the development of occupational fields for the profession.

Before the introduction of the Diploma Examination Regulations, or DPO (Diplom-Prüfungsordnung), in 1941, each newly created chair of psychology had to be justified for that particular university, which led on occasion to severe disputes. With the DPO, however, the need for new chairs resulted from a state regulation, since it was possible to train psychologists only at universities with a chair. This created a fundamentally new situation for the academic institutionalization of psychology, which I will consider after the treatment of the DPO in Chapter 7. However, for reasons of clarity, the quantitative considerations in this chapter include the period up to 1945.

Efforts to achieve the academic institutionalization of psychology are nearly as old as those to establish psychology as a science in its own right, and indeed this institutionalization was necessary if independent research and teaching were to be financially possible. I will therefore consider the period before 1933 in some detail and outline how a group of scientific psychologists had formed by that time, which members of the group held chairs, the efforts this group made to establish the new subject in the universities, and the successes and failures they experienced.

Psychologists as an academic group: the formation of a scientific community

Even before the Nazi period, a recognizable group of academic psychologists had formed. It would be virtually impossible to define the membership of this group on the basis of mutually accepted theoretical fundamentals or a common research program. Karl Bühler (1926), for example, began his book, *The Crisis of Psychology*, as follows: "So many psychologists side by side as there are today, so many individual approaches have never been together before." According to Woldemar Oskar Döring, author of *Main Trends in Recent Psychology*, anyone "demanding of a science that it present a unified and homogeneous image of itself would be disappointed by recent psychology" (1932, p. 7). The earlier, topical textbook-style accounts of psychology were superseded in the twenties by introductions presenting the reader above all with the variety of psychological schools and their numerous areas of application.[1] Nevertheless, there was a feeling of belonging to a group of researching psychologists. Perhaps this can roughly be characterized as the feeling that the problems of consciousness (*Bewusstsein*) or of mental life (*Seelenleben*) were to be solved by empirical methods. Expressed negatively, psychologists were united in their rejection of philosophical speculation.

Things are simpler and less ambiguous when one describes the group of academic psychologists from a sociological point of view. In the twenties, a common communication network had developed through scientific journals and congresses. With the Society for Experimental Psychology, which was to change its name in 1929 to the German Society for Psychology, a unified representative body had been created for scientific psychologists, which also formulated the academic interests of the group for outsiders (Tables 1 and 2). Whereas early journals had reflected the connections between psychology, philosophy, and physiology, the journals founded in the twenties were more identifiably psychological publications. Also significant is that in this period journals appeared that dealt with applications of psychology. The Society for Experimental Psychology was founded in 1904. Since at that time there were no operational criteria for defining psychologists, such as the scientific dissertation or later the diploma, the executive committee was left to decide

---

1 Most of these books are listed in Pongratz (1980, p. 83). To be added are Messer (1927) and Saupe (1928).

Table 1. *Founding of psychological journals and series*

1881	*Philosophische Studien* (from 1904 on, *Psychologische Studien*)
1890	*Zeitschrift für Psychologie und Physiologie der Sinnesorgane* (later, *Zeitschrift für Psychologie*)
1899	*Zeitschrift für pädagogische Psychologie* (five name changes; finally from 1933, *Zeitschrift für pädagogische Psychologie und Jugendkunde*)
1903	*Archiv für die gesamte Psychologie*
1903	*Beiträge zur Psychologie der Aussage,* until 1906
1908	*Zeitschrift für angewandte Psychologie und psychologische Sammelforschung* (continuation of *Beiträge zur Psychologie der Aussage*)
1910	*Untersuchungen zur Psychologie und Philosophie* (from 1925, *– und Pädagogik*) (series)
1912	*Fortschritte der Psychologie und ihrer Anwendungen*
1915	*Arbeiten zur Entwicklungspsychologie* (series)
1916	*Deutsche Psychologie: Zeitschrift für reine und angewandte Seelenkunde*
1918	*Schriften zur Psychologie der Berufseignung und des Wirtschaftslebens* (series)
1919	*Praktische Psychologie*
1921	*Psychologische Forschung*
1922	*Psychotechnische Rundschau*
1924	*Jahrbuch der Charakterologie*
1924	*Industrielle Psychotechnik*
1925	*Psychotechnische Zeitschrift*
1926	*Neue psychologische Studien* (series)
1931	*Vierteljahresschrift für Jugendkunde*

Table 2. *Membership of the Society for Experimental Psychology, 1904–45 (from 1929 on, the German Society for Psychology)*

Year/month	Membership	Year/month	Membership
1904/10	85	1927/4	247
1906/12	112	1930/4	278
1908/4	128	1932/1	339
1910/4	147	1934/1	288
1912/4	177	1936/8	231
1914/4	193	1937/4	276
1921/4	200	1941/12	Increase; no figures
1923/4	215	1943/1	Increase; no figures
1925/4	209		

Source: Congress reports; collated by Wolfgang Michaelis with additions by myself. For 1936, 1941, and 1943 in accordance with UAT 148. For 1943, and similarly 1941, reference is made in a newsletter to an encouraging upswing in membership. See also Geuter (1990) and Traxel (1983).

whether someone had shown himself to be a scientific psychologist. Section 2 of the statutes stated that a person "who has published a work of scientific value in the field of psychology or bordering areas" could become a member (Schumann 1904, p. xxi). At its founding, the society had eighty-five members, a number that was to grow continually over the following years. Perhaps more important than this quantitative growth was the fact that the composition of the membership changed too. Looking through the membership lists, which were published every two years in the congress reports, one gets the impression that the proportion of members from other disciplines or occupations declined steadily. At the same time, the proportion of active practical psychologists increased. In the opinion of Traxel, by 1939 the society had become "much more a professional association" (1983, p. 98). Every two years, except for disturbances caused by the First World War and economic crisis, the society held a scientific congress. Possibly the clearest indication that academic psychologists at the end of the twenties can be regarded as a relatively uniform academic group is the fact that they were active in the politics of their subject, presented themselves to others as psychologists, and demanded the academic representation of psychology in the universities as a subject in its own right – steps that established their separation from other academic groups.

Psychologists with chairs and psychological chairs: comments on terminology

Which chairholders belonged to this group? At the start of the century, there was no professorship defined as being "for psychology" (this first was in Jena in 1923); nor was it clear who could be regarded as a psychologist. If the institutionalization of psychology is to be examined in terms of the academic chairs, it is important to establish who can be counted as a psychologist at the university and what a chair for psychology is. The bureaucracy provides no clear criteria here. In January 1933, for example, full professors researching psychological questions and teaching the subject held chairs designated as follows:

Philosophy: Köhler, Berlin; Rothacker, Bonn; Stern, Hamburg; Krueger, Leipzig; Jaensch, Marburg
Philosophy, especially psycholoy: Wertheimer, Frankfurt; Gelb, Halle
Philosophy and psychology: Ach, Göttingen
Philosophy and pedagogy: Kafka, Dresden Technical College
Philosophy, aesthetics, and pedagogy: Marbe, Würzburg
Philosophy, pedagogy, and experimental psychology: Schultze, Königsberg
Pedagogy and experimental psychology: Katz, Rostock
Psychology: Herwig, Braunschweig Technical College; Peters, Jena
Pedagogy: Fischer, Munich
Educational sciences: Kroh, Tübingen

These professors did not share a common administrative designation of their activity, but they were linked by their psychological research and teaching and their efforts within the group of academic psychologists. They all spoke at its congresses, published in its journals, and were members of its Society at the beginning of 1933; and some were very active in its cause. Not least, contributions to psychology were expected of them. However, they frequently saw themselves not just as psychologists but, as was usual at that time, also as members of the communities of pedagogues and philosophers, which were not yet rigidly separated from that of psychologists.[2] As long as psychology was not an independent discipline, a full professor would also be expected to accept responsibility for philosophy or pedagogy, or even both. However, the criteria already listed seem a sensible starting point for deciding which chairs to include in this investigation.[3] This is not to say that the administrative designation is unimportant, but it is more a reflection of the gradual separation of psychology, philosophy, and pedagogy. For a philosophy chair it was possible to wish for a psychologist interested in philosophy, as the Königsberg faculty did when a successor to Narziss Ach had to be found in 1922; but one might end up with a philosopher interested in psychology, as happened in Bonn in 1929 with the successor to Gustav Störring. When a chair is explicitly designated for psychology, there is no longer a tug-of-war between the disciplines. This makes it important to trace how the definition of philosophical chairs gradually shifted in favor of psychology. The specification of a chair for psychology can be taken as a measure of institutionalization.

One further clarification is necessary: What is a chair? In addition to the full professors, there were honorary professors and budgeted posts or ad personam appointments for associate professors, who might be civil servants or not. Then there were instructors (*Dozenten*), who were paid at first from student lecture fees and were later civil servants. The distinction seems to be most appropriate between professors with budgeted posts, who were

2 This means that in a different context Rothacker or Krueger could also be classed as philosophers, and Kroh and Aloys Fischer as pedagogues.
3 The statistics on university teachers compiled by von Ferber (1959), which are often quoted, cannot be used for psychology. Von Ferber distinguishes between the subjects of philosophy (including philosophy and psychology or pedagogy), pedagogy, and psychology. But he arrives at figures very different from mine. For psychology he gives the following figures for 1931 and 1938 (after which there are no figures):

	1931	1938
Full professors	5	2
Associate professors (budgeted)	5	4

The figures for associate professors are almost comprehensible. They could refer to Wirth, Klemm, and Volkelt in Leipzig and Sander in Giessen. The latter post then became a full chair in 1934; an additional budgeted associate professorship was added in 1935 in Kiel. The numbers of full professorships is most problematical, however. For psychology alone there was in fact only one in 1931 (Jena) and two in 1938 (Jena and Halle), in addition to one at a technical college (Braunschweig). Cf. my statistics in Table 5.

appointed to a specified field and would have a successor in that post, and the ad personam professor, where the professorship ceased when the holder retired. All full professorships were budgeted; I also intend to count budgeted associate professorships as chairs of psychology.[4]

The institutionalization of psychology in universities and technical colleges before 1933

German psychology began as a subdiscipline of philosophy. Its institutional basis until World War I was philosophical chairs (Ash 1980, 1982). In 1892, there were thirty-nine professorships of philosophy, three of which were occupied by psychologists; in 1914, there were ten psychologists out of forty-four philosophy professors in all. But psychology could not develop freely within the framework of philosophy. A precondition for calling a psychological researcher to a chair for philosophy was the existence of at least two such chairs at the university in question, since the faculty gave priority to covering historical and systematic philosophy. The development of a new discipline within an existing one was, however, only one method of institutional establishment, and not necessarily the most usual one. Another possibility was for the new discipline to create an extraordinary or associate professorship, which would then later be transformed into a full professorship. In 1910, Prussian Minister of Science Trott zu Solz described this as the strategy of his ministry in a speech to the Prussian House of Representatives (Busch 1959, p. 111; cf. Ringer 1969, p. 53).

For psychology, the path via philosophical chairs seemed initially to be the most promising. Given the prospect that their new empirical methods could provide answers to philosophical, especially epistemological, questions, the philosophy faculties seemed to be the right place. However, conflict with traditional philosophy was inevitable. There were only a limited number of chairs in philosophy, so every additional psychologist meant one less classical philosopher. The conflict flared up when psychologist Erich Jaensch was called to the philosophical chair of Hermann Cohen in 1912 in Marburg, against the wishes of the local philosopher Paul Natorp. A total of 107 philosophers sent a declaration to university faculties and all education ministers in which they criticized the "serious damage" caused to philosophy by the "withdrawal of chairs dedicated solely to it"; they demanded that in the future experimental psychologists have their own chairs "and where old chairs of philosophy are occupied [by psychologists]... to see to it that new chairs of philosophy are created."[5] The academic institutionalization of

4 This corresponds to the use of the term *Ordinarien* by von Ferber (1959, p. 45). It also corresponds to the contemporary legal definition of a university teacher. The regulations of 10 June 1939 (RGBl 1939, I, pp. 1010–11) defined these as *"die beamteten ordentlichen* [= full] *und ausserordentlichen* [= associate]" professors of universities and technical colleges.
5 "Erklärung," in *Zeitschrift für Philosophie und philosophische Kritik*, 151 (1913), pp. 233–4. See Ash (1980).

psychology had clearly developed into a political question. Traditionally, the ministers responsible for the appointment of professors followed the recommendation of the faculty, which presented a short list of three names. In Marburg, the minister had solved the dilemma of not wishing to create a new chair in psychology, while at the same time wishing to see psychology represented, by calling a philosophically interested psychologist, Jaensch.

Thus, when investigating the professionalization of psychology, we are confronted from the start with the problem of the relationship between science and the state. Long before developing fields of professional activity and establishing methods of training, at the very first stage of academic institutionalization it was necessary for the subject's representatives to demonstrate to state officials that the discipline was useful and autonomous. Of course, recognition first had to be gained within the philosophical faculties, but when it came to budgeting or creating professorships or readerships, it was the ministry that had to be convinced that the expenditure was necessary and appropriate. For psychology, this meant that it not only had to prove its value for philosophy, but also had to demonstrate its value alongside and distinct from philosophy. This was already clear at the Fifth Congress of the Society for Experimental Psychology in Berlin, in 1912, where various speakers demanded wider representation of psychology in the universities, and the mayor of Berlin clearly expressed the expectation that psychology present usable results (Goldschmidt 1912). In the 1912–13 debate about where psychology should find a place at the university (involving such leading psychologists as Wundt, Stumpf, Ebbinghaus, Müller, Külpe, and Schumann), it was still decided that it should stay in philosophy, not least for tactical reasons. This secured the existence of psychology at that time (Ash 1980).

The use of psychologists during the First World War for selection testing in the military, industry, and schools changed the path of its academic institutionalization in the twenties. The most visible result of this activity was the establishment of psychological or psychotechnical professorships at six technical colleges between 1918 and 1927, and also the establishment of one further university institute and the provision of further teaching contracts over the same period (Table 3). Furthermore, in business and commercial colleges a few professorships were created or teaching contracts given. In 1919, Wilhelm Peters received the full professorship for philosophy and pedagogy and became head of the Institute for Psychology and Pedagogy at the Trade College in Mannheim. At the Business College of Berlin, Walter Moede taught "psychotechnical business studies" (Dorsch 1963, pp. 82–3); in 1934, Richard Müller-Freienfels became instructor there. At the Trade College of Nuremberg, Würzburg Professor Karl Marbe was responsible for philosophy and pedagogy from 1926 to 1931 (Kürschner 1950, col. 1284).

The first budgeted post for an associate professor of applied psychology at a university was established in Leipzig to begin on 1 October 1923.

Table 3. *Establishment of lectureships and chairs at technical colleges*

Aachen	*1927* Karl Gerhards appointed professor of philosophy and psychology *1928* (probably) Laboratory for Psychotechnics established, led by Adolf Wallichs until 1934
Berlin	*1928* Section for industrial psychotechnics at the Institute for Tool Machines and Production Technology, led by instructor Walter Moede *1921* Moede appointed nonbudgeted associate professor of industrial psychotechnics and work technology; the section is renamed "Institute for Industrial Psychotechnics"
Braunschweig	*1923* Bernhard Herwig lecturer in psychology and psychotechnics *1924* (probably) Institute for Philosophy, Psychology and Pedagogy founded with psychological-psychotechnical section *1929* Herwig appointed nonbudgeted associate professor of psychology and psychotechnics; 1932 full professor
Danzig (Free City)	*1922* Hans Henning appointed professor of philosophy, psychology, and pedagogy *1923* Institute for Psychology and Psychotechnics founded
Darmstadt	*1922* Psychotechnical Institute founded under Erwin Bramesfeld in the Department of Engineering *1925* Two associate professorships for philosophy, psychology, and pedagogy included in budget and a Psychological Institute founded *1927* Matthias Meier appointed to a third professorship for philosophy (pedagogy included in 1929; psychology in 1930)
Dresden	*1922* Psychotechnical Institute founded, headed by Ewald Sachsenberg *1930* Institute for Philosophy, Psychology, Pedagogy founded with a second section for psychology
Hannover	*1922–4* Psychotechnics represented by instructor Adolf M. Friedrich *1929* Wilhelm Hische instructor and lecturer for psychology and psychotechnics *1932* Hische appointed honorary professor in the Faculty for Machine Economics
Karlsruhe	*1918* Willy Hellpach becomes lecturer for psychology, including economic psychology and pedagogy *1920* Hellpach appointed budgeted associate professor of applied psychology and the Institute for Social Psychology founded *1924* Adolf M. Friedrich appointed budgeted professor of social psychology
Munich	*1922* Habilitation of Friedrich Seifert *1927* Seifert appointed nonbudgeted associate professor of philosophy and psychology
Stuttgart	*1923* Fritz Giese instructor and lecturer for psychology and general pedagogy *1924* Psychotechnical Laboratory established at the Institute for Industrial Science of the Engineering Department *1929* Giese appointed associate professor

Psychology had been represented there not only by Wundt, but also since 1908 with a further chair held by Associate Professor Wilhelm Wirth.[6] The faculty applied for the professorship to keep Professor Otto Klemm (whose professorship was not a budgetary post) in Leipzig, in response to the offer of a budgetary associate professorship at the Braunschweig Technical College (UAL PA 636, f. 29). Klemm stayed. The fact that the ministry granted such a request at the height of the inflation of 1923 to keep someone from moving away indicates the reputation that psychology had gained with the public following its use during the war and in industry in the years after. It is not possible to prove that this is the reason why financial considerations did not dominate in this case since the ministerial records in Saxony have been largely destroyed. However, the Prussian Ministry for Science, Arts, and Education and others, such as the Ministry for Labor and Industry, were interested in furthering applied psychology, as was the Reichswehr.[7] In Prussia, it was the general policy of the Ministry of Culture under the Social Democrat Konrad Haenisch (1919–21) and afterward under Carl Heinrich Becker to

6 Wirth had held a budgeted associate professorship for "Philosophy, especially Psychology" since July 1908 and was codirector of the Institute for Experimental Psychology. When Krueger was called to Leipzig as Wundt's successor in 1917, Wirth received the newly created psychophysical seminar; he was then to teach "philosophy, especially psychophysics" (Fritsche 1976).

7 On 28 May 1920 a meeting was held at the Prussian Ministry for Science, Arts, and Education to discuss support for applied psychology. In addition to the hosts there were a representative of the Trade, Labor, Economics and Post ministries as well as four representatives of the Reichswehr Ministry. Psychology was represented by the director of the Berlin Psychological Institute, Carl Stumpf, and by Hans Rupp and Otto Lipmann. Rupp represented industrial psychology at the Berlin Institute and Lipmann was the director of the Institute for Applied Psychology of the Society for Experimental Psychology. Occupational psychology, meaning the assessment of aptitude for occupations, was regarded as important at the discussions not only because of its success during the war, but also because it was to be used to deal with current economic difficulties, to combat the curse of quacks and dilettantes, and finally to overcome the "authoritarian state" in the present "democratic and socialist age," since it opened the way for the most skillful people. Stumpf recommended the creation of a department for applied psychology and a corresponding associate professorship at all institutes (Rep 76 Vb, Sect. 4, Tit. X, 53A Bd.I, ff. 113–24).

Notes to Table 3 (*cont.*)

Note: Since Wilhelm Sternberg was still teaching psychology at the Breslau Technical College, in the twenties either psychology or psychotechnics was being offered at all ten technical colleges in the Reich and in Danzig.

Sources: Aachen: Wer ist wer (1935, p. 1230). Berlin: Spur and Grage (1979); Ebert and Hausen (1979); Rep 76 V b, Sekt. 4, Tit. X, 53 A Bd. 1; telephone message from the Archive of the Technical University Berlin. Danzig: Information from the Wojewódzkie Archiwum Państwowe, Gdansk, 14 April 1981. Darmstadt: *Die technische Hochschule Darmstadt*...(1936, p. 203); Kipp and Miller (1978). Karlsruhe: *Z. päd. Ps.*, 19 (1918), p. 428. Munich: R21, app. 10019, f. 8936; NS 15 alt/33, f. 59918. Stuttgart: Giese (1933). Additional information was gathered from widespread sources.

reform the universities so that they had more contact with the life of the nation (Ringer 1969, p. 69).

Whereas applied psychology had hitherto been cultivated at extramural institutions, it was now finding its way increasingly into the universities. The Hamburg institute had had a strong practical bias since its founding as a university institute under William Stern in 1919, and was called on to conduct vocational aptitude testing (Stern 1922). At the Psychological Institute in Munich, Gustav Kafka received a contract to teach applied psychology at the recommendation of the Inspectorate of Military Aeronautics, which had built up a psychological training center with Kafka at nearby Schleissheim (Sen 310). The Berlin Institute founded a section for applied psychology under Hans Rupp in 1922. In Göttingen, such a section had already been created in 1920 under Walter Baade, an instructor.[8]

A further change in the situation was the introduction of professorships for pedagogy at universities in Prussia in the course of school reforms. Like psychology, pedagogy had been represented until then only within philosophy, mainly as the history of pedagogical ideas and as the theory of pedagogical norms and values. In 1917, the Prussian Ministry of Culture held a pedagogical conference. According to comments on the conference, pedagogy was to include empirical studies on youth (*Jugendkunde*) and to concentrate on the school system ("Ueber die künftige Pflege der Pädagogik," 1918). This was an innovation for university pedagogy. After the November revolution of 1918, the training of elementary-school teachers was to be put on a better scientific footing. Prussia admitted elementary- and middle-school teachers to further training at the universities and created pedagogical academies for their initial training, which led to professorships in pedagogy being set up in the universities at the beginning of the twenties. In 1922 the main committee of the Prussian Landtag passed a motion to set up chairs for educational science at all Prussian universities and colleges.[9] For psychologists, this meant more than the creation of a new rival; it also provided an opportunity, since in places the professorship was for psychology and pedagogy, and in view of the fact that there were hardly any representatives of empirical pedagogy, psychologists had a good chance of getting the professorship, which they could then give a distinct psychological bias. This happened, for example, in the case of David Katz, who was called to a new chair for pedagogy and psychology in Rostock in 1919. Katz was later to write that this was a product of the school reform movement (Katz 1953, p. 477).

The effort to reform teacher training was a decisive factor, along with the initiatives of industry and the Society for Experimental Psychology, for the creation in Jena of the first full professorship in Germany defined solely for

8 For Berlin see the entry on Rupp in *Wer ist's?* (1935); on Göttingen, see the report in
    Z. päd. Ps., 21 (1920), p. 365.
9 Report in *Z. päd. Ps.*, 23 (1922), pp. 313–14.

psychology, commencing 1 May 1923 (Eckardt 1973). The first step toward establishing this professorship was taken in 1920 by the Thuringia Teachers' League. The new post was severely attacked by local conservative philosophers Bruno Bauch and Max Wundt, but was welcomed by the pedagogue Rein. The Thuringian Ministry of Education called Wilhelm Peters to the chair against the wishes of the faculty. As he was later to write to the ministry, the research was to "concentrate less...on the problems of psychology of the senses, but all the more on the psychology of individuality...and its applications," and "among the applications the pedagogic-psychological" was to receive special attention (p. 546).

The establishment of a psychological professorship at a university was nothing that could be taken for granted at that time. Throughout the Weimar Republic, psychologists were struggling for representation of their subject. At the Seventh Congress of the Society for Experimental Psychology in 1921, Marbe used an argument for the institutionalization of psychology that was often repeated later. Marbe contrasted the "living conditions of psychology" at the university not only with its scientific relevance to other disciplines, but also with its practical importance. In the universities, it was necessary to emphasize the scientific relevance of psychology, which Messer (1914) had also tried to do, whereas with state governments one had to see to it that they would reach into their pockets for practical things (Marbe 1921, p. 204). Since psychology had proved its practical value in war and was viewed as being important in industry, education, and medicine, Marbe reckoned with more financial support. Following this talk, the society passed a resolution to send a circular to the state governments demanding more funds for existing institutes, and to petition both universities in which psychology was not represented and the relevant governments for the creation of budgeted professorial posts (p. 210).

The influence of Marbe's phrasing can be seen in the circular that the chairman of the society, Georg Elias Müller, sent to the heads of the universities on 19 December 1921. In Jena, this arrived in the middle of the confrontation between the ministry and the faculty about the psychological professorship. Müller gave three arguments for the relevance of the "independent positive special science" (cited from Eckardt 1973, p. 529): (1) its relevance for other sciences, (2) its achievements in war, and (3) its importance for industry, education, and the work of physicians and the legal profession. He called on the universities and technical colleges to pay due attention to the interests of psychology when creating new chairs and making appointments.

However, the initiative seemed to meet with no success. A look at further events shows that the situation was to get worse for psychology during the twenties, leading to lamentations that would only give way to hope in 1933. In 1922 Narziss Ach, an experimental psychologist with a chair in philosophy in Königsberg, went to Göttingen to succeed G. E. Müller. At the same time, Otto Schultze was called to Königsberg to a newly created chair for "Pedagogy with an obligation to teach Pedagogy and Philosophy as well as Experimental Psychology." Against the will of the faculty, the ministry now wanted a philosopher as Ach's successor. Thus, in Königsberg (in contrast to Rostock) a new pedagogy professorship had a negative influence on the position of psychology. Nor did the voices of the local teachers' associations change anything when they appealed for the "Chair for Psychology" (as Ach's philosophical professorship was generally referred to) to be preserved.[10]

A further, spectacular loss of a chair was required before the Society for Experimental Psychology raised its voice in public protest. Perhaps the Executive Committee of the society may also have been influenced by worries about finding positions for their scientific offspring when they agreed on a manifesto entitled "On the Support for Psychology at German Institutions of Higher Education" at the eleventh congress in Vienna in 1929. The decisive reason for the turmoil in the committee, however, was the fact that the successor to Gustav Störring – head of one of the most renowned seminars for experimental psychology at Bonn (the others being Berlin, Göttingen, Leipzig, and Würzburg) – was to be the philosopher Erich Rothacker. The ministry had initially considered the recommendation of the faculty that the post go to Erich Becher, but he became so ill in June 1928 that he could not enter into negotiations, and in fact died on 5 January 1929. It was initially intended that Rothacker be called to another vacancy as an associate professor of philosophy. However, when difficulties arose in finding a successor to Störring – Karl Jaspers and Moritz Schlick refused – the ministry went for a new solution; it appointed the Protestant Rothacker to the budgeted full professorship, commencing in July 1929. This also solved a problem posed by a statute of the university which required that the post serve the needs of the evangelical (i.e., Protestant) worldview, although it had traditionally been occupied by a psychologist since Benno Erdmann and Oswald Külpe. After a long disagreement, the associate professorship went, by way of a compromise, to the Catholic Siegfried Behn in October 1931. He was appointed "Personal Full Professor" for "Philosophy with Special Emphasis on Experimental Pedagogy" and also as director of the Psychological Institute, which then had two directors.

It is difficult to decide whether this to-ing and fro-ing about appointments really demonstrated an opposition to psychology on the part of the Prussian ministry under Becker, but this is certainly the way representatives of psychology saw it. A few days after Rothacker's nomination, Ach and Bühler lodged a complaint at the ministry as delegates of the Executive Committee

10 The entire matter is documented in Rep 76 Va, Sect. 11, Tit. IV, 21 Bd. 30. On 31 July 1923 the faculty applied once again for a chair for experimental psychology (f. 329). The files do not show that things went any further than this.

of the German Society for Psychology. A memorandum notes that they "complained about the failure to appoint a psychologist to Störring's chair" and expressed the opinion that "the Prussian teaching administration intends to gradually eliminate psychology." This was rejected by the ministry, which emphasized that the financial situation made it impossible to fulfill the demand for the creation of chairs for psychology at every university.[11]

A little earlier, when the Bonn conflict was still smoldering, the society had published its declaration (*Kundgebung*). We find the same arguments for the furthering of psychology as G. E. Müller had cited in 1921 in the circular already mentioned. In its first sentence, it complains that at a number of German-speaking universities "psychology was being ousted from its former position. Chairs that had previously been primarily for psychology were repeatedly given to representatives from the fields of pedagogy or pure philosophy" ("Kundgebung..." 1930, p. vii). Methodological developments in psychology were emphasized, but not only in the experimental sector. It was no coincidence that at this same congress the name of the society was changed to the German Society for Psychology (DGfPs), dropping the word "experimental." It was then emphasized that psychology had gained an "unanticipated importance for practical areas of culture" (p. ix), for education, jurisprudence, economics, and medicine. It was therefore the task of psychological chairs "to provide the scientific basis for the vital and productive practical application of psychology" (p. ix), and to train future practical psychologists. The entire committee – consisting of Bühler, Stern, Ach, Katz, Lindworsky, Poppelreuter, and Volkelt – signed the declaration.[12]

Not all academic psychologists wished to be fighters for the institutionalization of psychology in this way. In a letter to Max Wertheimer (4 August 1929), Wolfgang Köhler referred to Vienna as a "sort of rabble revolution." It was necessary to speak out against the manifesto. After the declaration, "which you will find as lamentable as I do," he no longer wished to remain a member of the society (though in fact he did not resign).[13]

Since no new teaching positions for psychology were created in the aftermath of the great economic crisis of 1929, we find the same complaints at the twelfth congress in Hamburg in 1931. The chairman of the society, Karl Bühler (1932), began his address with a reference to the desperate straits in which psychology found itself and the decline of its research establishments in the universities. At the same congress, Felix Krueger commented that psychology "now...had to fight for its right merely to survive" (1932, p. 70). The psychologists received support from the German Teachers'

11 Rep 76 Va, Sect. 3, Tit. IV, 55 Bd. 11 and 12, for the events in Bonn and the memorandum (Bd. 11, f. 412).
12 This declaration was published not only in the congress report but also in the important journals: *Z. Päd. Ps.*, 30 (1929), pp. 376–8; *Z. ang. Ps*, 33 (1929), pp. 550–2; *Ind. Pst*, 6 (1929), pp. 202–4; and *Psychologie und Medizin*, 4 (1930), pp. 222–4.
13 Wertheimer Correspondence box 1, Folder Letters received 1929, August–December.

Association, whose head office for educational science declared in 1930 that because of the "importance of psychology for the training of all types of teachers and educators," the demand for chairs for psychology was justified ("Die Stellung der Psychologie..." 1931, pp. 159–60). In a talk at the head office in 1930, William Stern had drawn a frightening picture of the situation. At some institutes, the resources were too meager; at some, their leadership "had been replaced by unsatisfactory compromise measures." At other universities, psychology was surviving "as an adjunct to a philosophical or pedagogical seminar." Finally, there were universities at which psychology was not represented at all. "The entire East, to the east of Berlin and Rostock, is without any independent full professor of psychology ... as is all Rhineland and Westphalia and the non-Prussian southwest Germany" (1931, pp. 74). Bühler's complaint at the twelfth congress in 1931 sounds much the same: "The entire east and west Prussia are today without properly staffed psychological institutes: Königsberg, Breslau, and Bonn have been lost" (1932, p. 3).

Since Stern left Breslau in 1916, the associate professors of pedagogy there had not been psychologists but philosophers (Richard Hönigswald, Siegfried Marck). Königsberg and Bonn have already been mentioned. "The entire East" referred to the universities of Breslau, Greifswald, and Königsberg. "All Rhineland and Westphalia" meant the universities of Bonn, Cologne, and Münster. "The non-Prussian southwest Germany" meant the universities of Freiburg, Heidelberg, and Tübingen. There were no budgeted posts for professors with responsibilities including psychology in Breslau, Greifswald, Cologne, Münster, Freiburg, Heidelberg, or Erlangen. In Königsberg and Tübingen, psychology was the responsibility of the professor of pedagogy (Schultze and Kroh) and in Bonn of the philosophical ordinarius. Munich, where the ordinarius for pedagogy, Aloys Fischer, lectured in psychology, could also have been included. Stern's complaint was that psychology was inadequately represented at nine of twenty-three universities in the German Reich. If Erlangen and Munich were included, the figure was eleven. At eight of these universities, the situation would improve during the Third Reich; at three there was no change; and at three, where Stern had felt that the representation in 1930 had been adequate, a deterioration had to be withstood. However, first the fascist civil service law was to take its toll.

The Nazis come to power: dismissals and the reaction in the discipline

On 7 April 1933 the Hitler government proclaimed the Law for the Restoration of the Professional Civil Service (RGBl, 1933, pp. 175ff.). According to paragraph 3 of this law, civil servants who were "not of Aryan descent" were to be retired; paragraph 4 stated that civil servants "who in their previous political activity have not provided the guarantee that they will unconditionally support the state at all times" were to be dismissed. The first wave of dismissals hit the psychologists hard. As a result of paragraph 3, Professor David Katz and Professor Wilhelm Peters were relieved of their duties in April 1933. Max Wertheimer had already left Germany in March, and on 28 August 1933 he was, in the euphemistic phrasing used, given leave of absence to take on a visiting professorship in New York. On 25 September 1933, he was retired under paragraph 3, followed on 10 October 1933 by Peters and on 1 January 1934 by Katz. In April, the Senate of Hamburg University requested that William Stern give no lectures that summer and dismissed him in the course of the semester. Adhemar Gelb in Halle received his dismissal on 7 September 1933. The lectures he had given in the summer semester had been filled to overflowing. Since Katz was covered by a special subsection (paragraph 3.2) under which war veterans could not be dismissed, the Mecklenburg state government came up with the idea of abolishing his professorship. The Nazi press had been vociferously demanding his dismissal.[14] Under the Nuremberg Racial Laws of 1935, the ancestry of wives also became a criterion for the suitability of civil servants and the wave of dismissals caught the Munich ordinarius Aloys Fischer, who was retired prematurely on 25 June 1937 (Schumak 1980, p. 337; Ebert 1979, p. 464). Karl Marbe would also have fallen victim to this regulation if a lowering of the age limit had not allowed him to become professor emeritus at the age of sixty-five on 1 April 1935. In the summer of 1935, he was still filling in for his own, unoccupied chair.[15] Aloys Fischer died in November 1937, the seriously ill Adhemar Gelb in August 1936. The other dismissed professors emigrated.

In 1933 chairs in psychology were occupied at fifteen universities (see Table 5). Thus, the persecution of Jews by the Nazis meant that psychology lost more than a third of its full professors. In comparison with the number of full professors dismissed overall, which Charles Hartshorne (1937) estimated at 10.9 percent – he later thought this too low (Beyerchen 1977, p. 47) – this seems to be a relatively high proportion.[16] On the basis of recent studies on émigré scholars and scientists, we find that the figure of one-third

14 Sources: On Gelb, UAH PA 6557; Bergius (1962). On Katz, Carlsen (1965, pp. 147); Grassel (1971, p. 157); Katz (1953, p. 480); Z, f. 173. On Peters, UAJ D 2246. On Stern, Letter SAH 21 April 1981; SAH C 20/4, Vol. 5, meetings of the University Senate 27 April and 28 July 1933. On Wertheimer, PA Wertheimer ff. 43–4.; AHAP Wertheimer Memoirs, Transcript p. 46.
15 Information from the Commission for the History of the Würzburg University, 27 May 1981. Wellek (1964, p. 239) is therefore wrong to say that Marbe and Kafka were affected "directly or via their wives." Kafka is discussed in more detail in a later paragraph.
16 Comparisons must still be based on Hartshorne (1937). More recent investigations are not known to me. Overall, Hartshorne (p. 93) estimates 14.34%, later 20%, for physicists; Beyerchen's figure is about 25% (1977, p. 47). On Hartshorne see Beyerchen (1977, p. 221, n. 20). For psychology Hartshorne has fifty-one dismissed lecturers, which seems too high. Merz (1960) talks of about 50% losses without exact figures. Ash (1984) researched members of the DGfPs in great detail and arrived at forty-five emigrants, of whom twenty-five taught psychology in universities.

roughly corresponds to the proportion of émigrés for the total teaching staff of German universities and colleges, but at the same time is higher than the proportion of émigrés among all German psychologists (Ash 1984).

A number of associate professors were also dismissed, such as Curt Bondy in Göttingen, Jonas Cohn in Freiburg, Richard Hellmuth Goldschmidt in Münster, Erich von Hornbostel in Berlin, Traugott E. K. Oesterreich in Tübingen, Erich Stern in Giessen, and Heinz Werner in Hamburg. In Berlin, Kurt Lewin resigned from his post; the assistants Karl Duncker, Otto von Lauenstein, and Hedwig von Restorff later had to give up their positions. Finally, a further loss attributable to politics was Köhler's departure from Berlin in 1935, when he was no longer prepared to tolerate the continual interventions of the administration in his institute. However, he was not dismissed; in fact, he had difficulties gaining approval for the application to leave state service that he had first made on 10 June 1933 (see Ash 1985; Geuter 1984; Henle 1978).

At technical colleges, there were two full professorships held by psychologists according to our criteria and a further two that also encompassed psychology. The new political developments left their marks on all but one of them. In Darmstadt, Matthias Meier and Hugo Dingler had full professorships for philosophy, pedagogy and psychology, although Dingler had little time for psychology, and Meier did not develop an interest until later. Meier was dismissed on 7 April 1933, then reinstated on 29 March 1934. Dingler received emeritus status prematurely at the age of 53, but after his support of "Aryan physics" in the thirties, he taught philosophy and history of science at Munich University from 1938 on. At the Technical College of Dresden, Gustav Kafka, a psychologist with a chair for philosophy and pedagogy, applied for premature emeritus status at the age of fifty in 1933, nominally for "health reasons," although the political situation was probably responsible. The application was approved in 1934. The associate professor of general and applied psychology in Dresden, Walter Blumenfeld, was dismissed and emigrated to Peru. Of the four full professors at technical colleges, only Bernhard Herwig in Braunschweig remained in his position without interruption. Also unaffected were Walter Moede (Berlin), Friedrich Seifert (Munich), Fritz Giese (Stuttgart), and Wilhelm Hische (Hannover), who represented psychology at their institutions. At the Technical College of the Free City of Danzig, where following the election victory of the NSDAP the Senate of Hermann Rauschning forbade the employment of Jews and members of the political opposition in public office, Professor Hans Henning was dismissed for political reasons. The only full professor of psychology at a trades college, Otto Selz in Mannheim, was dismissed on 7 April 1933. He was murdered ten years later in Auschwitz.[17]

---

17 Sources: On Henning, Letter from W. Ehrenstein to the curator of Frankfurt University 8 October 1934, in PA Wertheimer f. 52. On Meier, *Darmstädter Archivschriften* 3,

Thus, Nazi policy toward civil servants had a profound effect on psychology. I am not referring here to the qualitative losses that resulted from expelling professors of scientific standing, including almost all representatives of the internationally renowned school of Gestalt psychology, or the practical liquidation of an entire research institute in Hamburg – where, in addition to Stern, Heinz Werner left and their co-worker Martha Muchow committed suicide in September 1933 (see Geuter 1984). What was a tragic fate for the individuals as well as a scientific loss also led to some temporary institutional setbacks for academic psychology, although this was not to be characteristic of subsequent developments. Hamburg, Frankfurt, and Rostock were so affected by the dismissals that psychology there never recovered. In Rostock, the chair was abolished, in Hamburg, it probably went to the art historians; and in Frankfurt, the chair was obviously lost, although it is not clear where it went.[18] The State Teaching Authority of Hamburg and the Reich Ministry for Science and Education wanted to see psychology represented there again as soon as possible. However this was thwarted for a long time by the resistance of the faculty, which preferred to have a "race scientist."[19] Halle and Berlin were severely hit by the losses at first, although things were kept running with stand-ins and associate professors. At some institutes, such as Danzig, Dresden, and Jena, chairs vacated by dismissals were filled again immediately.

How did the leading representatives of the discipline react to the dismissals? Only Köhler protested publicly, in a newspaper article, to the dismissal of Jewish professors (Ash 1985; Henle 1978). The others directed their activities not toward the individuals who were dismissed, but toward the representation of the subject. Even more than previously, they attempted to emphasize its usefulness, though now with a different political emphasis. We will trace this later in the chapter when we consider the policies adopted for the academic institutionalization of psychology (see also Chapter 5).

Despite numerous dismissals, those remaining placed great hope in the new state. After four of the seven members of the Executive Committee of the German Society for Psychology resigned for political reasons, the new president, Felix Krueger, from the renowned Leipzig Chair, who as a German

Hochschullehrer (1977, p. 137). On Dingler, *Darmstädter Archivschriften* 3 (1977, p. 39); Beyerchen (1977, p. 180). On Kafka, Wehner (1964). On Selz, Seebohm (1970, p. 23). On Danzig in general, see Levine (1973 pp. 56ff., 102).

18 The Curatorium records spoke in 1941 of the chair being cut from the budget (alt Ve 6h). According to Rausch (Pongratz et al. 1979, p. 219) it was used for a philosopher; according to Thomae (Z, f. 88) it was for prehistory. The matter cannot be cleared up with data from existing faculty files.

19 The faculty wanted the professor for race and cultural biology, Walter Scheidt, to represent psychology, while the Hamburg teaching authority asked for a list of psychologists. The Reich Ministry tried to get psychology represented in Hamburg from 1936. No records could be found in Hamburg about the psychological chair from 1937 to 1941; for the later period see Chapter 7. For details and documentation of events in Hamburg see Geuter (1984).

Nationalist approved the 1933 coalition, summoned the thirteenth congress to Leipzig in Autumn 1933 (see Geuter 1979). Krueger commented optimistically on the situation in his opening address. The "cultural transition" in which "cultivated humanity" found itself corresponded to mental reality, and psychology could make a contribution to the current regeneration of the souls of the German people, which he felt he could perceive. Thus, there was hope for the "situation of the science of the soul (*Seele*) in contemporary Germany" (Krueger 1934). Now the only important thing left was to convince the state authorities of psychology's importance, because psychology was still inadequately represented, just as Krueger had complained in 1931. He did not mention the dismissals in this context, but instead attacked Becker's Prussian Education Ministry, which in the Weimer period had shown preference for "that conceptual acrobatics whose name begins with 'soci'..." (Klemm 1934, p. 8). But now there was "fortunately a self-purification of the sciences... from the whole" (p. 8). Psychology had been purified more from the outside, but Krueger did not hesitate to turn to the authorities responsible for this in the attempt to secure chairs for the field, rather than for individuals. Thus, he intervened after the dismissal of Stern in Hamburg, not for Stern, but in order that "at German universities and colleges teaching and research in psychology at least do not decrease to the detriment of the science" (SAH A 110.70 ff. 1–2). In the same text, he assured the Hamburg authorities that he had been able to prevent the discussion of the events in Hamburg at the congress in Leipzig, despite pressure from others.

On 8 December 1933, Krueger sent a circular to the members of the DGfPs Executive Committee informing them of talks he had held "with all responsible ministries of the Reich and of Prussia" as well as the Propaganda Ministry and the Foreign Office. The head of the personnel department at the Prussian Ministry for Science and Education had explained to him that the ministry did not intend to restrict psychology in any particular way and had no intention "of making appointments to chairs that had hitherto covered either solely or mainly psychology on a different basis than before" (NS 15 alt/17 f. 56948). Krueger informed the Rostock rector on 9 May 1934 that the Hamburg administration had no intention of edging out psychology either (UAR R6 H 27/2).

Psychology was obviously not going to be subjected to any "special treatment." After the founding of the Reich Ministry for Science and Education, Krueger notified the committee members on 2 July 1935: "As a result of my repeated supplications the deputy state secretary of the Reich Ministry for Science has given me permission to state: 'The Ministry, far from appointing other than proven psychologists to existing Chairs of Psychology, is inclined to gradually increase the posts for our science'" (f. 56951). This was no placatory phrase. Admittedly the losses caused by National Socialist civil service policies were considerable, but in terms of the number of chairs psychology's situation at the universities had improved by the beginning of the war, even when compared with 1932. The decisive reasons for this success were developments outside the universities. The expansion of military psychology and the resulting increased demand for academically trained psychologists were needed for the academic institutionalization of psychology to make headway.

New chairs and appointments, 1933–41

As already mentioned, up to 1941 there was no clearly defined training program for psychology that could be used to argue for a new chair or that an appointee was expected to fulfill. Therefore, reasons had to be given from case to case to show why psychology should be represented at a given university or a certain psychologist appointed to a chair. If they were written down at all, such considerations can be found among the papers of the institutions involved: the universities, the ministries, the Nazi Party, and occasionally the Wehrmacht. The language used was particularly plain when the topic was the establishment of new chairs.

During the Nazi period, political criteria were expressed relatively openly. But just as the persecution of the Jews in the universities had followed from earlier discrimination, so the influence of political considerations in appointments was nothing basically new.[20] However, in the Nazi period such considerations were expressed more explicitly, and they were given more weight. This was ensured by a new appointment procedure, which created a number of opportunities for the Reich government and the party to influence the results.

The appointment procedure

Before the Third Reich, the procedure for appointments was that the faculty recommended a successor when posts become vacant. Usually a "short list" of three was drawn up, but occasionally a single person was named, as when Krueger was appointed to succeed Wundt. The cultural minister of the state in question then had to decide whether to accept the list and offer the chair. Usually the first on the list would be selected, rarely, as in the case of Peters in Jena, it would be someone who was not on the list at all. Since the finances and administration of the universities were under state control, the ministers were able to do this. During the Third Reich, there were many more institutions involved. Kelly (1973) has described them in detail, also covering the changes made in the procedure and analyzing conflicts between the state and party organs involved. An outline of the procedure will be given here

20 See the call of Peters to Jena (Eckardt 1973) or Behn's statement for Erismann on 2 July 1931, that he was not as pro-Jewish, pro-Russian, pro-communist, or atheistic as others claimed (Rep 76 Va, Sect. 3, Tit. IV, 55 Bd. 22, ff. 30ff.). In 1919 Max Weber advised prospective Jewish private instructors "lasciate ogni speranza" (1922, p. 530).

to make clear the various levels at which different organs brought forward their arguments.

The starting point for all nominations in the Third Reich remained the recommendation of the faculty,[21] which was generally accepted as far as psychology was concerned. The faculty passed on its suggestion via the dean to the rector, who after consulting the academic senate formulated a statement of his own. This he attached to the recommendation, which was passed on via the state ministry or, in the case of Prussia, directly to the Reich Ministry for Science and Education (RMWEV). This ministry, created in May 1934 under Minister Bernhard Rust, was united with the Prussian ministry. If the university also had a head of administration he too was entitled to submit a statement. With the formation of the RMWEV, the state ministries lost all practical importance in this procedure.

The influence of the party was ensured by the fact that Hitler, as Reich chancellor, reserved the right to nominate all public servants personally (Broszat 1975, p. 310). In September 1935, the deputy of the Führer, Rudolf Hess, received formal entitlement to examine the appointments to higher levels of the state civil service (Kelly 1973, p. 328). The RMWEV therefore had to consult Hess's office in the Brown House in Munich before making a recommendation that then went to the Reich chancellery for final approval. Hess's office and, after his flight, the party chancellery under Martin Bormann based their judgment above all on the reports of the Reich leadership of the National Socialist Lecturers' League (NSD), which in turn gained its information from its local leadership. The NSD was formed in 1935 as a section of the party in its own right, having previously been a division of the Nazi Teachers' League. From 1938 on, the local "Führer" of the NSD was informed directly of the faculty recommendation and passed his own comments on to the Reich leadership.[22]

Officially, then, the only step in the procedure at which the party could exercise its influence was the political assessment of the persons recommended from the RMWEV by the deputy of the Führer. In view of the influence of the NSD on the statement of the deputy's office, however, the ministry sometimes contacted the NSD leadership directly. From 1938 on, the Amt Rosenberg also attempted to become an official participant in the procedure. A science office was established there in 1938 under Professor of Philosophy Alfred Baeumler that had its own informants and provided reports for the RMWEV and the deputy of the Führer. Reports for psychology were

21 Hirsch (1975, p. 345) claims that the faculty recommendation no longer played any role after 1933; the state ministries, and increasingly the Reich Ministry, made appointments in consultation with the party. Unfortunately the postulates of this study have been advanced in ignorance of the facts.

22 Account based on Kelly (1973) and the Guidelines of the RMWEV, 14 May 1938 (131/123, No. 96). See also Kasper et al. (1943, pp. 37ff.); a summary of the procedure is in Losemann (1977, pp. 52–61).

provided above all by Professor Hans Volkelt from Leipzig and also the lecturer, and subsequent professor in Halle, Kurt Wilde. A crucial influence, not necessarily evident from the files, may have been exerted by the party at the level of the local "Führer" of the NSD. As a member of the staff, he was often directly involved in the selection procedure, and one would expect that the faculty would attempt to secure his support, since his comments had a decisive influence on the judgment of the deputy of the Führer.

In the ministry, the section for universities and colleges (WI) was responsible for appointments until 1937. This was a section of the Science Office (Amt W) led by Theodor Vahlen. WI was led by the chemist Franz Bachér, at first in 1934 on a temporary basis, but from 1935 on full time. His deputy was theologian Eugen Mattiat, who in this period was obviously concerned mainly with appointments for the humanities. When the research office WII was dissolved in 1937 as rearmament began again and Vahlen left, a new organization plan was introduced in Amt W. The three sections that had existed were replaced on 21 January 1937 by nine sections, of which W6 to W8 concentrated on specific disciplines. The new head of Amt W was the cultural minister from Baden, Otto Wacke, who was replaced by Rudolf Mentzel in May 1939. Section W6 was responsible for the humanities, the subsection W6a under Mattiat for "philosophy including psychology and the pedagogical sciences," among other disciplines. Mattiat was responsible for handling personnel questions concerning professors. But during 1937, Heinrich Harmjanz, a folklorist from Königsberg and a member of the SS, joined the ministry and took over as head of W6. Subsection W6a was initially given over to Frey, who had hitherto shared responsibility for the subsection. Subsequently, Harmjanz took over and then "called the tune in appointment politics in the humanities for five years" (Heiber 1966, p. 648). Thus, first Mattiat and then Harmjanz were those at the ministry who had a decisive influence on appointment decisions in psychology.[23] Harmjanz also played a key role in drawing up the DPO of 1941. Because of the role of the SS man Mentzel and the important position of Harmjanz, who later as personal secretary to Rust (from April 1942) and head of the revived ministerial office could do what he liked, and in view of the numerous ties between the ministry and the SS and the initial emphasis on military research, Heiber referred to Amt W as a "domain of the Army and the SS" (1966, p. 655). This was modified by Losemann (1977, p. 4) in light of the work by Reinhard Bollmus and Michael Kater, since department heads did not just follow Himmler's every word.

Bearing in mind this strong influence of Nazi and SS members in the ministry, it is very difficult to separate the interests of party and administration in appointment policies. Kelly (1973, p. 322) states that the ministry

23 Details on the RMWEV from Heiber (1966) supplemented by a diagram of responsibilities of Amt W, 21 January 1937 (Rep. 4/180) and also REM 11887/2 and 3.

denied the party the right to make academic evaluations and wished to see its role restricted to political evaluations. Although the formal ruling was that the party would judge only political suitability, as in the new Reich regulations for obtaining the right to teach in 1939, in the cases investigated party members also put forward academic arguments. This makes the dichotomy postulated by Adam (1977) between a ministry that had won the right to stick to scientific criteria and a party that considered only political ones seem too simple.

According to Kelly (1973), the party made only limited use of its political influence; less than 10 percent of the assessments by the NSD leadership were negative, and these did not necessarily lead to refusals (p. 380). On the whole, the state administration gained the upper hand in science policy. The files studied show no case of an appointment to a chair in psychology failing due to intervention by the party, even though the Amt Rosenberg made massive attempts to stop Hans Wenke from getting the chair at Erlangen (see the subsection entitled "The Primacy of Practical Relevancy, 1937–40"). Here, however, the question is not only whether interventions of the NSDAP stopped or enabled appointments, rather it concerns the relationship between academic and political criteria as one element in the larger question of the reasons for establishing chairs or making specific appointments to them.

"The primacy of politics": self-recommendations and appointments, 1933–5

In 1933–4 dismissals meant that chairs in Danzig, Halle, Jena, Hamburg, Frankfurt, and Rostock[24] became vacant, although the last two were withdrawn so that no successors could be appointed (see section entitled "The Nazis Come to Power"). Friedrich Sander's rapid move from Giessen to Jena meant that the psychology professorship at Giessen was vacant, and Kafka's early retirement left the chair at Dresden empty in 1934. In 1935 Berlin, Königsberg, and Würzburg followed. The Königsberg chair, vacated when age led Otto Schultze to retire, was not reoccupied until 1940; that in

---

24 In Rostock the DGfPs and the faculty tried to get a representative for psychology after Katz's dismissal. In the fall of 1933 the faculty requested that the assistant and Associate Professor Hans Keller be given the responsibility: "The faculty has no doubt that the maintenance of a special representation of psychology, in view of the size, the wide connections and the importance of the objects of this science, is an irrefutable requirement for the teaching syllabus" (12 October 1933, UAR R 6H27 [1]). Keller was sent to Berlin in 1936 (UAR R VIII F 107), after which psychology was represented only by Hans Koch, who was budgeted assistant from November 1935, became instructor on 29 June 1937, qualified as lecturer in Würzburg 27 January 1939, and became head of the Rostock Psychological Institute on the same day (Miehe 1969; *WPsM* 1940, 2, H. 7; 42 a M 1, personnel news from 5 July 1940, p. 326; information from the University Archive of Wilhelm-Pieck University Rostock, 1 September 1981).

Dresden was reoccupied 1936. These two appointments will be considered in the next section.

It is not, perhaps, surprising that unoccupied chairs gave rise to a certain degree of opportunism and that the "old soldiers," who had been on the winning side and had now come out on top, should stake their claim to a professorship. On 8 October 1934, Walter Ehrenstein, who had just been appointed to succeed the dismissed Hans Henning in Danzig, recommended himself for the chair in Frankfurt in a letter to the curator. Not forgetting to mention that he was one of the longest-standing party members among his colleagues, he asked "to be included on the short list if the university decided to approach the ministry" (PA Wertheimer f. 52). The curator's answer was brief: the chair was not to be reoccupied. Fritz Giese, who boasted in 1933 (pp. 13, 40) that he had already taken *Mein Kampf* as the subject of psychological seminars before the "time of change" – like Poppelreuter (1934) – suggested to the Prussian ministry on 4 July 1933 that an "Institute for German Psychology" be set up in East Prussia, for which, as a pioneer in the field, he would offer his services.[25] It is not intended to imply that these two examples, found by chance, mean that the "Nazi fiends" in psychology have been unmasked. The behavior may indeed be understandable in individual cases.[26] The point here is that these self-recommendations, based on good political track records, had no influence at all on the universities and only little on the state administration.

The party was not able to put professors where it would have liked them either, and even the ministry was cautious in this respect. A conflict in Halle demonstrates this. After the dismissal of Gelb, the ministry tried, obviously for political reasons, to force the instructor and party member Heinrich Schole on Halle's philosophical faculty. After the faculty had presented a short list in July 1934 with four unlikely candidates, the ministry decided on an interim solution and appointed Otto Schultze to represent psychology in the chair of dismissed philosopher Emil Utitz; he met with massive rejection by the faculty and stayed only briefly in Halle. Eugen Mattiat from the ministry wrote on 15 March 1935 asking for a new list of three names, with a comment on Schole and Carl Jesinghaus, who had returned from Buenos Aires. Obviously the ministry was interested in finding professorships for the two. However, the faculty found them unsuitable. On 19 June 1935, Ehrenstein also tried to enter the debate with a letter to the curator. On 22 November 1935, the ministry called on Schole to carry out the duties of professor, so that the faculty could form an opinion. But this turned out to be negative. Whereas Gelb in 1931 had been accepted solely on the basis of his psychological writings, the argument was now that Schole was exclusively a psychologist, but the chair was for "scientific philosophy and psychology."

25 Rep 76 Va, Sect. 11, Tit. IV, 21 Bd. 34 ff. 147–8.
26 Ehrenstein, for example, had his sick mother in Frankfurt, as he reported in the same letter.

The implication is that this was a tactical argument because Schole was not wanted for other reasons – including political ones. The negative assessment of Schole's ability was supported by the rector. The ministry remained unmoved, and on 4 July 1936 it once again called on the faculty "to declare its position on Schole's final appointment." When the university stuck to its list, and even Erich Jaensch, the well-known National Socialist spokesman of the DGfPs, named four alternatives in a letter to Emil Abderhalden as president of the Leopoldina Academy in Halle, the ministry appears to have backed down. For the summer semester of 1937, the instructor Wolfgang Metzger from the faculty list was appointed as stand-in. The decision went against Schole's "sense of justice." He wrote to the rector on 18 December 1935 that he had been led by the ministry to expect appointment:

> A natural feeling of justice keeps the conviction alive that at the seats of higher education at which for 14 years I have struggled tenaciously against Jewishness, the decline of the *Volk*, liberalism, Marxism etc. – without hope of advancement! – I have earned the right to be fully integrated. My simple logic remains this: Here eight or nine unoccupied chairs, there the cry for three years: we have no reliable men! I claim to be reliable.[27]

His logic was not, however, the logic of the universities. Mattiat tried unsuccessfully to get Schole accepted in Hamburg, Berlin, and Königsberg. After the ministry had given him temporary posts in Königsberg, Göttingen, and Greifswald (which did not require faculty approval), he was finally made an associate professor (but not a civil servant) in Greifswald. He never did become a full professor.

Mattiat obviously could not rule against the universities. Nevertheless, the first appointments in the Third Reich seem to have been very dependent on political criteria. Jesinghaus, who could not remain a professor in Buenos Aires in 1933 and whom Mattiat had already backed against the faculties in Hamburg and Halle, was called to Karl Marbe's chair in Würzburg in 1935. Politics obviously played a role in Giessen. The philosophy faculty there had been thoroughly purged in 1933: Ernst von Aster, a Social Democrat, was dismissed, August Messer retired, and the Jewish professor Erich Stern (nonbudgeted) was also dismissed – all three were professors of philosophy and pedagogy. A new full professorship for psychology and pedagogy went to Gerhard Pfahler, who also became the rector at Giessen and a local leader of the SA (Chroust 1979, p. 30). Waldemar Lichtenberger, who worked at the Institute before Pfahler's appointment, reports that he had told him that he was a close friend of Rudolf Hess (Z, f. 111). In Jena in October 1933, Friedrich Sander took up the professorship from which

---

27 Rep 4/898; for the entire case also Rep 4/847 and 6/1368.

Wilhelm Peters had been dismissed. As Peters later reported, this was only because Sander was a more than nominal party member (UAJ D 941). Thus, against all custom the newly promoted full professor became dean of the mathematics and natural sciences faculty. It is more than disquieting to read that, after the war, Albert Wellek argued that Sander's appointment, to a chair empty only because of Nazi persecution of Jews, was the result of an "*orderly* recommendation procedure" in the faculty (1960, p. 180, n. 2).

A rapid rise could be followed by a nasty fall if mistakes were made or the past caught up with someone. Take the case of Johann Baptist Rieffert, who tried to become Köhler's successor in Berlin. The former head of Army Psychology was appointed associate professor of philosophy, particularly characterology while Köhler was still there, in October 1934. He immediately set about denouncing Köhler's assistants to the SS. When Köhler left in 1935, Rieffert was given temporary charge of the institute. He never got around to enacting his plans to meet the "psychological needs of the National Socialist state" at the institute, however. The ministry found that he had not declared his earlier membership in the Social Democratic Party (SPD). All his zealotry and political friends could not save him from being suspended and finally dismissed on 9 August 1937. Associate Professor Hans Keller took over the leadership of the institute, followed later by Professor Walter Malmsten Schering, who had specialized on the war philosophy of Clausewitz. Köhler's chair was not reoccupied until 1942. Rieffert fell on hard times before getting a job as a psychologist with the Borsig firm in 1940.[28]

The primacy of practical relevance, 1937–40

In the first years of the Nazi dictatorship, political considerations played an important role in appointments, and political opportunism could be rewarded – but this was to change. The most prominent criteria for chairs and appointments now became the practical relevance of scientific activity to psychological applications outside the university. This replaced pre-1933 criteria of interdisciplinary relevance or those of ideological relevance in the early part of the Third Reich. This is easy to say with hindsight; of course, in reality things were more complicated. Such a tendency can nonetheless be ascertained in the period 1937–40 and becomes clearer after the new examination regulations were established in 1941. Nevertheless, the rapid realignment of criteria was amazing.

In the twenties, it would be hard to find a case in which an appointment was influenced by practical professional needs. At the most we find an orientation to practical questions such as school reform, which was important to the ministry when it appointed Peters in Jena. When the Frankfurt faculty

28 Letter from Borsig GmbH, Berlin, 11 August 1983; cf. Geuter (1984) and Ash and Geuter (1985).

in 1928 discussed a successor for Friedrich Schumann, for example (in the end, Wertheimer was called), it attached a long foreword to the list on the value of experimental psychology for the "renewal of philosophical concepts." When in 1930 the faculty in Halle needed a successor for Theodor Ziehen, it placed "special emphasis on experimental-psychological activity and ability."[29] But the faculty expressed the same opinion again from 1933 to 1938 when a successor was to be found for Adhemar Gelb. At the end of 1933, the faculty could not see a suitable candidate to occupy a "chair for psychology oriented primarily to natural sciences." They requested that the ministry appoint an instructor as a temporary replacement and suggested Wolfgang Metzger, who had taken over Gelb's former assistant post in Frankfurt, because he was able, "with exact natural scientific methods, to set up experiments designed to bring the problems under investigation closer to a solution."[30]

The case of Halle thus demonstrates a certain continuity of criteria for appointments on the part of a faculty. But other criteria were introduced here as early as 1933. Not only that, Mattiat from the ministry tried to get Schole to Halle for political reasons, as we saw in the preceding section; in November 1933 the Reichswehr Ministry also got in on the act. Along with Metzger, the faculty had named Philipp Lersch and Heinrich Düker. The Psychotechnical Laboratory of the Army Ministry now wrote to the Prussian Ministry of Science in support of Lersch. The Dresden Institute, at which Lersch was an instructor, had recommended him for the chair in Halle. It was desirable that the problems of military psychology receive consideration in the universities, and Lersch was deemed suitable for this in view of his experience as an army psychologist.[31]

This was perhaps the first intervention by the Psychotechnical Laboratory, but it was not the last. Subsequent events, especially concerning new chairs in Breslau and Erlangen, show that the desire to give consideration to the problems of military psychology increasingly entered into university and ministerial deliberations. Table 4 shows vacancies and appointments for the years 1936–41. Documents are not available in all cases, but those examined indicate the arguments used, who was interested in which specific appointment, and who suffered as psychology gained in strength in this period.[32]

29 PA Wertheimer, letter from Natural Sciences Faculty, 30 July 1928. For Halle, Rep 6/1368, letter from Faculty, 27 July 1930.
30 Rep 76 Va, Sect. 8, Tit. IV, 48 Bd. 10 ff. 57–8.
31 Ibid., f. 59.
32 Documents do not show the reasons for Bollnow's call to Giessen, Petermann's to Göttingen, G. H. Fischer's to Marburg, or von Allesch's to Halle and Göttingen. Von Allesch had the experimental background for Halle, and was sent to Göttingen shortly before retirement when a new start was sought in Halle. Fischer succeeded Jaensch, who had been his teacher; Petermann was known for his criticism of Gestalt psychology and his work on race psychology. The inclusion of these cases might add detail to the overall picture, but would certainly not blur the trend of the primacy of practical relevance.

Table 4. *Vacancies, appointments, and substitutions*

Year	New vacancies	Chairs declined[a]	Appointments	Substitutions
1936			Dresden Technical College: Lersch	
1937	Dresden Technical College Göttingen Leipzig Munich		Breslau: Lersch Dresden Technical College: Straub	Halle: Metzger
1938	Giessen Greifswald Tübingen	Giessen: Klemm Leipzig: Sander	Halle: von Allesch Greifswald: Schole[b] Munich: Kroh Tübingen: Pfahler	Giessen: Bollnow Göttingen: Pfahler, Petermann Leipzig: Klemm
1939[c]	Breslau	Berlin: Sander Breslau: Straub, Gottschaldt	Giessen: Bollnow Göttingen: Petermann Leipzig: Lersch	Frankfurt: Metzger[d] Posen: Hippius
1940	Marburg		Breslau: Eckle Erlangen: Wenke Königsberg: Lorenz Marburg: G. H. Fischer	Rostock: Koch[d] Prague: Scola
1941	Göttingen Halle		Göttingen: von Allesch Strassburg: Bender	

[a] This column is not complete.
[b] First as an associate professor (without tenure), from 1942 as a nonbudgeted professorship. The other chairs were budgeted.
[c] From 1939 chairs in the occupied areas allocated under ministerial guidance are included.
[d] As nonbudgeted professor after the chair had been lost.

There were five out of eight chairs whose designation was specified more toward psychology in the period 1937–41: Halle, Königsberg, Marburg, Munich, and Tübingen. Two out of five newly created chairs were established in this period: Breslau and Erlangen. After the occupation of Austria and Czechoslovakia and the conquest of Poland, the ministry also became responsible for the universities of Graz, Innsbruck, Vienna, Prague and Posen. In Austria, Otto Tumlirz (Graz) and Theodor Erismann (Innsbruck) kept their posts, whereas Karl Bühler, who had once been close to the Social Democrats and had a Jewish wife, landed first in prison and then in exile (Ch. Bühler 1972). His chair in Vienna remained vacant till 1943. At the German University in Prague, Johannes Lindworsky was full professor of psychology from 1928 until his death on 9 September 1939. His successor was found under German rule; in December 1940, Franz Scola became associate (budgeted) professor (BA R 31/649). In Poland academic appointments quickly followed the military occupation. Among 119 Baltic-German scientists who came to Posen University at the end of 1939 (Kalisch and Voigt 1961, p. 192) was the psychologist Hippius (Z. f. 82). Following the official establishment of the "Reich University of Posen" on 27 April 1941 (Wroblewska 1980), a chair for psychology and pedagogy was established but not occupied until 1943.

When Oswald Kroh went to Munich in July 1938, his vacated chair for educational science in Tübingen was renamed a pedagogical-psychological chair. The Reich ministry had made it plain that it wanted Gerhard Pfahler to get the position. The faculty then listed him pari passu with another Nazi, Friedrich Berger, an SS man and professor of pedagogy at Braunschweig Technical College. Political-ideological motives were evident. While the faculty seemed to signal a preference for Berger, the rector recommended that the state ministry support Pfahler, since the Reich Führer of lecturers, Walter Schultze, felt it would be appropriate. In addition, Berger's work was not up to Pfahler's.[33] Pfahler was appointed on 1 October 1938. In 1939 the educational science seminar of which Pfahler had become director was renamed the Institute for Psychology and Educational Science.[34]

This appointment in 1938 was obviously still governed by the "primacy of politics." But it was not typical of this phase, which is why it is mentioned first. Even Erich Jaensch, second to none in his public efforts to create a psychology oriented to the "German movement" (Jaensch 1933), used a different criterion for appointment matters. He was asked by the rector at Tübingen to give a confidential report on Gert Heinz Fischer, who had qualified under Jaensch in Marburg and had also been named on the list. In his evaluation, Jaensch recalled Fischer's activity as an army psychologist, and commented that "a university teacher for psychology, who in future will have to prepare others for the profession of army psychologist and related professional fields, will not be able to do without this experience."[35]

33 205/96; Adam (1977, pp. 144ff.) also describes these events. He mentions a letter from Himmler to Heydrich on 5 September 1938 saying that Professor Hauer had asked for Berger's appointment. Adam thinks that the RMWEV followed the wishes of the leadership of the NSD.
34 131/124; see also *Psychologen-Taschenbuch* (1955, p. 92).
35 205/69, Jaensch to the Rector, 20 July 1938.

This argument came to play an increasingly important role. On the death of a professor of philosophy and pedagogy in 1937, the University of Erlangen applied to have the chair changed to one for psychology and pedagogy. In its petition of 1 December 1937, the faculty referred to the lack of a chair for psychology as "the most noticeable deficit ... in the entire range of chairs in its responsibility." In an age of "experimental psychology," this had been bearable, but not today; first, psychology "allows an augmentation and combination of all those sciences that refer to the racial soul (*Rassenseele*)" and, second, the career regulations for Wehrmacht Psychology from November 1937 assumed "that in the universities the opportunity existed to study psychology and to obtain a doctorate." But a specialist course in psychology was hitherto not possible in Erlangen.[36] The rector and the Bavarian State Ministry supported the petition, which was approved by the ministry.

When the faculty attempted to get Hans Wenke, a pupil of Eduard Spranger and Wehrmacht psychologist in Nuremberg, appointed to the chair, the Amt Rosenberg intervened. The three-year-long conflict casts some light on the rivalries between various parts of the Nazi Party over science policy, which have been dealt with extensively elsewhere (Bollmus 1970; Losemann 1977). In 1938 the faculty authorized Wenke to teach, and on 19 January 1939 the ministry awarded him a lectureship, after which he was to receive an associate professorship. The obligatory inquiry at the party chancellery met with the reply that there were no objections to Wenke (14 September 1940). The NSD also approved of Wenke, as did the security service of the SS. However, Baeumler, still head of the Science Office at the Amt Rosenberg, objected on 10 October. Ideologically, Wenke was a pupil of Spranger and was not dedicated to the National Socialist worldview. He was not to be entrusted with a chair for educational science. It is probably significant that in connection with his politically motivated call to Berlin University in the spring of 1933, Baeumler had been involved in a bitter feud with Spranger.

In December 1940, Wenke was made associate professor (budgeted). The conflict with the Science Office of the Amt Rosenberg, however, carried on until 1943, when Wenke was to be made a full professor, which required a further statement from the party chancellery. The correspondence reflects considerable tension between the party Science Office and the NSD. The Science Office made use of an argument that ran counter to the support otherwise extended to psychology in the Third Reich. Wenke was a representative of the "humanistic (*geisteswissenschaftlichen*) psychology that had pushed itself to the fore in connection with the growth of Army Psychology"; this psychology, particularly of Wenke's type, was still without exact

36 UAE T 11, I, 44 W, Faculty to the Rector, 1 December 1937. The first psychological dissertation was submitted in Erlangen at the same time as Wenke became associate professor (Letter UAE, 3 June 1981).

foundations.[37] It is hard to decide whether this was really a serious argument. The attacks launched against the Wehrmacht from the Amt Rosenberg (see Chapter 5) would suggest it was; the suspicion that Baeumler had other motives for his offensive against Wenke would suggest it was not. This illustrates again the problem of analyzing records. It is scarcely possible to distinguish real motives from pretexts. Anyway, in our case none of the objections was successful. The party chancellery judged Wenke on his political behavior in the Third Reich, and not on the ideological purity of his scientific activities. He became full professor in March 1943.

Events in Breslau resembled those in Erlangen in some ways. In 1936 three chairs were vacant: the old professorship for pedagogy once held by William Stern and two philosophical chairs, one of which was Roman Catholic by statute. This was to train future priests in accordance with the concordat between the government and the Holy See (July 20, 1933, RGBl II, 1933, p. 684). The ministry decided that one of the chairs would have to go and reduced the budget. This was the beginning of a long quarrel. In late 1936 the philosopher August Faust was appointed. The university wanted to have one secular philosopher in addition, in order to avoid having the Catholic philosopher as second examiner. Voices were raised at the university to save all three professorships, one of them for psychology and pedagogy. The faculty, the rector, and the new professor, Faust, drew attention to the needs of the Wehrmacht and the guidance of the young with the use of psychology. Faust pointed out that the labor service included in the Four-Year Plan made well-trained vocational counselors and industrial psychologists necessary; the Wehrmacht also needed well-trained military psychologists. The commander in chief of the army wrote to the ministry in 1937 confirming the arguments advanced by the rector "since the desolation of numerous chairs of psychology in German universities endangers the training of new recruits for Wehrmacht Psychology" (Rep 76/131, f. 164).

The ministry was in a tricky position, because it had introduced a policy of allowing only two chairs for philosophy, one for its history and systematics, the other for psychology and pedagogy. In this case, the second chair had to go to a Catholic professor, which duly happened early in 1937. The university continued to press for a third chair. The NSD also got involved, calling for the influence of theology to be reduced.[38] The ministry finally found a solution without having to request funds for a new chair from the Finance Ministry; a chair that became vacant in the Theological Faculty was declared superfluous, and this became a budgeted chair for an associate professor of psychology and pedagogy.

---

37 MA 116/17, Wenke. Information according to written communication from UAE.
38 These worries are understandable. In 1937 and 1938, attendance at the lectures of the Catholic philosopher Bernhard Rosenmöller was between three and eight times larger than that at the lectures of Faust and Lersch (Rep 76/131, f. 196). On the policy of two chairs for philosophy, see Rep 76/131, f. 160 and Rep. 76/202, f. 106.

There were two considerations at work here: one was to curtail theology, the other to meet the demand for professionally active psychologists in the labor service and the Wehrmacht. In this case, the latter tipped the balance. The faculty nominated Philipp Lersch at the top of its list, followed by Bruno Petermann and Otto Klemm jointly second. They made it plain that had the rules allowed, they would have named only Lersch; "he is going new ways, and ... these ways are of special importance for the national interest of the German *Volk* of the present, namely for its military capability" (Rep 76/131, f. 166). Oswald Kroh, asked his opinion of the list, also favored Lersch. He emphasized that of all university teachers, he was closest to the work of the Psychological Laboratory of the War Ministry (f. 169). Finally, Colonel Sehmsdorf of the Eighth Army Corps Testing Station in Breslau also spoke up for Lersch. The ministry appointed him. A handwritten note by Mattiat says that he "has the edge in all respects" among those recommended: "Among others, the Psychological Laboratory of the Reich War Ministry (Colonel von Voss) actively supported Lersch" (f. 177). Though Lersch had been associate professor for only a year, he was found worthy of a personal full professorship, though the post itself was budgeted at the associate level so that a successor could be classified differently if desired.

This is, in fact, exactly what happened when Lersch went to Leipzig in 1939. An instructor, Christian Eckle, from Tübingen was called to the chair in Breslau as associate professor. In this case we can again see how important the argument of practical relevance had become in psychology. At first the faculty had nominated Werner Straub, who had been Lersch's successor at the Dresden Institute. It argued that he worked along the same lines as Lersch and underlined the Wehrmacht's interest in continuity. The military head of Wehrmacht Psychology wrote directly to the rector in Breslau (21 October 1939) that the Inspectorate for Aptitude Testing would welcome Straub's being called. He was serving as army psychologist in Dresden, but would be posted to Breslau if appointed there. However, Straub rejected the offer. The dean deeply regretted this because he knew nobody else who was so at home with Wehrmacht Psychology as Straub (Rep 76/131, f. 232). The dean wished that the ministry would provide a full professorship to win Straub over. In his letter he argued solely in terms of the need to give educationalists and future psychologists in the Wehrmacht or employment offices a thorough training in psychology. Ten years earlier it would have been impossible to argue only on the basis of training requirements. A reference to the research work of the individual, previously the most important item, is missing in both this letter and the faculty's list. Instead, the rector pointed out that Straub was particularly suited for teaching because of his background in practical psychology (f. 234). The faculty was content to note that Straub was active in the same fields as Lersch and that he was particularly well acquainted with research on expression, which was important for Wehrmacht Psychology (f. 220; cf. Chapter 3).

Since Kurt Gottschaldt also refused the post and the ministry would not upgrade it, the faculty finally submitted a list mainly of younger instructors (23 February 1940): Eckle, Wellek, G. H. Fischer, Metzger, Keller, and Seifert. At a time when comments were beginning to get shorter and shorter, there was only one sentence on each candidate in this mixed bag – with the exception of Eckle. Precisely this brevity reveals the central arguments very well. Wellek was "too much of a specialist in musical aesthetics" (under different circumstances three years later he was to receive a call to Breslau; see Chapter 7). Only one short comment was made in favor of G. H. Fischer: he "is a leading army psychologist"; against him counted that he was not sufficiently "matured." Metzger and Keller were put down because of their experimental psychology, and Seifert leaned too much toward philosophy. Metzger had already been criticized for his narrow research interests in 1938 by both the ministry and the NSD in connection with the chair in Halle (Rep. 76/202, f. 105). Research in experimental psychology was no longer highly viewed. Eckle seemed appropriate for another reason: "He seems above all suited to maintain contacts between the chair and the Wehrmacht, the labor service, and vocational training, which we value highly" (Rep 76/131, f. 241).

One might think that the points about training Wehrmacht psychologists in Breslau arose due to some special local cooperation. But this is not backed up by the events described in Erlangen, nor by those in Munich. Here the Wehrmacht intervened after the educationalist professor Aloys Fischer had been dismissed under the new race laws discussed earlier. Initially the faculty wanted a full-fledged pedagogue, but it needed someone who would also take charge of the Psychological Institute, which Erich Becher had led until 1929. Only a month after Fischer had been forcibly retired, the Wehrmacht expressed its interest in cooperating with his successor. It favored Lersch, but he was going to Breslau. In the end, Kroh came, an educationalist who had worked in the field of psychology and had written on army matters (Kroh 1926). The new name of the chair, which Kroh took over in July 1938, was "Chair for Pedagogy and Psychology (with Special Consideration for Army Psychology)" (Schumak 1980, p. 340). This was unique in the Third Reich and reflects the importance of Wehrmacht Psychology.

Wehrmacht Psychology intervened in psychological matters at the universities in a number of ways. Its military head, Hans von Voss, was just as concerned to see psychology adequately represented as Krueger and Jaensch, his successor in 1936 as president of the DGfPs. On 7 May 1936 von Voss spoke with Werner Zschintzsch, the permanent secretary at the ministry. One topic was making use of psychological institutes for Wehrmacht Psychology research; a second was the preservation of psychological professorships. Von Voss argued that vacated chairs should not be transferred to other disciplines and felt that his views were understood (NS 15 alt/17, f. 56946). In 1937, he requested that the philosophical faculties allow doctorates in psychology (see Chapter 6). On 9 August 1938, he informed Baeumler at the Amt Rosenberg that as of April 1939, the number of Wehrmacht psychologists would probably increase, so that the "patriotic duty" of all university teachers of psychology was "to take care that there are new psychology recruits for the Wehrmacht" (NS 15 alt/22, f. 58844).

In June 1938, von Voss demanded a rapid replacement for the vacant chair in Leipzig.[39] The appointment of Lersch to this chair is a clear enough sign that the ministry was also encouraging research oriented toward Wehrmacht Psychology. The Leipzig chair of Wundt had the longest tradition in Germany, and together with the Berlin chair enjoyed the highest prestige. Because of this, the Faculty Commission wanted to have a scientist "who has shown with his publications and teaching that he is also willing to go his own new way." At first Sander headed the list, but he rejected the ministry's offer. An unpleasant row ensued as Volkelt tried to establish his claim to the chair. He had been named second on the list together with Klemm, but the faculty was unwilling to accept someone who closely followed Krueger's line "without fully reaching his characteristic scope." Lersch had originally been placed third on the list, because no one thought the ministry would appoint him. However, when Sander refused, it turned out that the ministry favored Lersch. It was also reported that "at the Psychology Conference in Bayreuth Lersch had been called the coming man in psychology." He was moved up to first place.[40]

Lersch had helped to build up psychology in the armed forces from 1925 to 1933. He had already received backing in 1933 in Halle, 1937 in Breslau, as we saw, and 1938 in Munich. As professor in Breslau, he cooperated with the testing station of the Wehrmacht and was the university examiner at assessor examinations in Wehrmacht Psychology (see Chapter 6). In 1932 Lersch had presented a theoretical evaluation of the experience with expression diagnostics in Wehrmacht Psychology in *Gesicht und Seele* (Face and soul). His new book, *Der Aufbau des Charakters* (The construction of character) appeared in 1938. His appointment was a recognition of the psychology of expression and characterology, as well as the importance of Wehrmacht Psychology. The sequence of Wundt, Krueger, and Lersch on the Leipzig chair mirrors the change of predominant concepts in the history of German psychology.

Working for Wehrmacht Psychology was a credential for aspiring professors. Wehrmacht Psychology clearly expressed its interest in university training, which supported the institutionalization of psychology as an independent discipline. In Germany under the kaisers, the status of a discipline and the extent of state support depended, according to Ash (1982), on the

---

39 *Leipziger Neueste Nachrichten*, 1 February 1939. Quoted in UAL PA 636, f. 69.
40 Details from a letter of the dean of the philosophical-historical section of the Philosophical Faculty to Minister of Education in Saxony, 10 January 1939 (UAL PA 62, ff. 89ff.). For Volkelt's efforts see Geuter (1984).

role it played in the system of elite education at the universities. As long as psychology was not specifically preparing for a profession, its formal role was limited to a basic training in philosophy for teachers, which meant it had no justification for an independent existence. The conflict over doctorates in psychology clearly demonstrates this (see Chapter 6). However, the demand for Wehrmacht psychologists and the employment of psychologists in industry, the labor administration, and the Labor Front established the need for professional training. That made it possible to demand professorships and also explains why practical experience became a criterion for qualification.

Two further factors were important for successful institutionalization. The less important of the two was the weakening of the main academic rival, philosophy. Before the Nazi period, the rivalry between "pure" philosophy and psychology had delayed the establishment of the latter, but now the balance of forces clearly shifted in favor of psychology. Philosophy was in direct competition with the pretensions of Nazi ideology. Even the right-wing German Philosophical Society (whose chairman from 1927 to 1934 was Felix Krueger, followed by Hans R. G. Günther, who went to Wehrmacht Psychology in 1936) had a harder time than the DGfPs in this period (cf. NS 15 alt/17). For all the problems of von Ferber's figures for the numbers of professorships in philosophy/psychology/pedagogy (see note 3), they are impressive for philosophy: fifty-six full professors in 1920 and 1931 compared with thirty-six in 1938 (von Ferber 1956, p. 207). In fact, some of these chairs had been taken over by psychologists. The new chairs for psychology were not newly budgeted posts (with the exception of Königsberg – but there an old chair was also lost), but were taken from other disciplines. The professorships in Erlangen and Giessen replaced philosophical teaching contracts with psychological ones. Both at Halle in 1938 and at Marburg in 1939, a chair was specified in the direction of psychology to the detriment of philosophy. In Tübingen and Munich, chairs for educational sciences were redefined for psychology and pedagogy; in Breslau a new chair was created after the abolition of the associate professorship for pedagogy at the expense of theology. Thus, psychology grew strong while other disciplines – theology, pedagogy, and above all philosophy – were weakened. The development outlined here for the years until 1941 was to continue after the DPO became established (see Chapter 7).

The more important factor, however, was the change in Nazi policy toward the universities. Initially, this had concentrated on ideological control and the political and anti-Jewish purging of lecturers. After the Four-Year Plan of 1936, science became oriented more and more toward effectiveness. Kelly writes that "it became apparent to the state officials that the universities must continue to be effective... in satisfying the technical needs for an industrial society, especially one preparing for war" (1973, p. 459). Practical relevance therefore seemed more important than ideology if psychology was to improve the deployment of people in industry and the armed forces.

Between 1937 and 1941, about half the psychologists appointed professor were party members, although in this period political criteria were not decisive for receiving appointments.[41] For example, Philipp Lersch was not a party member, but as we have seen he received various calls. Yet veteran party member Hans Volkelt, who regularly wrote reports for the Amt Rosenberg, was never offered a post through the entire Nazi period. An investigation of Frankfurt University by Zneimer (1978, p. 155) shows that by 1936 the party had ceased to be a vehicle for advancement at the universities. Kelly (1973, pp. 324–5) is of the opinion that as the shortage of university teachers grew worse the pressure on lecturers to join the party decreased. From 1940 to 1941 the ministry paid more attention to scientific qualifications.

It is difficult to discern direct party influence on the awarding of chairs for psychology; at least in none of the cases investigated was a candidate openly pushed into a post by the party. In Breslau, Halle (von Allesch), Leipzig, Erlangen, and Marburg, the ministry followed the recommendation of the faculty without objections being raised by the party. This was not so in Tübingen, but there two Nazis were competing against each other. About the other cases I found nothing. Of course, it is not possible to ascertain from the documents whether the faculties checked matters out in advance to avoid disagreement.

An additional comment is appropriate on Königsberg and Strassburg, which are to some extent exceptions in the period under consideration. In Königsberg, the Austrian zoologist Konrad Lorenz, whose ethological theory of decline through domestication and civilization augmented Nazi criticism of culture – he himself linked his views to race doctrine (Kalikow 1980) – was called to the chair in September 1940. According to Hans Thomae, this was to the "astonishment of the psychological world" (1977, p. 154) and occurred because of "the intervention of Minister Rust against the resistance of the faculty" (Z, f. 89). After the retirement of Schultze in 1935, the full professorship for "pedagogy, philosophy, and experimental psychology" had gone in 1937 to the folklorist Heinrich Harmjanz (who also became a senior official in the RMWEV).[42] Lorenz was awarded a new professorship for psychology on 1 January 1941 along with the directorship of an Institute for Comparative Psychology as a second section of the Philosophical Institute. The Königsberg records are totally destroyed, but from the statements of Thomae and Kalikow we can conclude that it was the biologistic orientation of Lorenz's work that led to the full professorship being made available for psychology.[43]

41 For details see Ash and Geuter (1985); for physics Beyerchen (1977). For a different approach see Adam (1977, p. 207).
42 Rep 6/1368, RMWEV to Curator, 5 August 1937.
43 Thomae mistakenly refers to a "philosophical chair" (1977, p. 154). Facts here correspond to the staff lists and lecture lists of the University of Königsberg. On the loss of records, the Wojewódzkie Archiwum Państwowe in Olsztyn, 6 May 1981. The Central State Archive in Potsdam also reported having no files on appointments to chairs. Possibly the ministry's faculty files would prove helpful.

As in Posen[44] the Reich University founded in Strassburg after the occupation was to include a chair for psychology. The plan at Strassburg was to link the natural sciences and humanities as "biological" faculties; there seemed to be the promise of large-scale financial support. Ernst Anrich, a historian from Bonn, worked out the plan, which included a full professorship for psychology and pedagogy and an associate professorship for medical psychology. This was intended to provide the link between psychology and clinical practice. On Anrich's recommendation, the latter went to the instructor Hans Bender, who had studied both medicine and psychology. It probably is a sign of the shortage of suitable candidates that the full professorship remained vacant.[45]

The strengthening of the discipline: quantitative aspects of institutionalization, 1933–45

The institutes in Berlin, Frankfurt, and Hamburg enjoyed an international reputation and attracted U.S. students in the twenties (Z, f. 178). These institutes were all severely affected by the Nazi persecution of Jews and political opponents. This must have left many North Americans with the impression that psychology had been devastated in Germany, which may have contributed to the common perception of psychology's continual decline throughout the Third Reich. Yet the subject did not meet the "distrust and aversion of the Nazis" (cf. p. 9) and was not persecuted, curtailed, or done away with as a subject, nor did such plans exist – as Wellek (1960) claimed. On the contrary, psychology (the lack of comparative studies does not yet allow us to say "above all, psychology") was able to recover institutionally from the initial blows, and indeed was able to improve its position in the universities relative to 1932.

This can be well demonstrated quantitatively through an overview of the chairs held by psychologists over the whole of the Third Reich (Table 5). I use the following criteria of improvement to assess whether psychology grew stronger over this period:

1. the establishment of new chairs, which include psychology as an area to be taught;
2. the transformation from a nonbudgeted to a budgeted professorship;
3. the upgrading from an associate to a full professorship; and
4. the specification of a chair to the advantage of psychology.

In each case, the converse is seen as a weakening of the subject. Points 3

---

44 In Posen, Hippius was temporarily in charge of the institute. The new chair for psychology and pedagogy, for which the Amt Rosenberg rejected H. R. G. Günther and L. F. Clauss in 1941 (MA 116/13), was empty until Eckle came in 1943.

45 Details from Z, f. 84, and contemporary letters from Bender to Pfahler, Anrich, and Kroh (UAT 148); see Chapter 7.

and 4 are weighted equally, so that a downgrading accompanied by a specification in favor of psychology is classed as "no change."

This overview takes no account of whether the designated teaching and research were actually carried out. This would make such a comparison impossible, especially during the war, when many professors were enlisted. Nor are the effects of general restrictions on science considered. The count covers the existence of budgeted professorships. Three budgeted associate professorships in Leipzig (Wirth, Klemm, and Volkelt), which existed alongside the full professorship and were all created before 1933, are not included. The comparison covers the period from 1932 to 1945.

Table 5 shows that at three universities the position of psychology worsened; at seven it was unchanged. At one a nonbudgeted post was transformed to a budgeted one; at one an associate professorship was made a full post; at six the chair was specified toward psychology (in two further cases a specialization occurred parallel to a downgrading of the professorship). A further five budgeted posts were set up in which teaching was to include psychology. According to our criteria, therefore, the situation at thirteen of the twenty-three universities was improved. An exception should perhaps be made in the case of Cologne, since the associate professorship set up in 1942 was never occupied, although the Institute for Experimental Psychology had been set up under Robert Heiss in 1938. It should also be noted that the comparison with 1932 does not take into consideration earlier setbacks. In Cologne, for example, Associate Professor Lindworsky taught psychology until 1928. We have already mentioned Bonn, Breslau, and Königsberg. Nevertheless, the overall picture is quite clear. Claims that psychology was nearly abolished in the Third Reich, or that its existence at the universities was threatened, are without foundation. On the contrary, its institutionalization advanced. In a speech in 1930 Stern (1931) had bemoaned the situation at nine universities: Bonn, Breslau, Freiburg, Greifswald, Heidelberg, Cologne, Königsberg, Münster, and Tübingen. In Bonn, the situation had already improved in 1931 with the appointment of Behn. There were improvements at all the other universities apart from Greifswald and Heidelberg during the Third Reich. When did these improvements occur?

For our purposes, the Third Reich can be fairly naturally divided into three parts by the start of the Four-Year Plan (1936) and by the introduction of the DPO in 1941. In the period 1932–6, the situation improved in Würzburg (through specification) and in Giessen and Kiel by upgrading from associate to full or from nonbudgeted to budgeted professorships. It should, however, be emphasized that the dismissals and vacancies meant that the overall situation in teaching and research was actually a good deal worse than before. From 1937 to 1941, four professorships were specified more toward psychology (in Halle, Königsberg, Munich, and Tübingen). In Marburg there was also a specification, but it was coupled with a downgrading. Two new professorships were created, in Breslau and Erlangen.

Table 5. *Psychological chairs at the universities, 1932–45*

Year	Berlin[f]	Bonn[a]	Breslau[b]	Cologne[b]	Erlangen[b]	Frankfurt[c]
1932	KÖHLER (Phil.)	ROTHACKER (Phil.)		Vacancy Head of institute		WERTHEIMER (Phil. (Psy.))
1933						Dismissed
1934						Metzger
1935	Rieffert					
1936	Keller		Heiss (Phil. Psy. Character.)			
1937			LERSCH (Psy. Ped.)			Metzger
1938	Schering					Madelung
1939			Vacant			Metzger (non-budgeted Psy.)
1940			ECKLE (Ass. Prof. Psy. Ped.)		WENKE	
1941				Vacant	Budgeted Ass. Prof. Psy. Ped.	
1942	KROH (Psy.)			Budgeted Associate Prof. Psy.		
1943			WELLEK			
1944						
1945	KROH	ROTHACKER	WELLEK	Vacant	WENKE	

Table 5. (*cont.*)

Freiburg[b]	Giessen[d]	Göttingen[a]	Greifswald[a,e]	Halle[f]	Hamburg[c]
●	●	●		●	●
Cohn (Ped. Phil.)	SANDER (Ass. Prof. Exp. Psy. Ped.)	ACH (Phil. Psy.)	*von Allesch* (Ass. Prof.)	GELB (Phil. (Psy,)) Dismissed	STERN (Phil.)
Dismissed	Vacant			Vacant	Dismissed
					Deuchler
Stieler (Phil.)	PFAHLER (Psy. Ped.)			SCHULTZE (subst.)	
				SCHOLE (subst.)	
		ACH Vacant		Vacant	
				Metzger (subst.)	
	PFAHLER	PFAHLER (subst.)	*von Allesch*		
	Vacant BOLLNOW (Psy. Ped.)	PETERMANN (Psy. Ped.)	*Schole* (Phil. Ped.)	VON ALLESCH (Psy.)	
		VON ALLESCH (Psy. Phil.)		Vacant	
HEISS (Phil. Psy.)				WILDE (Ass. Prof. Psy.)	ANSCHÜTZ (Ass. Prof. Psy.)
				WELLEK	
				WILDE	
HEISS	BOLLNOW	VON ALLESCH	*Schole*	WILDE	ANSCHÜTZ
●	●	●		●	●

Table 5. (cont.)

Heidelberg[a,e]	Jena[a]	Kiel[d]	Königsberg[f]	Leipzig[a]	Marburg[a]
●	●	●	●	●	●
Hellpach (Honorar Prof. Social. Psy.)	PETERS (Psy.) | Dismissed ┬ SANDER (Psy.)	*Wittmann* (Ass. Prof. Phil. Psy.) ● (Budgeted Ass. Prof. Psy.)	SCHULTZE (Ped. Phil. Exp. Psy.) Vacant ┬ Ipsen Vacant ┴ LORENZ (Psy.)	KRUEGER (Phil.) Klemm Volkelt ┴ LERSCH (Phil.) RUDERT (subst.)	JAENSCH (Phil.) G. H. FISCHER (Ass. Prof. Psy. Ped.)
Hellpach	SANDER	WITTMANN	LORENZ		G. H. FISCHER
●	●	●	●	●	●

Note: Filled circles indicate budgeted post. Capitalized names indicate chairholders or substitutes. Noncapitalized names indicate provisional representation of the subject or provisional direction of the institute. Italics indicate nonbudgeted professorships.

Table 5. (*cont.*)

Munich[f]	Münster[b]	Rostock[c]	Tübingen[f]	Würzburg[f]	Year
●		●	●	●	
A. FISCHER (Ped.)		KATZ (Ped. Psy.)	KROH (Education. science)	MARBE (Phil., Aesth. Ped.)	1932
		Dismissed			1933
		Keller (Ass. Prof.)			1934
		J. Ebbinghaus instructor		JESING-HAUS (Phil. Ped. Psy.)	1935
					1936
Vacant					1937
KROH (Psy. Ped.)			PFAHLER (Psy. Ped.)		1938
		Koch (assistant)			1939
					1940
					1941
LERSCH (Psy. Ped.)	METZGER (Psy.)				1942
					1943
					1944
					1945
				JESING-HAUS	
LERSCH	METZGER	Koch	PFAHLER	HAUS	
●	●		●	●	

[a] No change. [b] New chair. [c] Loss of chair.
[d] Improvement. [e] No chair at all.
[f] Specification to psychology.

After 1941 one chair was specified more clearly for psychology in Berlin. In Hamburg a new and more clearly specified chair was created, though this was not a full professorship, as Stern's chair had been eliminated. Three new chairs were created in Freiburg, Cologne, and Münster. In 1943 the professorship in Erlangen, which was only three years old, was upgraded. Thus, in the first phase improvements took place through upgrading of professorships. In the period 1937–41 chairs were mainly specified for psychology in connection with the importance of psychology for the training of practical psychologists. After 1941 the trend was more toward setting up new chairs as a consequence of the new examination regulations (see Chapter 7). Psychology was no longer an auxiliary of philosophy but a subject in its own right, one that prepared one for a profession.

The ten technical colleges in the German Reich (Prahl 1978, pp. 365–6) show no such marked developments. At the Breslau Technical College psychology was not represented during the whole Third Reich, though it was represented at the university there. The following information could be gathered about the other technical colleges.

Aachen: Adolf Wallichs and Herwart Opitz headed the Laboratory for Psychotechnics until 1942. In 1940 Joseph Mathieu was called as instructor for industrial psychology; on his appointment as nonbudgeted professor he set up the Institute for Work Leadership in 1942.

Berlin: Walter Moede remained and was made associate professor in May 1940.

Braunschweig: Bernhard Herwig remained full professor and set up the Institute for Psychology around 1934, which was renamed the Institute for Industrial Psychology around 1937.

Darmstadt: From 1934 until 1942 there was no institute. In 1942 the Institute for Philosophy and Psychology was founded under Matthias Meier, professor for philosophy, psychology, and pedagogy, who represented psychology at the college. The professorship of Hugo Dingler was eliminated.

Dresden: In 1936 Philipp Lersch came as professor of characterology, in 1937 Werner Straub as professor of psychology (1939: budgeted associate professor of psychology and philosophy). In 1938 Straub founded the Institute for Characterology and Philosophy, which was renamed the Institute for Psychology and Philosophy in 1939.

Hannover: Wilhelm Hische was honorary professor over the whole period.

Karlsruhe: In 1933, with the dissolution of the General Department, the Institute for Social Psychology was also lost.

Munich: Friedrich Seifert represented psychology throughout the period (see Table 3), but without a budgeted post or an institute.

Stuttgart: Fritz Giese was associate professor (not tenured) and died in

1935. It was not possible to determine whether there was a replacement (HStA Stuttgart, 6 April 1981), but it does not seem so.

At the German-language technical college in Danzig (a free city until the German occupation in 1939) Walter Ehrenstein was appointed professor of philosophy and psychology after the dismissal of Hans Henning in 1933. From 1938 he was tenured and held the post until the end of the war (I C, c.v. Walter Ehrenstein).

Thus, in contrast to what was happening at the universities, the situation at the technical colleges was relatively stable. Neither psychology nor psychotechnics was newly introduced at any college. The only new budgeted post was in Berlin, where Moede became a tenured civil servant. In Dresden Gustav Kafka's Chair for Philosophy and Pedagogy was redefined for philosophy and psychology. The Trade College of Mannheim, where Otto Selz had been full professor, was united with the nearby University of Heidelberg, so that there was no question of a direct successor on his dismissal (Bollmus 1973).

A more detailed investigation of the institutionalization of psychology could include further criteria, particularly questions concerning the independence of psychological institutes, as well as their staffing and equipment. Such material was not recorded systematically by administrators, so that it is difficult to investigate. Some details can be given by way of an example. Apart from new chairs, psychological institutes were founded – for example, the Institute for Psychological Anthropology under Jaensch in Marburg in 1933 (REM 2026, f. 132) and the Institute for Comparative Psychology under Lorenz in Königsberg in 1941. Some were reestablished – for example, the Institute for Experimental Psychology under Heiss in Cologne in 1938.[46] The assistant posts at some (though by no means all) universities were investigated, but hardly any losses were found. At Berlin University, the posts of senior assistant, as well as the two budgeted and one nonbudgeted assistants, were all retained. In 1937–8 they were held by Hans Rupp, who had been an assistant since 1907 and associate professor (not tenured) since 1919, together with Hans Keller as successor to Kurt Lewin, Hans-Joachim Firgau as successor to Karl Duncker, and Robert Beck as successor to Otto von Lauenstein. In 1938 Kurt Gottschaldt from the Kaiser Wilhelm Institute for Anthropology, Human Genetics, and Eugenics came as associate professor, and in 1936 Ludwig Ferdinand Clauss came as instructor for race psychology (though he had to take his leave in 1942 after party exclusion proceedings). In Leipzig the assistant posts were also retained. In Munich two posts seem to have been cut in 1937, but one assistant post was restored in 1938, and an additional post for a scientific auxiliary set up in 1943. In Freiburg an

---

46 For Königsberg, Kürschner ... 1940/1; Personal- und Vorlesungsverzeichnis, Summer 1942. For Cologne, UAF PA Heiss.

82   *The professionalization of psychology*

assistant post for psychology was created in November 1934, and in Jena with Sander's appointment in 1933 an auxiliary assistant post, which was upgraded to a budgeted post in 1940 in the course of negotiations to persuade Sander to stay. In Würzburg there was no change in the one post set up in 1904, nor in Frankfurt's two posts.[47] At the universities named, only the posts in Frankfurt remained unoccupied for a lengthy period. In Hamburg posts were abolished. In 1930 the institute had five assistants and one auxiliary; by 1939 there was only one lecturer and one assistant.[48] This institute, which had been second only to Leipzig in terms of personnel, was practically liquidated by the Nazis. In 1942 a new associate professorship was created, but still with only one assistant.

47 Collected from a wide range of sources. On Würzburg, Strunz (Z, f. 91) reports that Walter Jaide came to the institute as an assistant. But until 1942 there was only one post, which Strunz himself had occupied since 1937 (REM 821, f. 317); see Geuter (1986, p. 84).
48 Report of the *Z. ang. Ps.*, 35 (1930), p. 239; Personal- und Vorlesungsverzeichnisse, 1939–1941/42.

3

The potential of psychology for selecting workers and officers: diagnostics, character, and expression

The increased attention given to practical questions provided the major impulse for the institutionalization of psychology at the universities. It is possible for a discipline to enjoy such success for some time solely on the basis of fine words and promises. The social sciences live to some extent from the continual promise to develop new remedies for changing social problems. However, if the professionalization of psychology was to be successful in the long term, it could not just make promises, but would have to provide knowledge that could actually be applied.

Not all of the theories and methods of psychology were of professional use, and therefore of importance for its professionalization. Among the mixed bag of psychological wares, practical demand existed mainly for diagnostic assessment and the selection of personnel in industry and the armed forces, the two areas in which the professional development of psychology in Germany started (see Chapter 4). The need was for models and instruments that described and determined the abilities and personality traits of blue- and white-collar workers, soldiers, and officers. In fact, it can be said that in the period under examination the knowledge relevant to the profession was that of the selection and motivation of capable and conscientious workers and army specialists, and of strong-willed, self-controlled officer cadet applicants with leadership qualities, from among seventeen- to nineteen-year-old high school graduates. Psychotechnics, the psychology of expression, and characterology were therefore essential for the professionalization of psychology. In industry and labor management, the major area of psychological practice in the twenties, psychotechnics was expected to select individuals or assign work tasks according to abilities. Required later were also the determination of "work virtues" (*Arbeitstugenden*), such as a willingness to work, and the development of techniques for work motivation. In the armed forces, the psychotechnical examination of specialists, which had been carried out since

83

the First World War, involved the diagnosis of particular technical abilities, whereas the psychological selection of prospective officers, which began in 1927, was concerned with character traits. Here the psychology of expression and characterology were brought in.

The relevance of psychological knowledge was not restricted to industry and the armed forces. For example, independent of professional activities, psychological knowledge served to explain reality or could be used to provide empirical solutions to epistemological problems. Depending on one's political or scientific standpoint, one might have good reason to find either a socially critical psychology, or one that was relevant in terms of epistemology or worldview, or a practically oriented science that was not tailored to the dominant professional requirements to be more important. However, to discover factors of professionalization, it is necessary to determine the science used by professional experts to solve practical problems, the science for which there was a demand. This science could be described as being socio-technically relevant. When I speak of the development of scientific knowledge in psychology, I refer to this functional aspect of professional application. The critical analysis aims to determine the extent to which knowledge was connected theoretically, methodologically, and normatively to its applications in industry and the armed forces.

This chapter considers three questions. First, what theoretical and methodological developments were available in psychology to meet the external demands placed on it, and what was the practical function of these theories and methods? To answer this question, developments not in applied psychology in general, but in three, specific professionally relevant subfields will be considered: psychotechnics, expression studies (*Ausdruckskunde*), and characterology.[1]

Scientific knowledge relevant to the profession was produced not only in the universities. The largest employer of psychologists at that time, Wehrmacht Psychology, considered itself to be both a practical and a research institution. It produced, for example, several regular publications containing important results in expression psychology. The large number of standardized diagnostic examinations made it possible to amass huge amounts of empirical material for psychological research. In Wehrmacht Psychology, theories and methods of individual assessment were thus developed hand in hand with practical

1 The various developments concealed beneath the catchall phrase "applied psychology" are not considered. Originally intended by Stern for all practical applications of psychology, it was used variously in the twenties. Henning (1925) used it for everything except experimental psychology, from psychology of art through developmental psychology to forensic psychology. Messer (1927) preferred to distinguish between "practical psychology" and "applied psychology," which for him included branches concerned with the effects of the psyche on culture, such as psychology of art and religion. Wagner (1928) included applications both in cultural areas such as religion, ethics, and art and in areas such as education, industry, legal, and health services.

knowledge developed in the course of specific activities. This raises our second question, the extent to which the scientific content of expression studies and characterology – the two fields decisive for psychology in the Wehrmacht – was influenced by military considerations. More generally, the issue here is the extent to which psychological theories and methods themselves – and not merely their applications – were related to external requirements. This question has not been considered in previous historical work on characterology and expression studies in German psychology.

If scientific knowledge is to be used professionally, it must be possible to transfer it to specific procedures. It is a feature of professional activity that the professional no longer acts on the basis of individual experience but relies on the systematic knowledge of the discipline, which includes theoretical explanatory models and practical instruments for interpretation and guidance. The third question examined in this chapter therefore concerns the procedures provided by psychology for the professional activity of psychologists. Because of its paradigmatic character, the focus will be on Wehrmacht diagnostics.

In this chapter, the relationship between psychological knowledge and Nazi ideology will be considered only in passing. Nor will we consider the other institutions and instruments for human selection in the Third Reich – the racist laws and selection camps. Our attention is directed to that knowledge of psychology that was relevant to its professionalization and that was also used practically. This did not include that psychology in which the attempt was made to establish a link to Nazi ideology. In Chapter 5 we shall see how psychologists nevertheless attempted to prove the ideological relevance of psychology. Professionally relevant knowledge was developed in part before 1933, in part afterward. This account is therefore not divided into such periods, but according to the areas of knowledge and their relationship to the two major areas of application – industry and the armed services. The social problems that gave rise to the employment of this knowledge in these two fields is the subject of the next chapter.

General suitability for work and psychotechnical aptitude diagnostics

Two main areas of theoretical and methodological activity can be found in industrial psychology up to 1945: the determination of suitability for work and work performance, and the investigation of attitudes toward work and of the willingness to work. Approaches involving the social psychology of industrial organization, however, were only scarcely developed. The first area of activity mentioned marked the phase of "industrial psychotechnics" up to the twenties; the second was more prevalent in the following phase, that of the "psychology of the working individual" (Bornemann 1961). In the twenties and thirties there was a change of emphasis, replacing psychotechnics with characterology. The talk was no longer about external performance alone, but of the importance of the "working person," although psycho-technical selection according to performance was not abandoned. Both themes continued to be prominent in the Nazi period. However, methods of establishing attitudes toward work and stimulating the willingness to work were developed, and steps toward understanding and encouraging psychological leadership were pursued more and more. This will be considered in the next section.

Since the First World War, when practical application of psychology nearly merged with industrial psychology under the name of psychotechnics, the emphasis had been on the efficient use of labor power (Jaeger and Staeuble 1981). Time and motion studies, for example, led to an intensification of work. In addition to this, new technological and organizational developments in production also required new methods of evaluation, training, and deployment of labor power. As Moede wrote in 1920, machines had taken a physical load off the workers, but at the same time had placed greater demands on their senses and their attention, concentration, and determination (quoted by Wies 1979, pp. 88–9). Thus, the selection and training of workers focused more on such functions. The need for individual workers with better quali-fications gave rise to selection and training in terms of special abilities. Psycho-technicians saw their task as the provision of suitable scientific instruments for selecting, training, and deploying workers (Jaeger and Staeuble 1981, p. 54).

The methods and procedures developed for this purpose were decisive for the professionalization of psychology in the twenties. According to Jaeger and Staeuble, two theoretical concepts provided the basis for these developments (p. 59). The first was a model of people "as working machines, exhausting and regenerating themselves" (p. 60), developed by the psychi-atrist Emil Kraepelin. For Kraepelin, the ability to exercise, tendencies to tire or be distracted, and the like were basic categories of "mental work." This idea seems to underlie, above all, the early psychotechnical aptitude investigations. The second theoretical concept to be developed was that of "differential psychology." The authors trace this back to the requirement of classifying workers. A methodological framework and conceptual apparatus for differential psychology had been developed by Stern (1911).[2]

For science, models and a general theory of method were important for the solution of existing research problems. However, for professional practice, psychotechnics' practical procedures were valued more than its theoretical contributions. Contemporary textbooks show a corresponding practical bias

2 In *Differentielle Psychologie* Stern (1911) distinguished four basic tasks: the "variation theory" for testing the variation of features with many individuals, the "correlation theory" for testing the correlation between various features, the "psychography" for investigating the individual's characteristic features, and the "comparison theory" for comparing individuals.

(Giese 1927; Moede 1930; Weber 1927). Which methods were to be used for which problems?

Fritz Giese had laid out a wide range of problems confronting psychotechnics. He distinguished between "subject psychotechnics" and "object psychotechnics." The former was concerned with adapting people to working conditions, the latter by adapting working conditions and procedures to the "psychological nature of people" (Giese 1927, p. 6). Subject psychotechnics consisted of (a) vocational studies and career counseling, (b) selection and allocation of workers, (c) orientation and on-the-job training, and (d) personnel advising and guidance. Object psychotechnics consisted of (a) work psychology (time, motion, and fatigue studies), (b) rationalization of working conditions, (c) workplace illumination and accident prevention, and (d) advertising.[3] However, at the center of the methodological developments and practical work of psychotechnicians in the twenties were methods for determining individual abilities. Their basic principles can be clearly seen in the first psychological aptitude tests that Hugo Münsterberg developed in Boston to test street car drivers – at the instigation of the insurance companies, which wanted to see an end to street car accidents (Baumgarten 1955, p. 768). Münsterberg assumed that the ability to master new technology depended to a large extent on sensory performance and general mental functioning. He chose, therefore, to test simple sensory functions and nonspecific abilities. For the latter, for example, he examined people's reactions in situations where several stimuli had to be registered and responded to.

The prospect of using test equipment to make quantitative statements about sensory performance, attention, capability and reaction time was what attracted interest in wartime Germany for finding suitable pilots, drivers, and specialists in sound location (Rieffert 1921). The test developed by Moede for finding suitable drivers for the Reichswehr involved placing the subject on a testing stand, where he had to react to various signals with single or multiple actions of hands and feet, as in a vehicle. Reaction times, response distribution, and errors were measured. Tramm used similar methods for

---

3 The study of vocations was to provide job requirement profiles. This did not require an autonomous psychological method. Selection and training procedures are dealt with in this section, handling people in the next. Object psychotechnics was concerned primarily with increasing work output. The time and motion studies were undertaken chiefly by engineers and the Reichsausschuß für Arbeitszeitermittlung (REFA) (see Pechhold 1974). Object psychotechnics used technical methods to measure time and motion and physiological methods to test the effects of lighting or tools on performance, and particularly to test fatigue (after Giese). The aim was often to summarize experience, for example, when placing warning notices. Psychological attempts to sort out the accident-prone (e.g., by Marbe 1926) seem to have been less successful. However, the psychotechnical aptitude test was seen as a basic way of reducing accidents, since it kept unsuitable people away from certain jobs. The figures for a large metalworks show that technical measures, protection, and accident prevention campaigns had a much greater effect (Sonnenschein 1931).

streetcar drivers after the war. Such complex reaction experiments were to demonstrate the abilities required in emergencies.

In contrast to the United States, where paper-and-pencil diagnostics had expanded since the army tests of the First World War, German aptitude diagnostics were dominated by measurements with apparatus and work tests (Herrmann 1966, p. 263). In the twenties a number of practical test methods were developed, which were collected by Giese (1925, 1927). These included equipment to establish hearing acuity, and ability to judge speeds or assess sizes, to measure reaction speeds and the like. Various methods were used to establish general intelligence that were not yet collected in an overall intelligence test. In industry and in vocational guidance, practical-technical intelligence was tested with tasks related to technical comprehension or spatial imagery. An example of the latter was the use of Rybakov figures, where the subject had to divide given figures so that the resulting two parts could be rearranged to form a square.

Common selection procedure involved a person demonstrating specific skills. New to psychological testing was the ability to assess basic mental abilities generally. The most important features of the commodity labor power, such as sensory performance, attention, general intelligence, could be quantified and ordered, making possible comparisons among individuals, and between test results and indicators of subsequent performance. Questions of validity and reliability were often still judged by rule of thumb, but the correlation method introduced into German psychology at the turn of the century provided a mathematical tool for relating the results of psychological tests with quantifiable criteria of success.

The concentration on basic work skills can also be found in training methods, Giese's third field of subject psychotechnics. Firms faced the problem of training, for example, new street car drivers, crane operators, and typists. The aim was to provide effective, systematic instruction for untrained workers and apprentices. The idea that the essential thing was to provide training in basic abilities corresponded both with the qualification demands in question and with workers' increasing alienation from the content of their work. For example, in Saarbrücken a course for crane operators was based on a model of the engineer Adolf M. Friedrich. This involved various equipment on which general abilities such as attention, combinative ability, perception of motion sequences, judgment, and reaction capability were instilled (Bunk 1972, p. 112). Poppelreuter wanted to introduce a specifically psychological training method as part of his work for the German Institute for Technical Training (DINTA). The objective was to provide trainees with an understanding of the psychological laws governing their activity (p. 215). As a means of training in self-control, Poppelreuter constructed a performance display meter on which workers could observe their own performance (p. 210; Giese 1927, p. 206).

Work virtues, work motivation, characterology of the working individual, and psychological leadership

Psychology paid increasing attention to the personality traits of workers. The Danzig psychologist Hans Henning expressed employers' needs for information on character bluntly: "It doesn't matter what skills a worker has if he is a layabout or a bum" (Wies 1979, p. 145). The head of the psychotechnics section of the Reichsbahn, Richard Couvé, commented in 1928 that character studies were necessary where personal inadequacies "would have detrimental effects on the work process" (p. 145). The connection between this trend toward considering the working person on the one hand and actual economic and social developments in Germany on the other has not yet been investigated. One factor was workers' unexpectedly low level of social integration (see Jaeger and Staeuble 1981, p. 87). In the magazine *The Employer*, the limitations of psychotechnics were criticized and increased use of strategies of work motivation and leadership considered (Hinrichs 1981, p. 246). In the Third Reich a further factor of importance was that the entire individual was now held to be less responsible. Work was to be performed not in order to live, but as a willing service for the common good. At the same time the trend toward character testing was an international phenomenon (see Baumgarten, 1941).

The need for appropriate tools with which to study the working individual led to the dominance – or at the least to the discussion – of different terms and theoretical categories in psychotechnics. Talk was of the totality of the individual and the need for diagnostics of the entire personality. Some were more cautious; for example, Moede (1939) spoke of the need to augment the performance principle in aptitude testing with the expression principle. This change corresponded to developments in theoretical psychology. The limits imposed by classical experimental psychology, whose concentration on basic psychological functions had been appropriate for psychotechnical diagnostics, were no longer accepted. "Totality" was the central concept in Krueger's "holistic" psychology (*Ganzheitspsychologie*). The idea of the holistic perception of objects lay at the heart of Gestalt psychology. The switch in diagnostics from empirical to interpretive methods also corresponded to the fact that Spranger (1930) proposed a "psychology of understanding" based on Wilhelm Dilthey's (1924) critique of "analytical" psychology. The aim of this psychology, conceived as a human science rather than as a natural science, was to interpret the development of the individual from the standpoint of higher, objective mental structures. Such correspondences, however – with their widely varying contents and methods – existed only at a general level. The ideas developed in theoretical psychology did not gain direct entry into professional practice. Their importance was more diffuse, generating general modes of thought or, with Foucault, discourse. One such rule guiding perception and explanation was that the individual to be diagnosed was not to be seen as a collection of various functions, but rather viewed as a whole.

The individual was no longer seen as a machine giving its all, as with Kraepelin. Work character and will seemed to be more important than specific abilities. Good personality traits could compensate for a bad performance spectrum, but conversely, the performance spectrum "had little or no effect on the personality value" (Mathieu n.d., p. 63). As the will regulated the emotions and urges in Lersch's characterological model (see the following sections), so in the case of the "working individual" the will, as the "kernel of the personality" (Bornemann 1944, p. 39), had a decisive influence on actual performance. As Rieffert wrote after the war, the aim was to know something about the "the manner in which a man works: whether he is willing or not" (1945, p. 15). This is not to say, of course, that there was complete agreement about method. Above all, Moede, the advocate of an industrial psychology geared to maximum industrial efficiency, repeatedly praised the virtues of psychotechnical performance diagnostics over those of expression psychology and characterological interpretations.[4]

Richard Pauli's reinterpretation of Kraepelin's calculation test provides a clear example of the trend away from the pure study of performance to the study of the "working individual." When Kraepelin published his study on the work curve in 1902, he wanted to establish how the tendency to fatigue influenced individual work performance (Baumgarten 1955, p. 766). Since he saw fatigue as having a greater influence on performance than determination, Kraepelin (1921) wanted to increase the enjoyment of work with appropriate breaks. Pauli wished to add something to the experiment: the "energy of will" (1939, p. 120). In diagnostics the important thing was "not only a person's ability to perform, but also the will to do so" (p. 119). Pauli turned Kraepelin's method into a "characterological test procedure" (1943). In a mindless test, the subject had to add digits together for an hour. The quantity, quality, and consistency of the performance were not only indicators of ability, but also symptoms of underlying personality traits. True to the principles of expression psychology, Pauli saw each symptom as being ambiguous (see the subsection entitled "Expression Psychology"). For example, a high performance could be a sign of freshness, industry, stamina, strength of will, and the like or, alternatively, of narrow-mindedness, vanity, and ambition. Error-free results could indicate care and conscientiousness, but also timidity and pedantry. The psychologist had to interpret the symptoms negatively or positively from the general context. Each of quantitative indicators was assigned a range of possible qualitative psychological interpretations. According to Ernst Laepple (1940), the work curve could register "vital and temperamental aspects" as well as the "aspect of will and the

4 See Moede (1936, p. 293; 1939a; 1942, p. 179; 1942a; 1944, pp. 85–92).

actions of a personality"; for that reason he discussed its possible use for Wehrmacht Psychology (see also Arnold 1937–8).

Pauli's approach corresponded to Walter Poppelreuter's "symptomatic method." Already in 1923 Poppelreuter had named the determination of the "working personality" (p. 108) as the aim of vocational selection diagnostics, and had seen the behavior observed during a brief work test as a symptom of personality structure (p. 100). The interpretation of work curves in the course of psychological tests should cast light on the individual work type (pp. 113–14). The test did not result in a measurement, but in a personality assessment. With such a view of aptitude diagnostics, the functions of diagnostic tools changed. They were now seen as "situative occasions" that favored "observing subjects and developing an understanding of them" (Herrmann 1966, p. 267). Observing the subject in test situations, especially doing practical work, became the most important diagnostic procedure for determining "work character." Conclusions could then be drawn about personality traits. In rules governing the use of Giese's standard test method at the State Labor Office of Wuerttemburg in March 1925, point 4 states: "The main tasks of the tester are (a) observation of subject, and the behavior provoked by the test; (b) analysis of the results obtained. Numbers are not the important thing in psychology – the explanation of the numbers is given by observing the subject in order to obtain a picture of the person's inner situation" (after I C Dorsch).

Albert Huth (1936a) felt that the greatest opportunities for personality assessment in career counseling lay in "continual observation," where observation by psychologists was supplemented by that of parents, Hitler Youth leaders, and teachers, as well as by psychological interviews. Of course, if, as Huth believed, more extensive observation by Hitler Youth leaders was therefore more productive than psychological investigation itself, this would seem to cast doubt on psychology's claim to provide specific methods for judging personalities.

Here we see a real difficulty of psychotechnics at that time. It claimed to establish work character from behavior during tests, but this remained for the most part a postulate. Attempts such as Pauli's to establish a systematic relationship between observable performance and interpreted character traits remained the exception. More promising methods of determining personality, whose use lay solely in the hands of psychologists (and not other scientists, the lay public, or even Hitler Youth leaders) were those of expression psychology. These were developed primarily in the Wehrmacht, from where they partly found their way to industrial psychology.

Considering the mental preconditions of work performance not only yielded different diagnostic approaches. The problem of integrating and motivating workers also led to what Bornemann (1961, p. 347) has called "motivational maintenance." In view of the labor shortage at the end of the thirties and the intensification of the war economy (see Chapter 4), being willing to work and deriving pleasure from work became increasingly important. One way of increasing these factors was seen in vocational education. True to Nazi ideology, this should be "total education," as an article in *Industrielle Psychotechnik* put it. In addition to job training, "total education" should further comradeship and community spirit and make National Socialism an "experience renewed daily" at work stations (Anon. 1936, p. 159). To train "work virtues" (*Arbeitstugenden*), the Office for Vocational Training and Works Management of the Labor Front (which had taken over the training workshops of the DINTA; see Chapter 4) set up "Robinson courses," simple vocational courses developed by Poppelreuter during the Great Depression. Originally, unemployed young people were taught to work with wood, paper, cardboard, textiles, or metals without machines or special tools. The courses "Iron Educates" and "Wood Forms" became well known and were designed to instill a "stubborn and aggressive confrontation with the materials" (Mathieu 1935, p. 259) or "intention, hardness, and commitment" (Arnhold 1941, p.114). Such basic training was an excellent way of teaching young people not only work discipline but also cleanliness, care, loyalty, and a sense of duty (Arnhold 1937, p. 9). So not only selection but also vocational training concentrated more and more on general work attitudes. But in contrast to the methods of diagnostics, these new methods of vocational training can hardly be seen as a scientific contribution of psychology. They were a practical, social-technical contribution, based on experience, which obviously met the needs of economic bodies as well as the Labor Front.

This is also true of a social technological method designed to increase pleasure at work that was developed by industrial engineers, psychologists, and Labor Front functionaries: the internal suggestion system introduced in firms at the start of the war.[5] Hanns Benkert, director of the Society of German Engineers, a leading functionary in munitions production, manager of the Siemens Schuckert Works, and a confidant of Fritz Todt (Ludwig 1979, pp. 207–8, 396–7) wrote in 1943 about the attitudes of factory personnel toward innovations and outlined a system that had been introduced at Siemens to increase performance, produce "total mobilization," and create a "holy disquiet" (Benkert 1943, p. 20). The people must be motivated to think how to improve the work. Successful workers should be presented as examples for others to follow; women could be motivated by introducing a women's newspaper written by women. A panel at Siemens gave bonuses for good suggestions, which increased rapidly starting in 1937. Bornemann (1944a) saw a positive effect for the Hoesch firm too and underlined the effect of this system on the attitudes of employees and on the "spirit" of the firm. Labor Front functionary Herbert Steinwarz (1943) saw the suggestion system as a way of consolidating the community in the firm and creating an "organic leadership structure," since it indicated the able workers. One

5 See Benkert (1943), Bornemann (1944a), Michligk (1942, 1942a), and Steinwarz (1943).

can see here a transition to methods of industrial integration in Germany based on organizational psychology, the beginnings of which were the studies by Elton Mayo, F. J. Roethlisberger, and W. J. Dickson in the Hawthorne Works in Chicago in the late twenties and early thirties. Methods of personnel management and leadership training were also involved; these had been developed in the twenties and then refined in the thirties.[6] In the course of the war effort, the motivation of workers became more and more important (Arnhold 1941a), and the psychological training of managers and foremen (the so-called subordinate leader corps) was intensified (see Chapter 4). The methods employed in firms, such as increasing the work stimulus by the appropriate organization, encouraging cooperation by consultation, and the like, resemble those used in the Hawthorne Works.

The treatment of leadership personnel indicates that the differentiation between psychotechnical and characterological principles could also denote a differentiation between target groups. In practice the classical methods of psychotechnical aptitude diagnostics continued to be used to select workers for simple activities, and characterological methods were used for leadership selection. For the Borsig firm, Rieffert (1945, p. 12) established that personality should be investigated "where [it] is of particular importance for the job...for example in all cases where leadership is involved." Then methods of expression analysis, speech analysis, and graphology were employed. Ludwig Kroeber-Keneth reported for the Reemtsma firm that simple aptitude tests were sufficient for apprentices, but that graphological assessments were commissioned for higher employees (Z, f. 199). Poppelreuter felt that it was not appropriate to draw conclusions for the higher professions from performance in simple work tasks. He could use them to choose candidates for the police force, but refused to extend them to the selection of police inspectors (1923, p. 152). If characterological reports by Wehrmacht psychologists on prospective officers are compared with characterological assessments of apprentices at employment offices in the thirties (following I C Dorsch), it is noticeable that the latter include blatant value judgments and even defamations of sexual vitality (see subsection entitled "Characterological Diagnostics"). When the Labor Front's Institute for Work Psychology and Pedagogy selected deportees and prisoners of war for production work it did not use methods from expression psychology, but turned among other things to tests from the U.S. Army from the First World War (Schorn 1942; cf. Chapter 5).

6 In Giese's (1927) textbook the control of negotiations and meetings is only touched on. When Moede published two papers in *Industrielle Psychotechnik* (1930) and a popular essay on the handling of personnel by superiors, Otto Lipmann issued a strong rebuttal, and the committee of the Society for the Advancement of Practical Psychology also issued a statement, since Moede made no distinction between scientific methods and intrigues for getting rid of undesirable personnel (Lipmann 1930; Moede 1930a).

Expression psychology, characterology, and the selection of officers

The most decisive and typical contributions to the professionalization of psychology in Nazi Germany were made by expression psychology methods, the characterological system, and the situational tests of Wehrmacht Psychology. Wehrmacht Psychology not only drew on developments in (and often outside of) academic and professional psychology, but also developed new theories and methods itself. Most of the expression psychological investigations at that time drew their empirical material from tests on officer candidates. Just as U.S. psychology has often been accused of being based on research on college students, so expression psychology can be said to have been a psychology for seventeen- to nineteen-year-old males who had finished school and applied to be officer cadets. Characterology drew some of its research questions from Wehrmacht Psychology. A requirement of Wehrmacht evaluation reports was that they comment on a person's will-power, temperament, emotions, attitude toward his surroundings, and practical behavior, as well as on his special characteristics and strength of character (Kreipe 1937, p. 65; Masuhr 1937, p. 9). The integration of these various areas was undertaken by Philipp Lersch (1938) in his study on the structure of character. The first reference in his foreword was to his experience as a Wehrmacht psychologist. The available tools had proved inadequate to come to terms with the wide variety of human character types. "Again and again it was necessary to use concepts in the reports that were strange to scientific psychology (*Seelenkunde*) and had no place in its system.... Thus, my practical activity suggested the creation of a *Seelenkunde* that had absorbed at least as much of the vast psychological vocabulary as seemed necessary for the assessment of people" (p. v).

We have seen that characterological assessment was considered important for manager selection. Characterology and expression psychology promised to do the same for officers. Expression psychology methods were used to interpret expression as a manifestation of character; characterology provided a framework within which diagnostic observations could be interpreted. In the diagnostic procedures of Wehrmacht Psychology, standardized observation situations were created for studying features of expression and character.

I will consider expression psychology first followed by characterology, because the dominant characterological system of Lersch was created not least to provide a conceptual framework for the results of expression psychology diagnostics, and this framework in turn structured perception and guided actions. The needs met by the two systems will be examined and their practical value considered. Further sections will consider the link between ideals of the model soldier and psychological theories, and selection methods in the Wehrmacht.

Expression psychology

By observing people's expressions it was hoped that access to inner experience and to character could be gained.[7] The "media of expression" (Kirchhoff 1964) were handwriting, facial movements, body movements, gestures, physiognomy, and voice. This went a step beyond psychotechnical diagnostics, which restricted itself mainly to the performance of the "working person" and observation of work behavior.

Leaving aside some early systems not linked to the history of psychology as a discipline,[8] the foundation of expression psychology was laid by Ludwig Klages, who lived as an independent scholar in Switzerland after studying in Munich. Toward the end of the First World War, he published two books on expression. One of these, *Handschrift und Charakter* (Handwriting and character), had been reprinted fifteen times by 1932. Klages also used the interpretation of handwriting in a second book to explain his general theory of expression. In this book he formulated his law of body–mind relationship, that "each expressive body movement" is a realization of the "driving experience of the expressed feeling" or, in an alternative formulation, of the "form (*Gestalt*) of a mental excitation" (1923, pp. 22, 24). Controlled movements such as handwriting were also covered by this law. Handwriting is coordinated movement; by retracing these movements and interpreting the driving experiences they represent, it is possible to draw conclusions about the character of the writer. We can interpret and understand the mental contents of an expression when we are able to experience the mental excitation that lies behind a movement. Klages refers to interpretation and not explanation, because he is interested only in the mental aspect, not in the causes of movement. Here Klages follows his teacher Theodor Lipps, who felt that other minds should be understood, like works of art, by empathy (see Holzkamp 1964, p. 146).

Klages did not assign a fixed mental content to each expression as the old graphology of J. H. Michon and J. Crepieux-Jamin had done. Indeed, one of Klages's principles, which was later to govern expression psychology, was that every expressive feature had a double meaning. The experience expressing itself was "the result of the interaction of a driving force and a resistance in the soul" (1932, p. 32). The proper interpretation depended on whether the strength of the driving force or the weakness of the resistance dominated. For example, in Klages's graphology large handwriting indicated either enthusiasm or a lack of realism, small handwriting realism or a lack of enthusiasm; sloped writing indicated congeniality or rashness, vertical strokes

7 Similar matters crop up nowadays as "nonverbal communication" or "body language," in ignorance of the results and experience of the old expression psychology and of the criticism it received after the war. The fundamental difference from current concepts is the largely static view of the representation of a constant character through expressions.

8 Summarized by Bühler (1933).

rationality or aloofness. The choice of the positive or negative assessment depended on the general interpretation of what was called the "form-level" (*Formniveau*) of handwriting. This could be arrived at from the rhythm, depth, and richness of the script, and from the fullness of a person's life that it demonstrated. This was determined in several steps; however, according to Klages it could not be measured but only perceived. For graphology, whose other contemporary systems were compiled by Wenzl (1937), Klages's theory meant that it was now possible to interpret the whole character on the basis of handwriting, instead of compiling a picture of the personality from various features. This holistic method gained dominance.[9]

Lersch also helped to lay the foundations of German expression psychology in his study of the diagnostics of facial expression entitled *Gesicht und Seele* (Face and soul) (1932). For Lersch, this research formed part of the current of psychology centered on the individual as totality. In contrast to Klages, Lersch did not comment on the specific relationship between expression and psychological content. For him, the expressive phenomenon was "to be defined as an event taking place in the sphere of what we can perceive, in a coexistent polar relationship to a psychological event" (1932, p. 19). Therefore, in order to interpret expression it was not necessary to reach an empathetic comprehension of the underlying psychological stirrings. Rather, the psychologist should know the possible range of expressive phenomena as well as their objective determination and diagnostic significance. These were deduced by analogy from the functions of expressive organs in the relationship between person and world, generally using categories arising from everyday experience with facial expressions. For example, Lersch (p. 42) saw the eye as the organ for admitting the world into our consciousness. This then led to interpretations of opening the eye, direction of gaze, and glancing. A drooping eyelid implied a tendency to reduce the willingness to comprehend optically. Whether this was related to a dullness of mind or character or to a blasé attitude could be interpreted only from other facial expressions, for example, of the mouth, and their relation to that of the eye. Lersch wanted to use empirical material – films – to determine this relationship. The important advance beyond Klages's diagnostics was the attempt to objectify interpretation. However, when it came to establishing the characterological significance of facial expressions Lersch fell back on Klages's explanation that personal capacity to experience is the precondition for interpretation (p. 55), although he did not make this explicit. The hypothesis that the significance of an expression can be experienced or felt intuitively underlay all claims made for the possibility of interpretation by expression psychology.

Hermann Strehle (1935) was thus content to accept that the observer can feel the meaning of an expression intuitively, because it is an "externally

9 See, e.g., Heiss (1943), with minor modifications of Klages's theory.

perceptible manifestation regularly and uniformly connected and psychologically linked to a certain mental state or process, and that for the observer bears the direct character of that experience" (pp. 12; 2). He used concepts here that Rieffert developed in lectures at Berlin University, but did not publish. In his book on body movement and posture, Strehle described methods similar to those of Lersch, determining a connected experience by analogy from the psychological function of a movement. For instance, he viewed tense muscles as an expression of goal-oriented endeavor, and lack of tension as indicating a passive and moody state. Strehle saw not only kinetic forms of expression like Lersch, but also static forms.

The possibility of interpreting not only movements but also the form of the body in this way provided the starting point for Ludwig Eckstein in his *Sprache der menchlichen Leibeserscheinungen* (Language of human bodily appearances). Since body and mind are both parts of the same living being and governed by the same structural rules of "the living," then conclusions about psychological being could be drawn from the appearance of the body. The analogies Eckstein draws are crude; for example, since the muscle tension is the manifestation of the will, the muscles show *Willenscharakter* (1943, p. 65). They become embarrassing when Eckstein compares ethical stance with bearing; in the same way as the ethical stance resists attempts to degrade it, so muscle tonus resists gravity pulling down and establishes the bearing of the individual (p. 49). Later in this chapter we will consider the way in which this psychological theory pays reverence to the Prussian military tradition.

A further area of diagnostics used particularly in the army was voice analysis, although there are few publications and no books on this topic. In a paper at the twelfth congress of the DGfPs in 1931, Rieffert presented a system for establishing voice types using certain features. Rieffert (1932) distinguished between melos (the rise and fall of the voice) and rhythm. If the speech was characterized by melos, then it was more emotional; rhythm indicated will control. According to further characteristics (speed, force of expression), these two classes had subgroups, which were assigned to various character types. For example, the melos type – fast speaking and with soft timbre – was gregarious, whereas the rhythmic type – with tense rhythm and a harsh timbre – was a reserved, dominating person. Martin Keilhacker presented an extensive investigation in 1940 in which he claimed to draw conclusions on all aspects of a person's character, especially on his or her feelings, temperament, and will, from the way he or she read and phrased a text.

In the Wehrmacht, subjects were observed in various situations, and facial expressions and speech were registered relatively systematically. At employment offices or in industry, the observations made were more or less incidental or restricted to work situations. In many cases a fleeting impression of the subject's expression may have led to a characterological description. Both

Johann Sebastian Dach (1937) and Ludwig Eckstein (1937) carried out empirical studies on the reliability of such "first impressions." Both concluded that certain basic characteristics can be established by a first impression. Eckstein, who carried out his investigation in Army Psychology, was able to compare his results with extensive assessments made later.

The theoretical approach of Josef Schänzle (1936) differs from those of the other authors mentioned. In his investigation he studied films of the facial expressions of examinees who were thinking. According to Schänzle, expressive movement can be understood only from the relationship of the expressor to specific environmental situations. The analogy theory had to avoid the explanation of directed movements, since these could be explained only by this relationship (p. 13). Schänzle investigated the facial expressions associated with thinking postures, thinking movements, and the success or failure of reflection, believing that the experiences that accompanied the reflections were revealed in these expressions, though not the reflections themselves. Therefore, expression psychology could comment on thought temperament, but not on mental ability.

Apart from Klages's work and the investigation by Dach, all the studies considered in this section were carried out by Army Psychology, and all took their material from tests on prospective officers. Army Psychology showed such interest because expression psychology promised access to precisely those aspects of character that were of interest in these leaders of the future. Methods of expression psychology were developed further in the army, and found their primary application there. They were not designed as clinical-diagnostic procedures and were not used in a social-psychological or therapeutic context as current work on body language is. Diagnostic methods suited for clinical application had little impact in the twenties and thirties in Germany, since only a few psychologists worked in these fields. The Rorschach Test, in which the description of unstructured pictures was interpreted, and the Wartegg Test, in which drawings completed from a few initial elements were interpreted characterologically, were tried out in Wehrmacht Psychology but not included in the test procedure. Both were intended to provide access to the world of experience. But they were not designed to reveal the character of the "working person" or leadership qualities. The test of Ehrig Wartegg (1939) was, however, adopted in psychotherapy (Vetter 1940).

Characterology

In the twenties characterology was not yet a specialty of psychology, as it was later to become. Few significant contributions were made by academic psychologists. More psychiatrists and philosophers published in the *Jahrbuch der Charakterologie*, edited from 1924 by Emil Utitz (professor in Rostock and then ordinarius for philosophy in Halle) than did psychologists. In the foreword to the first volume, Utitz remarked that the field of characterology was not covered by representatives of any one discipline. However, Theodor Ziehen (1930), a philosopher and former psychiatrist who wished to base philosophy on psychology and the natural sciences, saw characterology as a subdiscipline of psychology, with a degree of independence. The problem of assigning characterology to psychology was linked to the subject matter claimed by each. Psychology was still concerned largely with the general laws governing mental functions. In contrast, characterology, according to Utitz (1928), dealt with the essence of character, which was seen as a personal striving. It thus differed from differential psychology, which sought only "psychological peculiarities" and did not describe the entire personality in terms of its striving. It could not therefore rely on developed systems from psychology or other sciences but needed an independent development.

The decisive theoretical impetus for the development of German characterology came from the work of Klages. Practical stimulation came from the psychology of expression and Army Psychology. The army psychologist Lersch was not the only one to come via expression psychology to the development of a conceptual system of characterology in order to fill the gap between practical character diagnostics and characterology as a science (Lersch 1934). Klages had also begun with expression studies and then found his way to characterology.

The philosophical basis for the two characterological systems that were most important at that time, those of Klages (1926) and Lersch (1938), lay in the German "philosophy of life" (*Lebensphilosophie*). Just as psychology was institutionally bound to philosophy, it was also scientifically dependent on it, since it hardly generated fundamentally new approaches from its own research. Experimental psychology had been linked with the nineteenth-century trend in philosophy toward questions of natural science and the attempt to provide empirical answers to philosophical questions. With the polemics of Friedrich Nietzsche, Eduard von Hartmann, and Wilhelm Dilthey against rationalist philosophy and the birth of "philosophy of life," psychology thus faced new problems. Important questions for von Hartmann (1930), such as the relationship between *Empfindung* (sensation) and *Gefühl* (feeling and emotion) or between feeling and will, came even more into the center of attention with Dilthey, whom Georg Lukács (1973) saw as the founder of German "philosophy of life." Lukács saw *Lebensphilosophie* as meeting the intelligentsia's need for a worldview in the crisis of imperialism following the fall of Bismarck. Dilthey had seen the key epistemological issues but had not really grasped the world as an objective reality, independent of consciousness. For Dilthey the ultimate source of knowledge was the experience of the world. If experience was the organon of knowledge of the world, then psychological work on emotional life and drives was of philosophical importance.[10]

10 The quest for worldview and meaning was on the whole decisive for changes of approach and theory in the twenties; see Bühler (1978 [1926], p. 123).

Already in Dilthey's *Ideen über eine beschreibende und zergliedernde Psychologie* (1894; Ideas concerning a descriptive and analytical psychology) ideas can be found that later reemerged in characterology. He opposed a psychology that explained the phenomena of mental life causally from its elements. The "understanding" psychology he proposed (*Strukturzusammenhänge*) approached mental life from structured, experienced original contexts. Dilthey used the concept of structure for the array of mental facts in an internal relationship that can be grasped only intuitively. Later in characterology, we find the related postulate of the "structural illumination" (Lersch) of a character as a basic principle of characterological diagnostics.

The reference to such intuitive experiences as the central organon of knowledge was a methodological postulate for Klages and Lersch – and after the war for people such as Albert Wellek. For Klages, systematic intuitive self-reflection makes it possible to gain access to the universe of others' character traits: "The number of differing characteristics we can find in ourselves is the number of properties of character we will be able to comprehend" (1926, p. 37). Klages's reference to intuition as the basis of knowledge forms part of an antirational philosophy in which vital *Seele* (soul) is opposed by destructive *Geist* (mind). In the cosmic natural state, body always contains living soul.[11] We can now see the roots of Klages's principle of expression in *Lebensphilosophie* – that *Seele* is the meaning of body, and body the manifestation of *Seele*. If the mental meaning of an external bodily appearance is accessible only via the inner experience that it expresses, then reexperiencing that oneself must be the only method of gaining access to another's expression.

For Lersch, intuitive experience is also the "prototype of adequate knowledge of reality" (1932a, p. 6). His book *Lebensphilosophie der Gegenwart* (Contemporary philosophy of life) begins with the comment that "in the search for new meaning and essential content of human existence typical of the present intellectual situation ... the concepts of life and intuitive experiencing have become the signature of numerous interrelated branches of intellectual endeavor." With this formulation he attempted to use *Lebensphilosophie* for the benefit of psychology. If intuitive experiencing and not rational knowing provides the main access to reality, and the question of epistemological validity is to be answered largely in the realm of the emotional-irrational (p. 7), then the theory of cognition presupposes a psychology of feelings, moods, emotions, drives, and the like.

Lersch and Klages shared this initial position, although Lersch did not follow the thesis of *Geist* opposing *Seele*. His more cautious formulation is that philosophy of life has uncovered a new level of reality and that the task is now to redetermine the position of the intellect within the totality of the person and within reality as a whole. Reason and experiencing are inter-

11 For this return to the cosmic-archaic see also the development in Jungian depth psychology; Bloch (1935, p. 330) linked Klages and Jung.

dependent; intuitive experiencing embodies disquiet, which keeps thought in action (Lersch 1932a, pp. 94ff.). In later work on the structure of character, we again meet this interdependence of reason and experiencing. The decisive difference between Lersch's characterology and Klages's arises from this point. In contrast to Klages, Lersch saw a stratification of the mental, with the will and intellect forming a superstructure over the world of the emotions.

Until the middle of the thirties, Klages was viewed, as Döring (1932, p. 86) wrote, as the "leader among present-day characterologists." Three editions of his *Prinzipien der Charakterologie* (Principles of characterology) were published in quick succession in the twenties, and the demand led Klages to prepare a fourth completely revised edition in 1926 entitled *Die Grundlagen der Charakterkunde* (The foundations of character science). In the foreword he wrote that in 1910 the name of characterology had been mocked, and now everyone had it on their lips. The same confident tone also marked the attacks on traditional psychology, which he criticized as restricting itself to work performance and rational, logically thinking people, acting usefully, but capitulating before the drives, passions, subconscious, and individuality. Klages approached his subject by starting from the experiential content of everyday psychological terms and sorting these names into categories, which he ordered to specific regions of inner life. He did not disclose, however, how he got from a given name to a given concept. The aim was to produce a system of concepts with which the character could be described. Klages distinguished three character zones: material, nature (*Artung*), and makeup (*Gefüge*). The materials of character are the abilities. "Nature" includes the facilities to utilize these materials, the moving forces (as he called the interests), the inner causes of strivings, the emotional dispositions. The third zone, the makeup of the character, consists of the predispositions of mental structure, the internal personal media in which experiential processes proceed. It has three structural properties: emotional arousal, will arousal, and expressive ability. He recommended that a fourth group of character features be studied further: those that relate to a "harmony or disharmony of the others" (p. 59), such as homogeneity, balance, disunity, and contradictoriness. If this framework is augmented with ways of presenting character externally, then a "network of basic concepts" is created with which one can "capture everything that can possibly be used to describe individual selves" (p. 61).

In the last chapter of his book, Klages developed a "system of driving forces." Roughly speaking, he distinguished among mental, personal, and sensory forces, which can be either binding or releasing. This is a reflection of his metaphysical distinction between *Geist* (mind) and life. From this metaphysics he drew his drive to self-assertion and the drive to self-abandonment (*Selbsthingebung*) (pp. 154ff.). By the latter he meant that the ego emotionally falls victim to life, whereas with self-assertion existence is maintained by the assertion of the ego over life. Thus, personal self-abandonment, which equals depth of feeling or a potential for passion, is a sign of separation from the ego; personal self-assertion, which equals egoism, is a sign of attachment to the ego.

Klages had thus provided a system for classifying character zones and moving forces with a vocabulary and grammar that could be of use in character assessments, and that offered concepts and structures for psycho-diagnostic perception. Thus, it was not his philosophy but his system that received the attention of practical psychology.[12]

In his book *Der Aufbau des Charakters* (The structure of character), Lersch (1938) dealt with what he felt to be the most important problem of character-ology, the question of structure, hoping to overcome Klages's weaknesses. In this book philosophical motives coincided with the practical needs of Wehrmacht psychologists, who wanted to gather all diagnostically deter-mined characteristics together in one conceptual model. "[We] seek a clear classification of the aspects under which we must examine the individual in order to determine the main features of his character and ascertain his individual peculiarities. We term this field of problems the *character structure*. This is the specific concern of this book" (Lersch 1938, p. 24).

Lersch's diagnostic starting point is equally plain in his definition of character: "The total structure of forms of experiencing (dispositions), which as relatively constant properties give a form to the reality of the person as a mental being (*seelisches Wesen*)" (p. 23). New to Lersch compared with Klages was the idea of character stratification. As in the ontology of Nicolai Hartmann, whom Lersch cited specifically, where layers of being rest one on the other, but are independent, human character has two layers according to Lersch, an endothymic ground and an independent superstructure. Endothymic states and processes take hold of the person; from them come mental energy and strength. By means of thinking and willing, the two zones of the superstructure, the person can, however, adopt an attitude to the "being taken hold of." A philosophical postulate explained why the endo-thymic base is overlaid by the superstructure: "The person wins his dignity, his liberty and responsibility from being able to make a stand against those endothymic experiences, restricting and suppressing some, but allowing others to have their full influence on his life. He does this through his *will* and his *thought*" (Lersch 1938, p. 188).

Even the basic structure of this model discloses a normative ideal. The thesis of the superstructure is justified by classic bourgeois ideals. Such an ideal is that will and thought govern the emotions and the senses. Vanquishing sensuality was a civilizing contribution of the bourgeoisie. The self-controlled individual who has internalized the external demands placed on the control of his or her emotions and who has subordinated his or her emotionality to

---

12 For the description of Klages in the psychological literature see Buser (1973, pp. 94–127) and Revers (1960, pp. 393–8).

the rationality of the age thus serves as a pattern for a general theory of character structure. Admittedly for Lersch, endothymic experiences nurture existence, but he does not uncover the dynamics of their rebellion against control as does Freud. Rather, it is the will that ultimately controls all mental dynamics. "By the strength of their will," according to his theory, people become the "bearers of responsibility" (p. 191). Referring to the aphorism in Nietzsche's *Wille zur Macht* that greatness of character consists in having emotions "to a terrifying degree," but keeping them reined in, Lersch sees the ideal relationship between will and the endothymic ground as a harmony between a rich endothymic life and the will's secure ability "to organize... experiences uniformly and to direct them toward an aim." (p. 182). The emphasis on the normative and regulatory function of the will in the overall structure of character is one of the features distinguishing Lersch's model from other structural and strata models, such as those of Freud and Rothacker, in which the ego is conceived as the integrating center or organizing layer of the personality. The extent to which military ideals influenced this model will be discussed in a later section.

Let us now briefly consider how Lersch envisaged this construction in detail. In the endothymic ground he distinguished four strata. The bottom layer is the life feeling (*Lebensgefühl*), where we internalize the fact of our living existence. Lersch divides this in turn into bodily feelings, moods, especially the basic mood of unhappiness or happiness, and the excited forms of life feeling, the emotions (affects) under which he includes states of upset such as ecstasy, rage, excitement, and angst. The second layer is self-awareness, the individual's feeling of standing out from the world around and about; this feeling is independent of the actually perceived world. The directed feelings in the third layer are then related to external influences. These include joy, sadness, worry, fear, hope, pity, and so on. In upper strata are the goal-oriented aspirations, including existential urges to activity, pleasure, and experience, as well as individual urges such as egoism, selflessness, self-interest, and domineeringness, and the transitive aspirations, which are directed to values and fill the person's being with meaning. Lersch includes here creativity, interests, ideal love for something (in which a person finds his or her real nature when staking his or her existence for an idea), normative aspirations for truth, justice, and duty, and the social aspiration for company and community.

In the superstructure, Lersch does not introduce strata, but has side by side the will and a noetic superstructure, thought, which are both differentiated into the various forms and abilities of thinking and willing. The endothymic ground and the superstructure form a natural whole. If this is disturbed, then the person is not authentic, as for example, when a certain expression comes from the endothymic ground but a different expression is created from the superstructure. Diagnostics in this case would require penetrating the layers of inauthenticity to reach the authentic kernel.

Characterological diagnostics

The few reports that are in the literature, or are otherwise accessible, show the function of characterological systems for practical psychological work. Most of these reports are from Wehrmacht Psychology. On the whole they are relatively freely formulated and not schematic. If one looks for common ideas from personality theory, it becomes clear that they all use a structural conception. Various parts of the character are described, without necessarily indicating interconnections between them. The concept of *Gefüge* (structure) is often used, and the term "format" frequently signifies a positive impression, reminiscent of Klages's use of the term "form level" (*Formniveau*) for the overall value of a character. Usually a dominant tendency will be registered; sometimes various aspects such as intellect and emotion are related to one another (see Masuhr 1939c). Finding dominating tendencies and structural relationships between characteristics corresponded to the suggestions made by Lersch at the end of his book. Characterological diagnosis had three steps: (1) establishing the dominant characteristic of a person; (2) staking out the characterological surroundings of this dominant – the theory could be used to formulate suppositions about further probable features that, if established, would be set in relation to the dominant; and (3) illuminating these features structurally, determining their superior or subordinate position and the position of the dominant in the totality of the character. This last was the most important step in the diagnosis (Lersch 1938, pp. 270–1).

It is noticeable that the reports for the Wehrmacht fit this model better than the model of typological classification. We find hardly any standard typological assessments. At the same time, the description of the special characteristic dominating the personality and its overall position is usually confined to a single general sentence: "He is a reliable conscientious character" (Masuhr 1939d, p. 27); "on the whole a cramped nature that can master new situations and orient himself in them only with difficulty" (Zielasko 1939, p. 48); "B. has above average abilities, reliable intellectual faculties, is quick on the uptake and has a clear mind" (I C Skawran 1, p. 49); "he has a firm, strong, and completely unspoiled nature" (p. 52); "he is of an ill-balanced nature and, with only moderate vital strength and freshness, is largely concerned to assert and distinguish himself" (Dirks 1940, p. 50). These descriptions, however, are far from Lersch's ideal of being "structurally illuminated." A further connection to characterological theory can be seen in the structure of some of the reports, which have "layers" from the endothymic base to the noetic superstructure and conclude with a general assessment of the characterological dominant (see Dirks 1940, pp. 57, 70). Other reports are structured in terms of intellect, emotion, practical-technical ability, and will (Masuhr 1939a, p. 51) or intellect, practical action, will, and social behavior (Zielasko 1939, p. 48). In the Wehrmacht there also was an attempt to concentrate the diagnostic perception of psychologists on the will and intellect. The guidelines for evaluating tests clearly focused more on this than on the diagnosis of social behavior, not to mention emotional make-up (II B, OKH, Anweisungen...).

The assessment form of the Labor Office in Stuttgart in the thirties provided for entries under the following headings: (1) surface/performance zone (intellectuality, predisposition to work); (2) deeper person, basic character structure; and (3) firmness of type (constitutional type, racial type, regional type, integration type). In the practical reports from the Labor Office available to me, the third category was never used. However, one feels much more strongly than in the Wehrmacht reports how the description of character was used to give a value judgment on the examinee. Terms are used for character such as "self-control," "commitment," "conscientiousness," "stiffening by the will and the character." "Vitality" was regarded as being primitive. The attitude toward vitality and sexuality seems to say more about the psychologists than about the examinees. Thus, a girl is reported to have "disturbances of drive and will" because she "already menstruated at the age of twelve" and "had read anatomical books during her schooling." Following several thefts, her strength of will was to be assessed. The report had no difficulty finding the reason for her weak will: "Fräulein...suspects that [it] results from her masturbation." The mixing of value judgments and description using characterological concepts is also shown by an extract from another report, dated 24 February 1939: "In summary, the youth can be described as being of modest value mentally (not particularly of low value). His basic mental structure is without a firm center and motivation, in particular he is selfish, unfeeling (and a liar), seeks admiration, is largely governed by drives and generally without bearing [*haltlos*]" (quoted from I C Dorsch).

The Prussian officer and psychology

Expression psychology and characterology were always closely linked with the Wehrmacht. The common self-image of psychologists would suggest that this military connection had no influence on theory and methods. In my opinion, however, psychologists were affected by military ideals and the military concept of the individual, which was one of the reasons they were able to meet the requirements placed on them. Support for this hypothesis is provided by a comparison of the military conception of the ideal soldier (or more particularly the ideal officer) with the categories used in psychological reports and theories.

The German military services felt that officers had a decisive influence on the fighting ability of the troops. They formulated relatively clear ideas about suitable "military leaders." In the "Supplementary Regulations for Officer Careers During the War" we read of the need for a "continual, conscious selection of leaders" and that, in the selection of officers, the "requisite leadership and character qualities" must be "clearly discernible as a predisposition and be capable of development" (H. Dv. 82/3b, p. 5). According to a colonel writing in *Soldatentum*, a journal published by the psychological headquarters of the Wehrmacht, the qualities involved were as follows:

> as far as will is concerned, strength of will, a steadfast character, self-awareness, purpose and constancy; for the intellect, logical thought, practical intelligence, sense of orientation and to a degree a worldview [*Weltanschauung*]. On the emotional side: warmth of feeling, tact, directness, social understanding, dedication and enthusiasm for ideal values such as Führer, *Volk*, and fatherland. (Schimrigk 1934, p. 142)

The arrangement of will, intellect, and emotion corresponds to Lersch's structure of character. But how were these terms used? Emphasis was given – not only by Schimrigk – to the fact that knowledge is not the decisive factor, because the will could compensate for any gaps. "Character qualities" counted for more than "intellect." In an article in a military weekly we read that war "puts the individual to the hardest test of his mental and physical resistance" and that therefore in war the "properties of character carry more weight than those of the intellect" (Schmidt 1938, p. 325). In an article by a captain in *Soldatentum*, character, including here the will, is described as being more important than reason (Brecht 1939). Thus, tests of character were more important to the military than tests of intellect. The assessment instructions for Wehrmacht officers were more precise about the character qualities required. Positive features named were honesty, complete frankness, being real men who did their duty unswervingly, making necessary decisions and carrying responsibility, placing the highest demands on themselves, showing a warm heart for subordinates, personifying military virtues and the ideas of National Socialism and soldiery, having a keen interest in the cause, being up to physical demands, with freshness, knowledge, and skill (H. Dv. 291, p. 5). The corresponding air force regulations also mention the importance of cutting a good figure and of being able to rouse people by personal example (RL 17/76). Almost the same characteristics, with the exception of ideological soundness, are mentioned by former Wehrmacht psychologists when they are questioned today about the type of officer they were looking for when testing (Z, ff. 33, 95).

This is apparent in the psychological reports. The official stance was that the psychologist only described, and an officer evaluated, military suitability (Masuhr 1937, p. 6), or as Simoneit (1938, p. 7) put it, the report was aimed "at the being and not the value of the being." The reports that are still available, however, show that positively diagnosed characteristics certainly did correspond to the military ideal. Positive characteristics in the reports are, above all, things like strength of will and self-control: the adjectives used are "firm," "strong," "purposive," "tough," and so on. A good report is given to someone who "has a good hold of themselves" (Masuhr 1939d, p. 27), someone of "manly, solid nature" (Masuhr 1939a, p. 50), or someone of whom the psychologist's first impression is "fresh, comes in [to the interview] erect, juicily tense (*saftig gespannt*)" (Eckstein 1937, p. 52). Spontaneity is good only as long as it is firm, clear, and manly. Sensuousness, sensibility, and empathy are not mentioned at all, unless in categories such as "feminine" or "lacking tension" that signify unsuitability as an officer. Such emotions were dull and at best dominated by will, as in Lersch's theory of character.

It is perhaps worth noting here that there also was military influence on the content of intelligence tests and the topics selected for the interview. An internal list of such topics used during the war had points such as "the performance of the infantry in the Polish campaign," "Why war against England?" and the candidate's attitude toward soldiering and the war. In the mental investigation, there were questions about battles, colonial history, the wars of Frederick the Great, and militarily relevant matters such as "What are the countries bordering Greater Germany?" "What are their capitals?" or "What is Murmansk?" (I C Munsch).

The linkage between diagnostic and military values should not be attributed to the political leanings of the leading psychologist involved. They are more an indication of the fact that the diagnostic instruments used for prognoses were always linked from their construction to the goals for which they were intended. In test construction this is now termed "external validation." In the thirties psychologists spoke of probative control. The criterion of external validation for psychological procedures in the Wehrmacht was always the subsequent military assessment.[13] In the case of fighter pilots the number of hits was also used as a criterion (I C Skawran 1). Military considerations had to be included in the psychological procedures and judgments if psychology was to establish its success as an institution.

A link between external criteria and diagnostic procedures may not be any more surprising than that between military ideals and reports, where the selection was oriented toward such ideals. But military models had their influence on the psychological concepts themselves. In expression psychology the influence of the Prussian military ideal can be seen in the attempt to use muscular tension as an indicator of the central (and militarily significant) characteristic of will. Strehle (1935) equated body tension (*Spannung*) with a determined will (*Anspannung*) and with willpower. He quoted a report (p. 153) in which the posture of a seated subject – "relaxed, bent back, open hands, loose neck muscles" – is judged to be a sign of an insufficiently "tense" will. In the test of shouting commands used in Wehrmacht diagnosis the sharpness and volume of the command were seen as indicators of determination and willpower. A weak-willed person is characterized thus: "Even the command to which the subject was strongly motivated was given without stiffening. The feet were together, but the pronounced 'hanging' of the arms

---

13 See the various contributions by Masuhr; see also Chapter 8.

clearly showed that the muscles remained relaxed" (p. 155). One has to have been a psychologist with the Wehrmacht to know that bringing the feet together when shouting commands is an indicator of willpower. In fact, what is being interpreted is whether the subject was able to personify authority and discipline voluntarily. Was an element of willpower also turning the body into armor against desires for pleasure?[14] Whether or not this was the case, suppressing "natural sings of displeasure" when stretching a chest expander or when receiving electric shocks were seen as "exemplary" (p. 163).

The relationship between mental willpower and muscular tension is an example of the principle from Klages's expression psychology that the *Seele* is the signification of the body and the body is the manifestation of the *Seele*. For Klages (1932) even handwriting, as long as it does not result from atrophy of the disposition, is seen as an indicator of a willed ordering of the world, a suppression of alteration and change. For Lersch there is "in the case of active preparedness and orientation in the environment a certain overall tension in the entire muscular system" (1932, p. 51). Or again: "The greater the preparedness for action of an individual, the more the entire body musculature is in a state of continual tension, like a bow that has been drawn to speed away its arrow at any time. On the other hand the lower the general preparedness, the lower... the tone of the general musculature" (pp. 50–1). Eckstein (1943) also saw a similar connection between the muscles and the will. Posture and inner tension were indicated for him by continual taughtness of the muscles. The prepared body is not in a state of rest but is tense: "The state of the musculature can be seen as the corporal side of a latent inner tension of attitude and preparedness" (p. 50). He saw in the musculature an instrument of the will, expressing its strength, nature, and orientation (pp. 63–4).

According to Strehle (1935), "acts of striving" manifest themselves in muscle tension: "A sharp rise of the tension graph marks a sudden, strong input of the will; a gradually rising curve would mark gradually increasing effort, a constant level shows a will attitude that can be described as comparatively persistent" (pp. 38–9). An eager, strong-willed person, ready to act, could thus be identified by his tense muscles. There is no way of mobilizing forces by relaxation in this procedure. Concentration cannot be achieved by relaxing the muscles as much as possible, as in meditation, but is registered by the tension of the muscles in the upper half of the face (pp. 35–6). Will is not knowing what you want from an inner, calm conviction, but rather the ability to get a grip on yourself mentally and physically and thus to create

14 The function of clicking the heels is open to other interpretations, since it also involves contracting the entire muscular system of the hips and seat. According to the theories of Wilhelm Reich, a body armor was thus created against sexual stirrings. Tightening the muscles of the seat is like gritting one's teeth, closing off an opening against the outside world, symbolically a sign of a refusal to open oneself, militarily an act of "self-control," or the muscular training for this.

a determined impression. This is a reproduction in expression psychology of a soldier's body under Prussian military coercion, conditioned and cramped. With the analogical method of interpreting expression, the content of the analogy determines the type of interpretation. It was not the Will that expression psychology was examining, but a certain concept of will. Will was attitude – and thus physical attitude, or bearing. Will was self-control, domination of the body, the ability to create a straight impression. In short, we see a Prussian militarized concept of will. This normative element in the theory of expression psychology in no way conflicted with a political theory that saw self-conditioning as a precursor of conditioning by others and that believed it could demand the impossible from people, if they only made it part of their own will.

It is not surprising then to find that in Eckstein's theory the Prussian attitude is equated with human nature, or in his words, "the Prussian form of regimentation" represents "the inner freedom and naturalness of mankind" (1943, p. 318). If the militarily enslaved body of the Prussian officer already represented the natural, beautiful, and strong-willed body for expression psychology, the psychologist did not need to have an explicit military model in mind when selecting applicants; it was already contained in the categories of the theory. Eckstein said of the Prussian officer: "The Prussian appearance indicates a well-developed musculature. But this is by no means used solely for movement. An important part of the energy is dedicated to controlling and reining in impulses from within.... The muscles are collected, efficiently tensed, and slimly bundled.... The Prussian appearance emphasizes the expression of will" (p. 317).

The military conception of the good officer also found a place in Lersch's characterological theory, in the ideal of the controlled and self-controlled person, whose will guards the house of his character. Treading the narrow path of duty, placing the highest demands on oneself, being decisive, showing ambition in the interests of the cause (or of values) – these ideal characteristics named by officers and quoted above, can all be found in Lersch's theory of the composition of character. Here, will dominated over the endothymic base and governed the emotions, since its function was to decide "which endothymic conditions and processes should determine the conduct and pattern of life" and to enforce "this decision *against all resistance*," including, of course, one's own emotions. It should "impose those [drives] of higher value over those of lower value" (Lersch 1938, pp. 189, 191; emphasis mine). The subjugation of emotions by the will corresponds to the subjugation to commands in military drill. Perhaps this is not the expression only of a specifically military ideal but of a more general political ideal of German character. In the stratification theory of the psychiatrist Hoffmann (1935), for example, the lower strata of vital drives and the middle strata of oriented emotions are covered by an upper layer of conscious will, rational thought, self-control, and the sense of duty.

An aside: changing metaphors and psychological theory

With the concept of strata, Lersch attempted to provide a new theoretical model for the relationship of emotion, thought, and will; he saw this as the big advantage of his system over Klages's. But his was not the only use of stratified or hierarchical systems. Although Lersch was clearly influenced by military ideals, such systems were common at that time, including Hartmann's conception of ontological strata and some layer theories in psychiatry (Hoffmann 1935; Thiele 1940). These cannot be explained by orientation to any specific image of the person.

Psychology has made use of various metaphors to grasp the psyche, which after all cannot be perceived with the senses. The succession of dominant metaphors is a sign of the attempt to do better justice to the object of investigation and to overcome the weaknesses of previous images (see Boyd 1979, p. 358). Metaphors function not only in relation to their original literal reference, but also with a whole field of associated implications about their implied object (see Black 1968, pp. 25–6). The choice of metaphors for psychological theories appears to be dependent on these implications and therefore, in my view, to be dependent on general intellectual, political, and social developments. The metaphors used in other disciplines, such as history, have already been investigated (Demandt 1978). A corresponding discussion for psychology has also begun.[15] If one wishes to examine the hypothesis that the changes of dominant metaphors depend on the intellectual and social context in a given period, it is necessary to look more generally at the changing imageries of an age and at the metaphors dominant in other sciences. As Lepenies (1978a, pp. 169–70) has demonstrated for the transfer of views on normality and abnormality between physiology and sociology in the nineteenth century, natural and human sciences frequently exchanged imageries.

A feature of psychology since the twenties was that biological metaphors replaced earlier technical or physical ones. For Kraepelin the machine provided the basis for his model of human behavior. Physical models dominated Freud's early views of mental life and Wundt sought causal explanations of psychological facts. But in the twenties – also in the psychoanalysis of the later Freud – we find a predominance of biological-organic and strata theories.

In the thirties the tasks of practical psychology were viewed more in bodily terms and less in mechanical ones. The call in the twenties to put the right man in the right place can still be interpreted in terms of assembling the

---

15 At the 91st Annual Meeting of the American Psychological Association (APA) in August 1983 in Toronto, the History of Psychology Section held a symposium entitled "Metaphors in the History of Psychology." See also David E. Leary, "Psyche's Muse: The Role of Metaphor in the History of Psychology," in *Metaphors in the History of Psychology*, ed. David E. Leary (New York, 1990). This book came out too late to be considered here.

parts of machines. But later the assignment of careers was described with a corporeal metaphor. Each person was to occupy a place appropriate to his or her abilities as a part or "member" of the body of the folk as a whole (*Volksganzen*). In pedagogical psychology, Oswald Kroh wrote that every "member of the *Volk*" should receive the education "that is suitable and necessary in view of his function in the whole" (1933, p. 309); pedagogical psychology should become a "pedagogical organology" (p. 327). The biological view of the "*Volks*-body" (*Volkskörper*) not only viewed the individual as an organ serving the body, but also included the possibility that the *Volk* could sicken and recover (p. 322). Here the biological metaphor met Nazi ideology.

Morphological thought repressed causality. Klages, who explicitly referred to the "importance of imageries as a starting point for the study of the soul," held that this must "in the first place be a *morphology*" (1926, p. 12). In Dilthey one can already see the replacement of causality with morphology, when he postulated conflicting basic types of philosophical worldview – naturalism (materialism, positivism), liberalism, and objective idealism – and propounded reason, will, and emotion as the psychological bases of these systems (see Lukács 1974).[16] Whereas Dilthey and Spranger (1930) based their systems of types on differing orientations toward mental content – in other words nonbiologically – Ernst Kretschmer, Erich Jaensch, and Gerhard Pfahler attempted to provide biological foundations for their typological systems.

It is interesting in this context to consider the reaction in characterology to Freud, who was perceived in the thirties as a "character morphologist." While the strata theoreticians took Freud's topographical model as a strata theory, they drew its sting by abandoning the instinct theory, robbing the model of its inner dynamic. Nor did they accept the superego, since this would have involved confronting the potential for social criticism in the model. Rothacker (1938) emphasizes the control of the id by the personal stratum (*Personschicht*) and the control by the ego, but no roles are played by the superego, the instinct dynamic, or the regulatory function of the ego between the pleasure principle and frustrating reality. Lersch, who dealt more explicitly with Freud, spliced elements of the Freudian conception onto his structure of the character – for example, narcissism (and Adler's inferiority complex) into the feeling of self-esteem. Endothymic experiences have for Lersch (1938, p. 190) the character of the id, but the important thing about this id, the role of the instincts, is abandoned. Strivings could not be traced back to a "single basic driving force" (p. 181). Lersch did not venture into the field of sexuality, and took flight at the mention of eroticism into the metaphoric language of poets (pp. 111–12). Nor did Klages (1926) include the "animal instincts" (p. 61) in his character model; he only wished to

16 This triad reappears in Lersch's characterology.

consider them as a vital substratum of character in the system of driving forces.

The stratification concept of the psyche was developed primarily by Rothacker (1938). He distinguished (1) the vital stratum, an autonomic and emotional layer of the fundamentally animalistic deep person; (2) the deep person (*Tiefenperson*), or the id, a layer of image-affective intuition and of the emotions; (3) the personal stratum organized by life experience; and (4) the uppermost stratum, the ego, as the psychic center and controller of the whole person. Rothacker drew an analogy between the layering of cortical person and deep person and the relationship of the cerebrum and the cerebellum, to indicate why a layer image was used. In the light of evolution it was not surprising that conceptions of mental structures were modeled on the strata of natural history. When Robert Heiss (1936) went beyond a structural consideration of character and saw each layer as a stage of ontogenetic development, this represented the transfer of Ernst Haeckel's biogenetic law to the theory of character stratification.

Since the end of the nineteenth century, the theory of evolution had been considered in the social sciences mainly in its social Darwinist variant, with its emphasis on selection and the transfer of that concept to politics and society (see Bayertz 1982). In the situation referred to by Edmund Husserl as the "crisis of European science," various sciences sought to achieve a new worldview via biology. Historians have linked this need for a new worldview to the disorientation of the intelligentsia in the age of industrialization (see Ringer 1969). The state of intellectual insecurity, and following the First World War that of political insecurity, created skepticism about the ideas of progress and causality. Since the beginning of the century, irrationalism had spread; for the Weimar period Peter Gay (1969) refers to an intellectual hunger for wholeness and a craving for community, resulting from a fear of the modern. Thus, the category of totality (*Ganzheit*), formulated by Hans Driesch purely for the scientific study of life, gained an importance beyond biological research. Rejecting an earlier dead and mechanical age, in various sciences a fundamental renewal of thought was sought from biology. It was not only in psychology that the opposition of organic and mechanical thought was postulated. This polarization was also found in political thought, and formed for Sontheimer (1978) a part of the "antidemocratic thought" of the Weimar period. Just as Karl Bühler (1926) hoped to resolve the conflict of psychology as a human and natural science by basing it in biology, so progress was hoped for in political theory from biological-organic conceptions of the state (e.g., by Krannhals [1928], whose work had an influence on the Nazis).

Perhaps these indications are enough to suggest the value of examining the changes in psychological theories up to the Nazi period against a broader historical background, as has begun in physics (Forman 1971) or political thought (Sontheimer 1978). Also worthy of investigation is the question of whether psychological theories contain a metaphor of the state. It has been shown that during the Bismarck period the theory of brain localization, with its regulated organization of individual centers and horizontal and vertical structures, corresponded to a model of the Prussian bureaucracy (Pauly 1983). In Lersch's character model, the vital impulses come from below, from the endothymic base, and these are then evaluated higher up, resulting in decisions involving the use of thought and will. Without the endothymic substrate, the superstructure would be an empty shell; but without the superstructure, chaos would reign. The integrated person manages to balance higher and lower strata, succumbing neither to wild emotions nor to cold rationalism. We find such a description in the psychological assessments of the Wehrmacht. It is worth considering whether this sort of internal hierarchy corresponded to an external social hierarchy and a concept of order and balance between social strata, or between the people and the state power.

Finally, if one wishes to determine why a certain metaphor of the psychical dominated at any period, it seems important to consider the connection between the choice of imagery and the practical function of the theory. Lersch's theory basically consisted of everyday character descriptions brought into a systematic form, which included concepts held to be diagnostically important and ethically correct, but which excluded questions raised by psychoanalysis concerning sexuality, aggression, and the protective armor of the character. The focus on the structure of character, involving architectural images of a building containing various floors with rooms on each floor, corresponded to a diagnostic approach that was interested in a person's state and its usefulness at a certain time. It did not include the question of the origin, the dynamics and the history of character, or of the connection between social relations and character structure, which was an approach adopted by the Frankfurt school, especially by Erich Fromm with the concept of social character. The requirement was rather for the conceptual systematization of information, and Lersch provided a model. In the final sentence of his book he wrote that his system could find its ultimate application as "an ordering of all questions that a person can be asked in order to investigate his essential psychic type (*seelische Wesensart*) with the many-sidedness, intertwining, and stratification of its constitutive features" (1983, p. 272). Characterology did not offer anything besides this ordering. When Lersch spoke about angst he described the states of excitement that we describe as angst, but not where it comes from.

Diagnostic procedures in the Wehrmacht and the situation tests

At the beginning of this chapter I mentioned the importance of being able to transfer professional knowledge to procedures that practitioners could apply systematically. Wehrmacht Psychology performed two services in this respect. First, it developed new procedures for the systematic registration of expression and situational procedures for the characterological evaluation of actions, all combined in a large-scale testing procedure for officer cadet applicants. Second, the "Instructions for Aptitude Testing" from the Inspectorate, edited by the Army High Command (OKH), provided a precise and uniform guide for the use and interpretation of all psychological tests, giving scientific guidelines for every Wehrmacht psychologist.

The development of the methods began in the twenties under Johann Baptist Rieffert, the first head of Army Psychotechnics (which was later named Wehrmacht Psychology). The new methods of characterological assessment were conceived primarily for the selection of officers, although parts were subsequently used for the selection of specialists such as airmen, range finders, and sonar operators, who were also subjected to other tests (see Chapter 4). The method used to select officers was central to Wehrmacht diagnostics and provides virtually a paradigm of characterological personality testing.

At the height of Wehrmacht psychology at the end of the thirties, the psychological test for officer applicants took two and a half days. As Rieffert's successor Max Simoneit put it, the aim was "to create observation situations" (1938, p. 14), placing the subjects in relatively complex, "lifelike" situations, in which the abilities and characteristics expected of an officer were called for. The psychological test was only a substitute for a "natural situation" (p. 14), and as soon as selection could be made in the most "natural situation" for an army – in war – psychological selection became superfluous (see Chapter 8). In the psychological test the performance, behavior, and expression of a subject were carefully observed, registered, described, and finally interpreted. The observations were seen as symptoms of characteristics; using a common term, Simoneit referred to the "symptomatological method."

For practical diagnostics in the Wehrmacht, Simoneit distinguished four areas of psychological analysis: biography, expression, intelligence, and behavior, all of which were seen as areas accessible to observation (1933, p. 46). Numerous methods were adopted and developed for the analysis of these four areas and to test special skills such as reactions and sensory performance. Here I shall concentrate on describing the innovations, based mainly on the loose-leaf collection of Army instructions already mentioned.[17]

First, an investigation contained a biographical analysis. After 1940, when the word "*Analyse*" had been expunged and Wehrmacht Psychology had adopted investigations of race and genealogy, this was known as the "biographical and lineage investigation" (*Sippenuntersuchung*). The psychologists were shown an applicant's curriculum vitae, personnel and genealogical forms, school reports, and health book. Additional information could be

17 Overviews of Wehrmacht Psychology diagnostics can be found in Kreipe (1937a), Mierke (1944), and Simoneit (1933, pp. 43–57). For the navy see esp. Feder et al. (1948); for the Luftwaffe see Gerathewohl (1950).

gathered in an initial fifteen-minute talk, following which the psychologist had to write down his first observations. Further questions could be cleared up in the main talk, or exploration, which formed part of the "mental (intelligence) analysis."

The second element of the examination was the expression investigation. This was felt to be very important, because the "involuntary stirrings and expressions of the person" could be observed (OKH Anw., C1). The applicant was observed throughout the test series, but special tasks were set for the investigation of voice and facial expression. Handwriting was to be evaluated at least generally, whether it was a personal or schoolboy style, natural or artificial, smooth or awkward. The graphological systems of Klages and Max Pulver served as orientations, though they were not binding (C4).

Speech investigations and mimetic recordings were developments created exclusively by Wehrmacht Psychology. In speech investigation the subject had to give a speech, and then to repeat this loudly, as though he were in front of a whole company of soldiers. The phonetic features were then evaluated characterologically. The subject was also required to give a loud command during the action analysis. A special test station was used to record facial expressions. The subject had to use a chest expander, be given an electric shock, prepare himself in front of a mirror for a photograph, look at a newspaper hanging on the wall, and then close his eyes. In all these situations he was filmed or photographed secretly. The interpretation of the facial expressions was made along the lines laid out by Lersch (1932), whose book *Gesicht und Seele* (Face and soul) was based on observations at these stations.

The third section, the so-called mental investigation, included a series of tests to establish intelligence, and the main talk. The tests and the diagnostic aims cannot simply be compared with current intelligence tests, because not only intellectual performance was important; the specific intellectual style and the relationship between thinking and character also had significance. Many of the methods used were adopted from psychotechnics, such as memory and arithmetic tests, Rybakov figures, technical work samples, and the explanation of technical drawings or films. The tasks related specifically to intelligence included the defining of terms, written organization tasks, writing essays, or an applied thinking test in which thirty-two bodies, which could be sorted into two equal groups using any one of five criteria, were placed in a series according to certain rules. The essays written about concepts, pictures, certain problems, or other topics were evaluated in terms of both intellectual performance and the personal features expressed. Aspects considered might include the relationship between abstract, conceptual, and concrete dimensions and whether the essay was lively or tame, powerful or feeble, thorough or superficial, ordered or loose, comprehensive or segmented, more subjective or factual, and so on (see II B Inspektion..., Atlas D, pp. 4–5). The description of picture cards was used to demonstrate "emotional and cognitive responsiveness" (OKH Anw., D17). The detective task, in which a possible solution had to be given for a "case," was intended to show intuition, decisiveness, and freedom from prejudice (D21).[18] One station providing an opportunity to observe both intellectual capacities (such as combinatorial ability) and social behavior was the newly developed battle game. Two parties of three applicants opposed each other in a board game, the aim being to eliminate the opponents and bring figures to a goal. The groups were in different rooms and informed each other of their moves by telephone. Psychologists, of course, were also present to observe.

The "great exploration" (see Kröber 1942) and a discussion circle with all applicants concluded the examination, though this was classified formally as part of the mental investigation. In the exploration the psychologist was to check and verify his findings. This could include questions that had not arisen, such as issues of "worldview, moral and political attitude (*Gesinnung*), and personal destiny" (OKH Anw., D50). The content was not defined in more detail, but the psychologist was expected to ask only such questions as he would be able to justify to the applicant's family. In the discussion circle the group of applicants discussed the "shared test experience" (D51). The instructions warned against any discussion of issues of the day or political and religious confrontations, although here as elsewhere the written material does not show whether the tests always went according to regulations.

The fourth area of investigation was the action analysis, which included a number of specifically characterological diagnostic methods. Two or four subjects were observed in situations in which they had to solve practical tasks. The aim was "to establish the type and strength of the present individual behavior of the young applicant, in order to conclude the overall psychological investigation by examining the military suitability of these established characteristics" (E1A). This was the only instruction that explicitly expressed the relevance of a certain test for military suitability. The evaluation of the observations was to provide characterological answers to the following question: "What is the degree of development of the practical senses, presence or absence of willingness for active conduct, degree of influence by success or failure, strength and importance of will in the total personality of the examinee?" (E1A).

The three tasks, "command series," "action series," and "leadership test," were methodological creations of Wehrmacht Psychology. In the command series the subjects had to get ready for a march and then carry out a number of tasks, such as stabilizing a damaged beam with a rope without falling off of it, making a bridge with a plank, and working with sandbags on a climbing wall. These tasks were then varied, with a different subject leading the other three. The action series involved work on a movable platform while standing on it, with shooting before and afterward. In the leadership test the subject

18 For the test essays see also Wohlfahrt (1939, 1939a).

had to teach the others for a short while about (a) a chosen, approved topic, (b) the description of a picture, (c) instructions for an organizational task (e.g., what to do in case of fire), or (d) instructions on a practical task. This was to establish the "winning contact, capacity for stimulation, assertiveness and participation, the suitability for leadership of people in terms of will and spirit" as well as to observe "mental and emotional responsiveness" (E3). Since the entire action investigation was directed toward the diagnosis of potential for activity and will, it was only logical that this included the Kraepelin Calculation Test as a characterological test – at least in the instructions (see the section entitled "Work Virtues, Work Motivation, Characterology of the Working Individual, and Psychological Leadership").

When selecting specialist personnel, the psychological testing of functions was more central. But procedures for characterological diagnostics were used to various degrees, and a description of the personality was generally required in the report, even though explicit personality diagnostic methods were not used. In Luftwaffe Psychology, for example, one had a general idea of the characteristics required of a pilot. Skawran (I C, 1, p. 9) named enthusiasm and commitment as important preconditions for the successful airman. Whereas the pilot needed to be aware and have practical intelligence, "activity, dynamism, and penetration" (I C Ge 4, p. 12), the bomber needed a calmer temperament and a sober and rational attitude, and tail gunners an adventurous nature (I C Ge 1 and 4). Good range finders needed practical skill with technical equipment, a sense of responsibility, high general reliability, constancy in dangerous situations, concentration, and a calm, reflective overall attitude (Kreipe 1936; see. Feder et al. 1948, pp. 52ff.). Radio operators required similar traits, such as levelheadedness, self-control, calmness, constancy, stamina, patience, and determination (Mierke 1941). For the army psychologist Gustav Nass, the ideal tank driver was a daredevil who did not make the "mistake" of thinking about his own life. He should show "firm character, selfless dedication to the task, willingness to sacrifice and ... strength of will," in short, "he must himself produce the will to give of himself" (1938, pp. 138–9).

Bearing in mind that the aim was the optimal allocation of individuals in a murderous war, it is difficult to reach a sober judgment on the psychological methodology involved. The attempt to match the personality profile of an applicant to a characterological "job profile" involved occupations for which drawing up a list of character requirements was a direct moral problem. This point will not be dealt with further here (see Chapter 9), since we are considering only the available psychological models and methods that were functional for practical demands.

For the selection of specialist army personnel, these were above all methods for the diagnosis of special skills. These were partly adopted from industrial psychology and partly the result of the development of new apparatus or tests. However, in terms of method they were not innovative. Basically they were simple tests of sensory and reaction performance as well as motor and sensorimotor abilities. For example, radio operators were tested for their ability to register and repeat Morse signals, for their ability to distinguish pitch, and for their writing speed. Some tasks from the mental investigation as well as the command series and the exploration, were also included. Special tests for the drivers' corps included field orientation, a test driving stand as designed by Moede, a technical-constructive work test (reassembling a tachometer), and laying a water pipeline. Importance was attached to vision tests for the range finders. To check depth of vision, a horoptoscope was used that Jaensch had employed to establish types: two hanging threads had to be adjusted to various levels. To select gas tracers, a testing box was used. The subject had to remember various smells and pick them out later from a number of bottles.

In the navy there were a number of specialist positions in crews for which psychological selection was employed. In a relay procedure (Mierke 1940) all applicants were subjected to a basic test (including intelligence, sport, education, and biography sections), following which they were divided into technical and nontechnical groups. The subjects were then allocated according to the results of the functional tests that followed. These were for technical comprehension, depth of vision, and suitability as a radio operator, as well as for concentration and reactions on an apparatus adapted by Mierke from Moede's test on drivers (see Feder et al. 1948, pp. 35ff.).

Luftwaffe Psychology developed new psychotechnical methods to establish the general sense of orientation and the ability to orientate independent of vestibular sensations, which after the war received the attention of NASA in the United States. Three methods should be mentioned here: the rotating-chair test, the orientation tent, and the Rhön wheel.[19] The rotating-chair test was used to determine the sense of location. The subject sat blindfolded on the chair and had to give his impressions of his situation when the chair was rotating or still. He then had to determine his spatial location from the sound of a metronome. Finally, a cabin was lowered over his head with a direction indicator showing the movement of the chair. When the light in the cabin was switched on, the subject had to say whether the impressions he had just described (e.g., the feeling of turning in the opposite direction after stopping) were correct. For the test in the orientation tent, the subject had to memorize a landscape that contained a tent. He then had to enter the tent and walk around inside of it while it was rotated. After thus losing his bearings he was given an orientation point relative to which he had to determine the other points. With the Rhön wheel, the subject was strapped onto a vertically mounted wheel two meters in diameter and rolled around sixteen times while

19 In an illustrated report on psychological tests at the Neubiberberg Airstrip, there are pictures of the rotating chair and the Rhön wheel (I C Däumling); reproductions in Geuter (1987). The instructions are in OKH (IIB); a summary of the methods is in Gerathewohl (1950).

he solved sums. These procedures clearly demonstrate that psychotechnical selection research was being continued, and new methods being dreamed up, despite official talk of the dominance of characterology.

Will, courage, and the limits of psychological prognosis

Soldiers, especially officers, were expected to show strength of will and courage. Wehrmacht psychologist Nass quoted an early comment by Hitler: "Sacrifice, commitment and courage are demanded, and those *Volk* will respond who call these virtues their own" (Nass 1938, p. 148). That those who responded to the call would necessarily possess these "virtues" was not, however, widely believed in the armed services: that was what psychological tests were needed for. But how were will and courage to be diagnosed? Diagnosing will had already proved more than psychology could swallow; with courage psychologists did not even want to take a bite.

The military importance of the prognosis of strength of will or of willed behavior was clearly seen in Wehrmacht Psychology. Simoneit referred to it as the "central problem of aptitude tests. If Wehrmacht Psychology fails in its efforts to solve this problem, it will lose decisively in military importance" (Simoneit 1937, p. 5). Diagnosing features of will had been a problem for psychotechnics in the twenties. For example, in connection with the work curve (see the section entitled "Work Virtues, Work Motivation, Characterology of the Working Individual, and Psychological Leadership"), determination had to be investigated. The will to achieve was regarded as being more important than actual performance. Theodor Valentiner had therefore attempted a diagnosis of will for career work with young people (*Die erste Tagung . . .* 1923, p. 396). In psychotechnics, however, the opinion was that it was hardly possible to diagnose the quality of will. Giese (1925, pp. 209–10) felt that courage, fearlessness, and endurance could not be determined experimentally – at most, one could observe their symptoms (e.g., by reaction tests) – and that one therefore had to be content with questions about life-style or the observation of work behavior. For the symptomatologists of the Wehrmacht, of course, the problem was the exact relationship between visible behavior and the factor of will. There was a practical reason for wanting to see progress here, since it was in the assessment of "will disposition" that most mistakes were still made (Simoneit 1937, p. 6). But they were not able to solve the problem. In a specially prepared investigation entitled *Wehrpsychologische Willensuntersuchungen* (Military-psychological studies of will) Simoneit defined the will as "that mental phenomenon . . . which, accompanied by the consciousness of freedom and of one's own strength, leads to the guidance and regulation of thoughts, behavior and actions" (p. 6). For him the difficulty lay in the fact that the will was a complicated psychic composite. To determine the will it was necessary to consider the components of the ability to will, willingness (wanting to will),

"will-method" (the capacity to use one's will), and willpower. This led him to the methodological platitude that an assessment of will can be based only on the entire psychical life. In Wehrmacht diagnostics, biographical analysis, expression analysis, mental analysis, and action analysis were all to be drawn on (pp. 10–11; see also Simoneit 1938, p. 17).

In 1933 Simoneit (p. 54) had written that the "will side" of the person was at the forefront of the entire action analysis. In practice, however, it was mainly the command series as part of the action analysis from which one expected "observation of practical bodily action, and especially of willed action" (OKH Anw., E1). But the evaluation of command series did not seem to offer a solution to the problem of how to make deductions about will from the behavior observed. Following an inspection tour, Simoneit (1940) commented that psychologists seemed unable to proceed beyond a mere description of performance or to make more than impressionistic judgments in their evaluation of the features of will.

"Will research" and the resulting adaptation of the command series obviously formed a central aspect of research by Wehrmacht Psychology in the following years (see Renthe-Fink 1941; Simoneit 1942). Renthe-Fink (1942) feared that attempts to make the command procedure "more lifelike," for example, by exchanging the subjects' athletic clothes for paramilitary uniforms, would cause these procedures to lose more and more of their character as scientific experimental test situations. He recommended returning to a precise, clear test situation as before. Flik (1940) wanted to assess will from the applicants' choice of sports, since more difficult sports required more willpower and hardness of will. A sports test itself was not seen as a viable way of diagnosing will (Masuhr 1937, p. 23). But all these considerations could not come to grips with the methodological problems or the dissatisfaction of the military.

The inability of Wehrmacht Psychology to diagnose courage had been a decisive reason for its rejection by some of the military in the early stages. But as Hesse (1930) wrote, the dangers involved prohibited the use of real tests of courage. The Inspectorate for Aptitude Testing officially rejected tests of courage as impractical "because courage – of the sort required in war – is activated only when vital values are endangered" (OKH Anw., E12). But the report as a whole should show "whether the preconditions for courageous deeds in war are present in the personality of the subject, or whether his general nature makes courageous behavior improbable" (OKH Anw., E12).[20] A comment on this matter was supposed to be included in the conclusions to assessments, but none could be found.

In the journal *Deutsche Infanterie*, Simoneit (1939a) asked rhetorically if courage could be tested or learned, and answered that it would not, because this would presuppose an "ability for courage," whereas courageous deeds

20 See also Masuhr (1937, pp. 24–5).

"were the result of all the mental, spiritual, and moral properties, abilities and strengths of the human personality" (p. 18). The possibility of transferring courage in sport or artistry to courage in war posed problems. Military demands thus revealed the methodological limitations of psychology. Even more so than will, the categories of fear, cowardice, and courage became practicably observable in war in a way that the diagnosis of "character" traits could hardly match (see Chapter 8).

The dysfunctionality of race psychology and typology for practical psychology

The Nazis had ways of selecting people other than characterological or psychotechnical methods. There was selection of the "inferior" according to political requirements or selection in terms of worldview. New forms, such as the selection camp, were also available. But these methods were not among those offered by psychology for its main practical purpose. It may seem surprising that I have not yet mentioned typology and race psychology, although these were so obviously linked to the Third Reich. In fact, both areas were of little practical relevance for professional psychology. If we consider their contribution to the professionalization of psychology, this would seem to lie mainly in the political and ideological legitimation of psychology discussed in Chapter 5.

Since Kretschmer's work, typologies were highly favored among psychologists. When Hubert Rohracher first published his *Kleine Einführung in die Charakterkunde* (Short introduction to character studies; 1934), he considered typological systems, favoring Kretschmer's position since this used the most general properties for classification and provided a doctrine of constitution to explain scientifically why a given person possessed a certain characteristic (p. 97). Rohracher's book was so popular that it had been through four editions by 1940. In addition to biologically based type systems, psychology also had systems of formal functional types oriented toward features of perceptional style, such as Jaensch's, which he described at first as biologically and later as racially based types (see, e.g., Jaensch 1938). Alternatively, there were formal "dispositions" such as emotional responsiveness and the general perception and assimilation of stimuli that Pfahler took as his starting point. He was later to link his types to the racial types of Hans F. K. Günther. Ach had a system of formal functional types classified by forms of will and awareness, without a biological foundation; Krueger classified according to basic forms of experience, Spranger according to value orientations.

For us, it is not worth discussing all these type systems in further detail, except perhaps to consider one point. What possibilities were seen for the practical application of typological knowledge in psychology? All the typologies mentioned, with the exception of Krueger's and Spranger's, probably assumed that a "type" was some sort of radical of the personality, from which various characteristics could be deduced and behavior predicted. The type was to be deduced in turn from various symptoms. The task of diagnostics was thus to provide a method for establishing the type rapidly and accurately from appearances. But what practical use could it be for psychology to determine the type when the pure type was usually a theoretical construct that would be found only in relatively extreme cases?

This practical problem meant that there were more supporters of typologies among academic psychologists than among industrial or military psychologists. Here the typological (and race-typological) systems proved to be dysfunctional, even though typological or racial studies were carried out on people by the Wehrmacht or various labor offices. For example, the arbitrary way in which performance was used to decide "racial type" did not permit a corollary in which "race" could be used to predict performance. To show this, I want to look at three investigations whose results are nothing more than racist apologia. Friedrich Becker, who studied under Jaensch, and Walter Schulz, head of the Rhenish Provincial Institute for Labor and Vocational Research in Düsseldorf, used Jaensch's typology in their studies. Becker (1938) studied the "types" of workers in various regions. The "internally integrated type" was to be found mainly in the Ruhr, an area of heavy industry, whereas the "externally integrated type" was more predominant in the precision engineering districts around Wetzlar. For the "theory of industrial locations," it was important to know whether each "type" lived where the work suited him best, because it was not work that shaped humans, but they who shaped the work (p. 75). Luckily, Becker found the right "types" in the right places, so everything could stay as it was. The work of Schulz (1937) on selection in the iron industry was without practical relevance, since he used the same circular arguments as Becker.[21] According to Schulz, behavior was genetically determined and a feature of race. If people were to be employed according to their ability, it was necessary to establish the racially determined forms of disposition. To do this Schulz used the simple trick of declaring the results of psychological function tests to be results about dispositions. Since dispositions are distributed according to *Volk* and descent, and the "tribes" know the sort of occupations suitable for them (p. 1140), the result was that the people working in the iron and steel industry were found to have the right origins for the job. In a further study Schulz (1936) tried to use Jaensch's typology to draw up a biological classification of type-appropriate (*artgemäss*) occupations. Ehrhardt and Klemm (1937) evaluated records of the Leipzig Labor Office (basing their classification on the racial doctrine of Hans F. K. Günther) and came to the conclusion that the "Nordic" and "West" races were superior, especially in working with their hands, whereas the "Ostian"

21 Schulz published similar essays between 1934 and 1936 in the journal *Die Rheinprovinz*.

and "Dinarian" races performed less well on average. Such racist stupidity was published by our scientific ancestors in scientific journals.

The results of these studies and the race typologies on which they were based were useless for practical aptitude testing in firms and labor offices. Thus, the use of Jaensch's typology was viewed with reserve by labor psychologists. Joseph Mathieu (n.d., p. 41) wrote that for practical reports typologies could be used "in part not at all, and in part very restrictedly." Otto Graf (1936) criticized Jaensch explicitly. As mentioned already, the typological entry in some reports of the Labor Office in Stuttgart had been left empty. Giese (1935) supported the "tribe psychological" evaluation in theory, but in practice the thing that counted was the effectiveness of selection for industry. There was no objection to typology in principle – as long as it lived up to these demands. Only a few pages on, however, Mathieu defended the diagnosis of *work* (not personality) types on the basis of work-duration tests based on Poppelreuter (Mathieu n.d., p. 47).

The divide between a psychology directed to ideological opportunism and one reflecting practical usefulness (see Chapter 5) was also apparent in Wehrmacht Psychology. The army psychologist and SS officer Ludwig Eckstein talked of the "racial soul" in the "racial body" in a pamphlet of the Reichsführer SS, as well as "spatial political aspects" of the "basic bio-dynamic laws" during the attack on the Soviet Union. He also warned Germans against subversion by the "east European races," and in the face of real genocide saw "selection and eradication processes" as natural (Eckstein, n.d.).[22] As well, he criticized in his *Psychologie des ersten Eindrucks* (Psychology of first impressions) the inappropriateness of the typologies of Jaensch, Jung, and Kretschmer for the psychological assessment of personality (1937, p. 92).[23] Simoneit also doubted the practical value of contemporary typologies: "The honest psychological researcher must admit that in quite a few cases nature made nonsense of his typological rules" (1934, p. 58). However, it was sensible to identify certain types of people who regularly experience a "failure of deployability" (p. 60).

When "types" were mentioned by Wehrmacht Psychology, this did not imply psychological classification systems but, as Simoneit (1941) expressed it, typical cases. According to Simoneit the type could be determined inductively from the diagnostic material using the criteria of distinctiveness and similarity. Instead of proceeding from the laws of mental functions proposed by general psychology to establish the "individual soul," he suggested the opposite for characterological-typological research, determining the "structural moments of the soul" from individual cases (p. 12). Some of

22 Since Eckstein is referred to as "Dr. phil. habil." this must have been published after his habilitation in 1942.

23 This despite an apparent disclaimer in the following sentence, in which he talks of the "irreplaceable value" of typologies for psychological work, which seems more an attempt to placate those he has just criticized.

124   *The professionalization of psychology*

the characterological types of Wehrmacht Psychology were the intellectual, the deep person, the fanatic, the swag, the unreliable one, the lazybones, the sensitive one, the man, the gentleman, and the unlucky one (Menschenformen 1941; see, also Junius 1943).

In the reports of Wehrmacht Psychology it is hard to find any psychological typology characterizations. At the end of the thirties racial investigations were introduced, and they were envisaged in the regulations of 1940. Physical appearance was to be used in the evaluation, though there were no guidelines about how this was to be done. The subject was also to state the German folk lineage (*Volksstamm*) to which he felt he belonged; this was supposed to show "tribal consciousness" (OKH Anw., G1). The instructions do not indicate what this had to do with aptitude criteria. Racial and "tribal" psychology seem to have served primarily to demonstrate the ideological orthodoxy of psychology (see Chapter 5). As with the psychological theory of types, they did not fit into the pragmatic criteria for successful psychological selection. Thus, racial psychology and typology, by providing an ideological legitimation for the discipline, contributed only indirectly to the professionalization of psychology in Nazi Germany.

4

Psychologists at work: the start of new professional activities in industry and the army and their expansion in the war economy

The first profession for psychologists was that of academic researcher and lecturer. Here we are interested in practical psychology as an occupation outside the university. What were the reasons for the employment of expert "psychologists" in various areas of German society? How were corresponding professional roles institutionalized, and what was the demand for psychologists? We will also consider the problem of competition from other professional groups, a point emphasized in the introductory model.

The development of psychology as an occupation cannot necessarily be deduced from the problems considered by psychology as a science. Problems taken up by psychology before Wundt did not simply become fields of applied psychology. Questions concerning the self-image of the citizen, for example, a psychological topic around 1800 (Jaeger and Staeuble 1978), did not find any place in institutionalized psychology. Problems of social integration (see Brückner 1974, p. 15) were considered only much later; they were left to educational, judicial, and punitive institutions. In applied psychology attention was focused primarily on questions of training and deploying the work force. Pedagogical and industrial psychology were the two central elements. The fact that the originally broader term "applied psychology," or "psychotechnics" as it was widely known, came to mean "industrial psychology" is a reflection of psychology's selection of problem areas in the twenties. A further selection occurred in the course of transition from a field of scientific reflection to a field of practical professional activity. This poses the question of why the labor administration and the Wehrmacht were the first institutions in Germany to employ full-time psychologists.[1] Why wasn't this the case in

---

1 Similarly, in Britain and Finland psychologists were mainly active in industry before the Second World War. In Belgium, Holland, Italy, Switzerland, and Spain, it was education, career advice, and industry; in Japan, first education, then industry. After the First World

125

clinical psychology, which had been central for the professionalization of psychology in the United States since the twenties (Napoli 1981), and why not in the field of forensic psychology? It is conceivable that the reason lies in the traditionally strong position of the medical and legal fraternities in Germany. But, then, why should psychologists have been able to take on tasks in the military formerly carried out by officers, who surely did not have a weaker social position?

If we wish to study the uses made of professional experts in psychology, the following questions seem important:

For which problems were psychological services sought?
Which institutions placed demands on psychology?
Why were specialist psychologists employed instead of psychologically trained
    members of other professions?
What form did the institutionalization of the professional role take, and,
    particularly, was a career created specifically for psychologists?
What was the demand for qualified psychologists?

For the time when there was no formal qualification for "psychologists" I will use the term after Giese (1922, p. 6) and Marbe (1921, p. 202) to refer to someone with a doctorate in psychology and active full time in a psychological field.[2] The questions just listed are considered for the various fields in which psychology was employed, for which it produced relevant findings, or in which psychologists were active. This includes industry, labor administration, education, psychotherapy, and the military. The break in the description of labor psychology following the transition to the war economy and the growth of Wehrmacht Psychology after the introduction of general conscription in 1935 follows the hypothesis that the war economy and war preparations boosted the expansion of professional psychological activity.

First concepts and footholds for the profession of psychologist

In 1913 Aloys Fischer, pedagogy professor in Munich, wrote an essay entitled "The Practical Psychologist – A New Profession" in which he outlined the possible fields of activity of future practical psychologists. He felt that the science was ripe for application and saw in the existence of various counseling surrogates a clear sign of the need for psychology arising from the differentiation of modern life. Fischer recommended that psychologists engage them-

---

War clinical psychology was obviously important only in the United States and Belgium. See contributions in *Psychology in Europe* (1956) and in Sexton and Misiak (1976); see also Ferrari (1939), Chleusebairgue (1939), and Napoli (1981). On the role of psychologists in the armies of other countries see Chapter 10, note 12.
2 On the question of doctorates in psychology see Chapter 6.

selves as counselors in education, the choice of careers, the choice of spouses, adult education, and the treatment of mental conflicts. They should take the place of doctors, counselors, fortune tellers, and relatives.

When Fischer wrote this essay, a program of psychotechnics had been outlined but had not yet been carried out (Jaeger and Staeuble 1981, p. 66). The First World War led to a breakthrough. Psychotechnical methods were used on a wide scale to select drivers, other military specialists, and highly talented pupils. Psychologists were active in brain-damage wards and in test stations for false limbs for the war wounded. Dorsch (1963, p. 81) saw the war as "setting the pace" for psychotechnics in Europe. Jaeger and Staeuble (1981, p. 74) felt that the war encouraged psychotechnics to take on many tasks pragmatically without theoretical reflection.

The problems of assessing vocational aptitude, selecting the ablest, and diagnosing and treating brain damage and head wounds continued to form the focus of psychotechnics in the twenties (see Piorkowski 1918). They were taken up by professors who worked in industry as a sideline or who founded communal institutes (see Dorsch 1963; Jaeger and Staeuble 1981; Wies 1979). These psychologists also raised the question of professional standing for practical psychologists. In 1921 Walther Poppelreuter, who had served in a war hospital for head wounds in Cologne and since 1919 had built up the Institute for Applied Psychology in Bonn, recommended training practical psychologists. They should study medicine and should then take a specialist follow-up course at an institute for the brain-damaged. Medical psychologists were also needed in career guidance and industry, since here the concern was mental and physical aptitude. Poppelreuter felt that psychologists "recruited from the usual faculties, with their links to philosophy" were unsuited for practical psychology (1923, p. 78). Fritz Giese, writing in a "Guide for Studies and Vocation" in 1922, lists opportunities for psychologists with a doctorate at institutes for practical psychology, at certain medical laboratories, and at career offices and advisory centers.

At that time there were not many psychologists in Germany. Poppelreuter (1921, p. 1262) spoke of twenty practicing psychologists. When Karl Marbe (1921, p. 150) claimed that "the call from the administration for able, practically qualified psychologists grows ever louder," this may well have been wishful thinking. There is certainly no sign of a rapid expansion in the number of publicly employed psychologists during the twenties. Giese (1922) seems nearer the mark with his opinion that economists and engineers who had dabbled in psychology had much better prospects with the authorities or with industry. Psychology was more likely to be applied by people who had been trained in short courses (see Poppelreuter 1921a, p. 10). Practical psychology in the initial phases was a matter for university staff active extramurally, career advisors, production engineers, and teachers. Only the last of Fischer's recommendations was realized at that stage: the use of psychologists in career guidance.

The twenties: opportunities in labor administration
and industry

After World War I, demobilized soldiers had to be given civilian occupations. This provided an opportunity for the application of psychological methods in job allocation and career guidance, to a degree that seemed large then but would seem smaller today. Psychotechnical methods were also used for the rationalization of individual firms. During the war, production had been converted to meet military needs; the legacy of war was large debts, loss of territory, and heavy reparation payments. This in turn caused raging inflation until the currency reform of 1923, which enabled big industry and the cartels to consolidate.[3] Whereas production had expanded during the inflation, the only option now available to industry, according to Sohn-Rethel (1973, pp. 41–52), was the intensive rationalization of the work process by increasing the pace of production, or as the contemporary phrase had it, "rationalizing from the person." It was here that psychotechnics was also to make a contribution.

As far as career guidance is concerned, the fundamental problem lay in the transition from an age in which the individual was usually assigned one of a small range of occupations, often at birth and for life, to an industrial age with a much wider range of vocations and relative freedom of choice. The individual was, of course, free but was forced to choose (Hesse 1972, p. 15), and the state and industry were interested in an optimal choice. Career advice had been given in some form since the start of this century, but it was the reintegration of millions of soldiers after the war that gave it increased importance. The shortage of employees – until 1923 unemployment was very low – made a centralized allocation of labor seem sensible. In the course of the establishment of labor exchanges and career advice institutions, we find the first full-time use of psychologists.[4]

A decree of the Office for Demobilization (9 December 1918) directed the individual German states to set up central offices and take steps to provide career advice and act as agents for arranging apprenticeships. Bavaria had already introduced vocational offices in 1917. In Prussia the tone was set by a decree on 18 March 1919 (see Hartwig 1959, p. 38; Meiser 1978, p. 15). For the first time the use of psychologists was recommended for public duties by a state organ. Paragraph 7 states that "for career advice, whenever possible, a doctor should be called upon; in larger career offices the cooperation of a psychologist also appears advisable" (Reichsarbeitsblatt [RABL] 1919, 17,

3 See, e.g., Hallgarten and Radkau (1981, pp. 140–80), Hinrichs (1981, pp. 107ff.), and Stolper et al. (1966, pp. 87ff.).
4 For a history of career advice see *Berufsberatung, gestern–heute–morgen* (1959); Dorsch (1963, pp. 112ff.), Hartwig (1947, 1948, 1959), Homann (1932), Meiser (1978, pp. 13–19), Neubert (1977), Stets (1963), and Wies (1979).

p. 329). The participation of a psychologist was also expressly recommended in Saxony; other states rejected it (Müller 1920–1). In 1922 a law led to public labor offices being set up over the whole Reich, as well as labor exchange administrations at national and state levels (RGBL I, 1922, pp. 657–71). On the basis of this law the "General Regulations for Vocational Advice and the Arangement of Apprenticeships by Employment Exchange Officers" were issued in 1923; these envisaged the establishment of advisory boards that were to include psychologists (RABL 1923, pp. 309–10). The 1927 law on employment opportunities and unemployment insurance, which for the first time defined career advice and arrangement of apprenticeships as obligations of the employment administration, no longer referred to these boards (RGBL I, 1927, p. 193f.). From 1923 until 1927 only the offices at the state level were obliged to provide career advice and to arrange apprenticeships. It is mainly there that we find the psychologists. Individual state employment offices carried out aptitude tests in conjunction with communal or provincial psychological institutes, which in the cases of H. Langenberg (Düsseldorf) and Wilhelm Hische (Hannover) involved one person acting in two capacities. Elsewhere, such as in Stuttgart, cooperation was with the university. There were not, as yet, budgeted posts for psychologists; they were positioned in employment offices either as consultants (*Referent*) or as career advisers. According to a survey by Stets (1923), there were twenty-one state employment exchanges and 592 local career advice centers in 1922 with 160 full-time counselors.[5] These included three male and two female psychologists. My own, incomplete lists show thirteen qualified psychologists with positions at employment exchanges in the period 1920–30.[6] For the year 1925 Rupp counted seventeen psychologists at employment exchanges (Wies 1979, p. 203); for 1930 fifteen can be determined (see the next section). The demand for psychologists in this field was thus only moderate.

Not until 1953 did psychologists became a recognized career group in the labor administration; until this time they were in competition with other professional groups. When the Leipzig City Council advertised for a career counselor, it sought a pedagogue, psychologist, or economist (*Praktische Psychologie*, 1920–1, 2, p. 382). There were no specific qualifications for such a post. The "General Regulations" of 1923 mention only a minimum age of twenty-eight and at least five years of work experience (RABL 1923, p. 310). Recruits came mainly from the social professions; in second place were teachers (Stets 1923), who as a group had initially opposed independent career guidance (Homann 1932; Müller 1920–1). In 1925 this conflict was removed by a decree that regulated the cooperation of the schools (see

5 The state exchanges are listed in Liebenberg (1925, pp. 158ff.).
6 Dorsch, Stuttgart; Handrick, Chemnitz, 1920; Hische, Hannover; Huth, Munich, 1928; Jacobsen, Hamburg; Langenberg, Düsseldorf until 1927; Moers, Bonn, 1921–8; Roos, Halle, 1929; Schorn, Aachen; Schulz, Düsseldorf from 1927. According to Roos (Z, f. 160) in 1929 H. Krüger, Petzold, and Wallau were also working in Halle.

Hische 1931, pp. 88–98). The competition with teachers in the early years of career guidance seems to have been of significance for psychologists. Former teachers such as Hellmuth Bogen, Paul Knoff, and Richard Liebenberg were involved practically and theoretically in expanding the use made of psychology in labor exchanges. A further problem for psychologists was their relationship with career counselors. It was the task of state employment exchanges to train career advisers in knowledge of human nature and developmental psychology and to teach them psychological methods of aptitude testing. Courses were held, and these were sometimes viewed as the responsibility of psychologists (see Becker 1930; Knoff 1926, 1928). However, the use of psychological methods by career advisers posed legal problems. Liebenberg (1925, p. 46) reported that central labor offices agreed to pay the cost of aptitude tests only when they had been carried out by "suitable persons." But there was no regulation about who was suitable until the fifties, when the Psychological Service of the Federal Labor Office managed to establish the application of psychological methods as its professional task. In the twenties psychologists cast doubt on the qualifications of career advisers. But these – for example, the consultant for career advice at the Ministry of Labor, Walter Stets – argued that psychology was in danger of being out of touch with the world and that psychologists should therefore never advise on careers, but only assist the career adviser (1939, pp. 35–6).

In industry the new psychological methods were applied primarily by engineers. Psychotechnical methods had been used in German industry since 1917. After World War I there developed a real psychotechnical movement, which brought with it a whole series of psychotechnical test facilities in factories.[7] Chestnut (1972) speaks of 170 industrial test stations in 1922; Jaeger and Staeuble (1981, p. 79) count 110 industrial firms (excluding subsidiaries) that carried out psychotechnical tests in 1926. The wide application of psychological knowledge in industry did not, however, mean that a professional perspective for psychologists had developed in industry. Wies (1979) has studied this contradiction, especially emphasizing the competition between engineers and psychologists. In the course of the rationalization of the twenties, a new occupation of production engineer arose; these engineers were responsible for the deployment of the labor force and often attended psychotechnical courses. They were generally the ones who undertook psychotechnical activities.

A count of industrial psychotechnical test stations in Germany, based on figures in *Industrielle Psychotechnik* for 1926 (pp. 246–53), showed sixty-three firms with their own test stations and eighteen that delegated testing to universities or communal institutes; twenty-four firms gave no further details. Forty-one of the test stations were headed by an engineer, two by a company

---

7 See Dorsch (1963, pp. 84–5); Jaeger and Staeuble (1981, pp. 74ff.), and Wies (1979, pp. 173ff.). Hinrich's data (1981, pp. 225–30) are partly imprecise or wrong.

director, five by other personnel, and one by a teacher. In thirteen cases the profession of the head of testing was not clear. The list does not include a single psychologist. Schulhof (1922–3) does mention a psychologist as head of testing in the evaluation of a survey of thirteen large firms and two public organizations by the Institute for Industrial Psychotechnics, but it is not clear whether this was a full-time post.

Thus, in industry in the twenties there was a discrepancy between the application of knowledge and the degree of institutionalization of the corresponding professional role. Perhaps the body of knowledge seemed so meager that a short course was felt to be sufficient to understand it. The center for such courses, attended primarily by engineers, was the Institute for Industrial Psychotechnics at the Technical College of Berlin-Charlottenburg under Walter Moede, according to whom "thousands" attended the courses during the twenties and thirties (Wies 1979, p. 82). Although DINTA, founded in 1925 on the initiative of the industrialist Albert Vögler, used psychologists (e.g., Walther Poppelreuter) in the training of engineers, it did not encourage the work of psychologists in the training centers and assessment offices set up by the institute.[8]

After a course at the Charlottenburg Institute there was an open conflict in the *Zeitschrift für angewandte Psychologie* about the relationship between psychologists and engineers (Z. ang. Ps., 16, [1920], pp. 391–3). Hans Paul Roloff, a teacher from Hamburg, wrote a report about the course criticizing the way the engineers were shown in forty minutes how to test intelligence. The industrial scientist Georg Schlesinger, who had helped to set up the institute, returned the fire, stating that the engineers were no laymen. The psychologist William Stern felt obliged to defend his colleague Roloff, and Giese replied with basic recommendations for establishing the professional status of psychologists. In 1923 Schlesinger then proposed a mutually beneficial solution; the engineers would not interfere in the core area of the psychologists, and those who really wished to be psychotechnicians would have to give up their former occupation and dedicate themselves entirely to the science (*Die erste Tagung...*, 1922–3, p. 30; see also Wies 1979, pp. 50, 80).

While academic representatives continued to argue over who should practice psychotechnics, a clear practical answer was being provided in the firms. Erwin Bramesfeld, professor of psychotechnics but a trained engineer, felt that engineers trained in psychotechnics rather than economists or psychologists were best suited to teach the subject in technical colleges (1926, p. 11). The engineer K. A. Tramm, well known among psychotechnicians, opposed the idea that aptitude tests should be conducted only by psychologists. For such applications of psychological methods the slogan must be "Make way for the best" (Tramm 1932). The only psychologist to support the idea

---

8 See the annual reports of DINTA, published as a supplement to the journal *Arbeitsschulung* (1929).

of entrusting specially trained engineers with practical psychotechnics was Moede, whose institute had an outstanding position in the psychotechnical training of engineers. Other psychologists criticized the dilettantism of part-time psychologists.[9] Their demand was that psychology be practiced only be people who had completed studies in the subject. The controversy was then reflected in differing ideas for the training needed by psychologists (see Chapter 6).

Academically trained engineers were not the only ones to apply psychology in industry. In the Reichspost, for example, staff members were given twenty-day courses to learn to use psychotechnical test procedures (Jaeger and Staeuble 1981, p. 80). The fact that it was not necessary to have studied psychology in order to apply a simple test procedure was recognized by Giese. Therefore, in 1919 he suggested introducing "psychological laboratory assistants," to carry out mechanical, unproductive mass tests (Giese 1919). Such assistants also seem to have been trained in industrial firms, although only one case is known to me.[10] In 1922 Giese referred to those personnel variously as lab assistants, psychotechnical assistants, or just psychotechnicians. For their training he envisaged a practical and theoretical apprenticeship at nonuniversity institutes.

The employment structure around 1930 and the development of industrial psychology with the Four-Year Plan of 1936

In a newsletter of the Academic Information Office, Bramesfeld published a short report in 1932 on the professional opportunities open to practical psychologists, painting an almost hopeless picture. Perhaps this was influenced by his own attitude, since he was in favor of psychotechnics remaining in the hands of engineers. He was not aware of a single case of a psychologist being employed in industry; there were a few possibilities in scientific institutes. Pure psychologists would have little chance in career advice either. He saw a path free of "obstructions" only if a psychologist wished to establish himself in a free practice as "counseling psychologist" (1932, p. 307).

Communal and provincial psychotechnical institutes were the only institutions to draw mainly on psychologists. They existed in Barmen, Bremen,

---

9 See Giese (1922a), Weber (1927, pp. 377–8), and Wies (1979, pp. 81ff.). Weber writes acidly: "All too often we find an engineer working with psychotechnical methods. He works mainly with numbers, makes a pretty equation and produces a neat solution with no loose ends. But the living, the constantly changing is not recognized. To attempt to capture the human soul with mechanical and other constructed fictions means the ruin of practical psychology" (p. 378).

10 In a short note in the *Psychotechnische Zeitschrift*, 7 (1932), p. 120, Dellwig and Poppelreuter report the examination of a psychotechnical laboratory assistant at the Vereinigte Stahlwerke A. G., Schalker Verein in Gelsenkirchen.

Dresden, Düsseldorf, Frankfurt, Gelsenkirchen, Halle, Hannover, Cologne, Münster, and Nuremberg (Dorsch 1963, p. 85; Jaeger and Staeuble 1981, p. 81). Heads of these institutes included the psychologists Theodor Valentiner (Bremen), Walter Schulz (Düsseldorf), Georg Clostermann (Gelsenkirchen), Wilhelm Hische (Hannover), and Josef Weber (Münster). Details of the legal status of these institutes are not known to me. In part they financed their activities with industrial contracts to test apprentice applicants or personnel. The Rhenish Provincial Institute for Labor and Vocational Research received one-third of its funds from the Düsseldorf authorities (Stets 1963, p. 7). The fact that the national auditor's report on psychotechnical facilities does not mention the Institute for Youth Studies in Bremen, for example, suggests that this was a private institute.

This report, which covers all test facilities in the bureaucracy in which psychological methods were applied, casts light on the employment structure for psychologists in public administration around 1930 (*Die behördlichen psychotechnischen Einrichtungen...*, 1930). According to the report, the railways and the Reichspost (two major supporters of psychotechnical testing), the National Statistical Office, and the police test centers did not employ a single psychologist among them. In individual cases they did cooperate with university psychologists. Three provincial institutes and their heads are named, of which two were definitely psychologists (whether the third was I do not know). Of the nineteen communal institutions, two or three had a psychologist; the rest employed teachers or engineers. At the labor offices the report mentions twenty-one academics acitve as experts in career counseling. Though this is not clear, it seems that fourteen (or possibly thirteen) psychologists were employed. There were definite concentration points, for example, the State Labor Exchange in Erfurt employed no less than five psychologists in Halle and one each in Erfurt and Magdeburg. In Hamburg and Kiel there were also psychologists at individual labor offices. In other states they had the position of consultant on career counseling at the state labor office. Thus, around 1930 the best career chances for psychologists were in career counseling.[11] The report also counts twelve psychologists in the Reichswehr, so that all in all some thirty psychologists were active in public institutions outside the universities around 1930. In comparison the number of academic psychologists at the twenty-three universities and ten technical colleges was many times higher. Bearing in mind that this ratio would be reversed over the next ten years, it becomes apparent just what a rapid expansion the development of Wehrmacht Psychology was to bring. When the Academic Information Office issued a pamphlet on psychology

11 Jaeger and Staeuble (1981) erroneously suppose the industrial opportunities to be greater, a conclusion they draw from their correct observation that, among the various conflicting tendencies in the twenties, psychotechnics oriented to industrial effectiveness gained the upper hand.

in its series *Die akademischen Berufe* in 1941, it was entitled "The Psychologist in the Wehrmacht." Instead of warnings about almost hopeless prospects (as in 1932), we now find appeals for trainees. But before considering developments in the Wehrmacht, let us look at the fields of labor management and industry up to 1938.

The economic upswing under the Nazis presented labor psychology with new tasks, especially when the Four-Year Plan of 1936 geared up the arms industry in preparation for war (see Mason 1978, pp. 210–11) at a time when the reintroduction of conscription took labor off the market. The labor offices were now expected to control the distribution of pupils after leaving school and workers to the various occupations.[12] The so-called monopoly law passed on 5 November 1935 gave the Reichsanstalt für Arbeitsvermittlung und Arbeitslosenversicherung (Reich Office for Labor Supply and Unemployment Insurance) a monopoly as agent for employment, apprenticeships, and career counseling (RGBL I, 1935, pp. 1281–2). A series of regulatory measures was aimed at ensuring that firms were provided with apprentices according to plan (Hartwig 1959, pp. 51–2). Beginning on 1 March 1938 registration was obligatory for pupils who had left school and labor office approval was required for apprenticeships (Loebner 1942, p. 511). This meant that it became the rule for graduates to consult career counselors. Labor offices even had "the right to summon young people choosing a vocation and their parents" (Stets 1941, p. 6). In 1940–1 some 90 percent of all graduates turned to the career counseling of the labor offices, according to Erich Schulz (1942, p. 291). Not all of those seeking advice underwent aptitude tests – Huth (1937, p. 91) speaks of some 20 to 40 percent – but it is safe to assume that the number of psychological examinations greatly increased. From mid-1932 until mid-1933, 45,041 men and 22,831 women are said to have been tested for aptitude; this represents 20 percent of the men and 13 percent of the women seeking work (Ulich-Beil 1935, p. 587). For Bavaria, Huth (1936) cites the incredible figure of 1,038,266 investigations carried out on 94,059 young people aged thirteen to fifteen between 1925 and 1935. Perhaps every psychological procedure was counted individually. A ministry official stated that the Reichsanstalt had carried out some 190,000 aptitude tests in 1938 (Stets 1939).

Psychological selection was also used at the labor offices during the establishment of the Luftwaffe. Recruits were needed for the new professions of metal aircraft builder and aircraft motor mechanic. From 1936 on flight technology training schools were set up for which the labor offices had to

12 Regrettably I came across an extensive study by Theo Wolsing (1977) only after completing the manuscript. In the second chapter he describes in detail official measures for training and vocational guidance. He also includes figures on career counseling (pp. 115–16): July 1933 to June 1934, 507,742 young people; 1934–5, 739,035; and 1935–6, 941,892. On the role of aptitude tests we read: "Aptitude tests ... developed ... in the course of the Third Reich, in view of the shortage of qualified labor, to a permanent element of career counseling" (p. 110).

recommend youths on the basis of psychological tests and a report from the Hitler Youth; the final choice was then made by the Aviation Ministry (Kipp 1980; see Z, f. 139–40).

In 1938 there was no longer a reserve army of the unemployed. On the contrary, the Minister of Labor estimated the labor requirement of the Reich to be 1 million (Mason 1978, p. 215). The lack of skilled workers was particularly acute. In 1939 the number of registered training and apprenticeship opportunities exceeded the number of graduates (Stets 1941, p. 4). The slogan was no longer "Make way for the best" or "The right man in the right place," as it had been for psychotechnics in the twenties, but "A workplace for everyone" (see Syrup 1938, p. 68). Career counseling was legally obliged to channel youngsters to suit the demands of the labor force (Stets 1941). Psychotechnical selection in large firms based on competition was no longer functional in the new situation, whereas the allocation of workers by labor offices according to their qualifications was. Waldemar Engelmann (1938) therefore recommended that all individual tests in industry be concentrated in the labor offices.

The increased use of psychological aptitude tests to control the labor market ought to have resulted in more opportunities for psychologists in labor offices. There are signs that this was indeed the case, but I have no documentary proof. The available archives of the Reich Labor Ministry provide no information. Proof could possibly be found in the files of the individual labor offices or other sources. A sign of the stronger position of psychology in career counseling may be the appointment in October 1934 of the first psychologist, Johannes Handrick, as chief consultant (*Hauptreferent*) for career counseling at the National Labor Exchange (*Ind. Pst.*, 11 [1934], p. 29). Handrick held this post until 1938, when the Reichsanstalt was taken over by the Labor Ministry. Career counseling was then headed by the newly appointed economist Walter Stets. I am aware of four cases in which psychologists were working at labor offices in the thirties without being mentioned in the 1930 national auditors lists. These were Walter Jacobsen, who started in Rostock in 1934; Wilhelm Lejeune, who worked in Bielefeld (1934–5) and Emden (late 1936); Maria Schorn, who worked at the Aachen Vocational Office; and Adolf Ehrhardt, who became head of the Department for Career Counseling at the Labor Office in Leipzig. Psychologists also worked in academic career counseling at the academic information offices of various universities (see *Kalender*... 1935–6, 116, pp. v–vi).

In industry no new employment developments can be determined for psychologists in the early thirties. According to Pechhold (1938), psychotechnical facilities that had managed to ride out the economic crisis were usually broadly based and not restricted to the selection of apprentices. As "the result of the perceptible shortage of labor" (p. 41), training now had a higher prestige than aptitude testing. Vocational training for skilled positions was thus a focal point of scientific activity in the field of labor in the thirties. It was centrally propagated by the Labor Front, which had absorbed DINTA (see Seubert 1977). However, this did not provide new positions for psychologists until the Institute for Work Psychology and Pedagogy was founded by the Labor Front (see later in this chapter). It was only toward the end of the decade that psychologists were employed full time in firms. The exception was Siemens, which employed the psychologist Adolf Zeddies from 1934 onward at the "Siemens Society for Practical Psychology" in Bad Homburg, founded in 1904 by Otto Siemens.

Other related areas of psychological activity were advertising and the Labor Service. Anitra Karsten reports that she started working as a freelance advertising psychologist in Berlin in 1929. In 1936 she gained the first lecturer's post for advertising psychology at the newly founded College for Advertising, but returned to her native Finland in 1939 (Pongratz et al. 1979, p. 91).[13] The National Labor Service also took on psychologists. Wilhelm Hische held courses at the Reichsschule für Arbeitstechnik (Reich School for Work Technique) of the Labor Service (Hische 1935). From 1933 until 1935, when the Labor Service was still voluntary, Ludwig Kroeber-Keneth was at its head office, responsible for the work of the "Führer Schools" and the choice of their directors (Z, f. 199; Kroeber-Keneth 1976, p. 154). Strictly speaking, Kroeber-Keneth was not a qualified psychologist, but he had lectured on expression studies, characterology, and graphology at the Berlin Technical College and had worked in Reichswehr Psychology. I will therefore refer to him as a psychologist.

The first years under the Nazis presented industrial psychology with new tasks. But the demand for psychology in the economic sphere, especially for the theories and methods of labor psychology, did not create a shortage of professional psychologists. This arose in the middle of the thirties in Wehrmacht Psychology with the massive rearmament program, although it had already become the most important sector for practical psychologists before the reintroduction of general conscription in 1935.

The occupation of army psychologist before the reintroduction of conscription in 1935

From the mid-twenties psychologists gained more and more responsibility for Reichswehr selection matters. Psychotechnical selection had already been used in the kaiser's army during the war, primarily to select able but inexperienced soldiers as drivers and so on.[14] University psychologists were involved in this selection, but at that time no full-time army psychologists were

13 Apparently there were no psychologists at the Society for Consumer Research, founded in 1935. But there was close cooperation with the German Institute for Psychological Research and Psychotherapy (Bergler 1960).
14 See Rieffert (1921), Dorsch (1963), Jaeger and Staeuble (1981), and Geuter and Kroner (1981).

employed. Often psychotechnical procedures were applied by regular or medical officers. The latter, the doctors of the Reichswehr, opposed the introduction of psychologists, and at their instigation the War Ministry ordered in 1917 that psychotechnical testing be controlled by troop physicians with special training in psychotechnics.

The Treaty of Versailles limited the German army to 100,000 men. Cooperation with scientific institutions, such as had existed with the Department of Experimental Psychology at Münster's Philosophical Seminar under Professor Richard Hellmuth Goldschmidt, was prohibited. Nevertheless, the Inspectorate for Weapons and Equipment undertook the evaluation of psychotechnics during the war. The work was assigned to Johann Baptist Rieffert, an instructor in Berlin, who began in 1920 and was to head Army Psychology until 1931. Since the Army High Command felt the results were positive, regular and medical officers were to recommence testing prospective drivers in 1922. However, testing was not felt to be among the real duties of officers, and the necessary special training would keep them away from those duties for too long. This feeling may have resulted from the fact that the Treaty of Versailles restricted the number of officers to four thousand, so that the army was inclined to delegate tasks to civilian staff members who were not included in the tally. In any event, in 1923 the Finance Ministry was requested to approve the establishment of psychological testing stations with qualified staff. Psychology lecturers recommended young psychologists for this work, and in 1925 the first posts assigned only for psychologists in any German public institutions were established and occupied by Ehmke, Lersch, Neuhaus, Sassenfeld, and Rudert.[15]

At that time the Reichswehr recruited volunteers for twelve years of service. Because of the economic situation, there were so many applicants that Otto Gessler, Reichswehr minister until 1928, wrote that "scarcely 5 percent of able-bodied applicants could be accepted" (1928, p. 96). In 1927 between 8,000 and 13,000 were accepted out of 250,000 applicants (Caspar 1959, p. 58, n. 45). The psychological aptitude test could thus serve to select the best. An instruction in 1925 ordered that every applicant be inspected by a commission consisting of a line officer, a medical officer, and a psychologist. The psychologist formulated an overall assessment, and the commission as a whole judged aptitude. The officer could declare a candidate unsuitable, but unity was needed for a verdict of suitable (RH 12–2/37, ff. 44ff.). This placed the psychologist in a fairly strong position, which is of interest in terms of professional rivalry. When in late 1926 the first psychological examination of officer applicants was carried out, the commission consisted of two officers, a medical officer, and two psychologists, Rieffert and Rudert. Rieffert cast the deciding vote. Later this ruling was not adhered to.

15 This account is based on an unpublished manuscript by Pieper (see sources I B), an essay by Simoneit (1940), and an evaluation of the file RH 12-2/37. See also my essay (Geuter 1985b) composed after this book was written.

Officers signed up for twenty-five years. At the start of the Weimar Republic there were fewer applicants than posts. The situation seems to have changed in the mid-twenties. By 1927 the vacancies had been filled (Hürten 1980, p. 238). In April 1927 obligatory testing was introduced for all would-be officers. A commission conducted a two-tiered examination. This was no longer aimed at finding specialist skills, but at assessing qualities of leadership. Obviously the psychologists were felt to be sufficiently qualified to judge this on the basis of their experience with character diagnosis. Their position relative to the medical officers was becoming stronger and stronger. After some alterations, a final form was found for the commission, which included two psychologists but only one medical officer, who was responsible only for neuropsychiatric aspects. With the introduction of the Officer Candidate Examination, the conflict between psychologists and officers over the competence to assess was to be of greater importance than the earlier conflict with the doctors.

The Officer Corps traditionally followed a policy of self-recruitment from "socially desirable circles" (Bald 1981). This was supported by Gessler and the commander in chief (1920–6), General Hans von Seeckt. Until 1927 the responsible division commander alone decided about future officers. One could suppose that psychological tests introduced an element of selection according to qualification in contrast to elite recruitment. These tests were indeed introduced at a time when the liberalization of officer recruitment was being debated in the Reichstag and when the predominance of nobility was being criticized, above all by the Social Democrats (Bald 1981, pp. 23–4; Caspar 1959, pp. 49–50). Despite this discussion and the introduction of psychological tests, however, the percentage of nobility and the sons of officers in the Officer Corps had actually increased by the end of the twenties (see Bald 1981; Doepner 1973). The more egalitarian psychological testing procedures did not do away with the feudalism of the Officer Corps; this was first achieved during the expansion of the army under the Nazis. Nevertheless, there was an obvious conflict between the feudal principle of self-recruitment and the objective principle of selection by civilian experts. In 1926 the president of the Reichstag, Paul Löbe, recommended that applicants be considered in the order in which they applied and appointments be overseen by two civilian parliamentary commissioners (Caspar 1959, p. 58). In all it is perhaps not surprising that psychologists and officers would tend to disagree in the commissions just when officers' sons were being interviewed.[16]

With privileges to defend, nobody was going to give ground too easily. The aptitude tests for officer candidates were a subject of controversy in the Reichswehr. After World War II former Major General Hellmuth Reinhardt commented that soldiers viewed the introduction of the tests with suspicion; they had opposed the fact that "here certain persons were passing

16 This was confirmed by all Wehrmacht psychologists interviewed.

judgments, e.g., on the acceptance or rejection of officer applicants, who had nothing to do with the training or command of the officer corps, had no experience in these matters, and did not have to bear the consequences of false decisions" (I B, *Die Anwendung...*, p. 7). Former psychologists in the Wehrmacht differ in their accounts of officers' attitudes toward psychology and psychologists. Some feel that psychology had a generally good image, others that the officers were convinced by the psychologists' good work; some speak of the suspicion that met the psychologists and their methods in the officers' mess (Z, f. 25, 33, 53–4; Pieper 1976, pp. 167–8). However, the unanimous opinion was that psychologists and officers were very much in agreement about their decisions. This would indicate that both sides had similar ideas of the suitable officer.

The dispute was thus more about who should decide on aptitude. Between 1930 and 1939 various articles appeared in the journal *Military Weekly* (*Militärwochenblatt*) in which military personnel commented on psychological aptitude testing.[17] Most articles were positive. One argument put forward was that by taking part in the tests the opponents of psychology became its supporters (Griessbach 1932) or that contact with psychologists could be of great benefit to officers (Anon. 1932a). It is noticeable that the supporters of Army Psychology, apparently to reassure the doubters, emphasized that psychology only described aptitude, but the officers evaluated it (Anon. 1932a) or that psychological findings were only an "ideal standard" to which the officer must add the military view (Hesse 1930).

The first attack on psychology was published in 1934 under the pseudonym *Tenax* Latin for ("tenacious"). The author tenaciously defended the tradition of selection by officers. The psychological task in the army of "seeing through somebody" was an "established military skill" (p. 45). If a psychologist had the ability, then it was by chance and not because of his science.

The conflict found an elegant institutional solution. The "Guidelines for Psychological Test Stations" (H.Dv. 26), issued in 1930 and revised in 1936, determined that the psychologist formulated the psychological findings and the whole commission passed judgment on aptitude. The officer chairing the commission then decided the degree of aptitude (Simoneit 1940, pp. 34–5). The report and the aptitude judgment formally retained their character as recommendations and were not binding for the appointing commander. The final decision thus remained in military hands, but according to reports in *Wehrpsychologische Mitteilungen*, the commission's judgment was usually followed. Furthermore, the traditional voting in of a new officer by the other officers of his company remained unaffected. This arrangement allowed the psychologists to go about their business and gave the officers the freedom

17 Hesse (1930), Anon. (1932), Anon. (1932a), Griessbach (1932), Scholz-Roesner (1933), Tenax (1934), Anon. (1935), Schack (1935), Schmidt (1936, 1938), Marx (1939), and Walzer (1939).

Table 6. *Number of Wehrmacht Psychology
reports, 1930–8*

1930	2,940	1935	39,654
1931	3,485	1936	81,640
1932	6,838	1937	115,825
1933	9,652	1938	152,015
1934	19,254		

Source: RH 19 III/494; Simoneit (1940).

to decide about new recruits to their ranks. It made possible a balancing
act between the rival officers and psychologists. Other rules were valid for
specialist selection, but this was less controversial.

Most of the soldiers in the Reichswehr had signed on for twelve years at
the beginning of the Weimar Republic, so that a renewal wave began for the
troops about 1930. Most of the officers, who served twenty-five years, had
been taken on from the old army, and between 150 and 200 were appointed
yearly. When in 1933 the army was expanded beyond the limits laid down
in the Treaty of Versailles, the number of officer candidates increased tenfold
to about 2,000 (Bald 1981, p. 43). This in turn stimulated a growing demand
for aptitude tests, which virtually doubled every year in the period from
1933 to 1936 (Table 6).

The increase in the number of tests led to more test stations and more
work for psychologists. In 1925 there had been six test stations, each with
one psychologist. In 1927 the navy received its own test station in Kiel; in
1931 a seventh army test station was set up. Around 1930 a second psychologist was appointed for each station, and then in 1931 a third. Reichswehr
Psychology was expanded in 1933, with two test stations for the army, one
for the navy, and a fourth psychologist for every station. When general
conscription was introduced in 1935, there were ten test stations for the army
and two for the navy, all with five psychologists.[18] From 1935 onward the
increase in test stations corresponded to the number of regional commands
(*Wehrkreise*). If the staff at the Psychological Laboratory is included, the
number employed as psychologists increased from six in 1925 to fourteen
in 1929, thirty-three in May 1933, and sixty-nine in July 1935. Not all had
been trained as psychologists. The sudden increase in demand led to many
former teachers, philologists, and philosophers being recruited. In the
Psychological Laboratory of the Reich War Ministry, the center for Army
Psychology at this time, the staff in 1934 included mathematician Heinz
Masuhr, philologist Gotthilf Flik, former officer Dietsch, psychologist Erich
Zilian, and teachers (later psychologists) Paul Metz, Karl Kreipe, and

18 Simoneit (1940, p. 13), RH 53-5/41, and RH 12-2/101.

Max Simoneit (Z, f. 149). The last had succeeded Rieffert as head of the laboratory in 1931. Constant pleas went out to psychology faculties for new recruits.[19]

The demand for psychologists that came from the Wehrmacht in the first half of the thirties was without parallel in any other sector. Thus, in a short time the occupational group (*Berufsstand*) of Wehrmacht psychologists (Simoneit 1940, p. 39) became virtually synonymous with the body of practical psychologists. The expansion of psychological activity in the Wehrmacht also posed legal problems that had to be solved. Until 1927 psychologists had worked on the basis of private contracts; after 1927 they worked as employees on the basis of standard salaries. In 1933 the first army psychologist was made a civil servant. This then became the trend, requiring a continual increase in the number of budgeted posts. The law also required a career regulation to govern appointments, promotions and so on, and after three years in the making this was decreed in October 1937. It will be considered in detail when we look at the period after 1935.

Professional activities in pedagogical and medical fields

In his opening address to the eleventh congress of the DGfPs in 1929, Stern named the following fields of application for psychology (Volkelt 1930, pp. xiff.):

education, training, youth work,
economic life,
the legal and penal system, and
the care and treatment of the sick.

But these were areas in which psychology claimed only to have potential and to be able to present scientific research. They were still far from being areas of professional activity. This is particularly apparent in the fields of law and pedagogy. The scientific journals give the impression that forensic and pedagogical psychology were major areas of psychological practice; in fact, scientific endeavors in these areas had almost no institutional impact in terms of establishing a professional role.[20]

---

19 See UAL Phil B I, 14[37], I, f. 25. By their own accounts Arnold came to Wehrmacht Psychology through the recommendation of Aloys Fischer, Schänzle through Kroh, Munsch through Lersch, and Flik through Max Dessoir (Z. ff. 70, 94, 103, 148).
20 Information on professional psychological activity in the fields of education and medicine is harder to come by than that in the Wehrmacht or career counseling. It would be necessary to review files from numerous institutions. There are no documents of the DGfPs or the Association of German Practical Psychologists indicating the activities of their members. The information in the members register of the congress reports is not precise enough for evaluation. Studies of further material, such as the files or lecture lists of teacher training institutions, would allow more detailed statements to be made on these points.

In the early stages, pedagogical psychology was mainly a psychology of school education. The expansion and differentiation of schooling had made it more difficult to steer pupils' progress. Just as in industry, the aim was to have "the right man in the right place," so pedagogical psychology was expected to put "the right pupil in the right school." Ability was to be the sole criterion for deciding which secondary school a child attended. Efforts were made to employ psychology for this purpose, but despite discussion about their usefulness school psychologists were not introduced until the fifties. Stern had demanded in 1907 that the post of school psychologist be created, but the only place where this took effect was in Mannheim (Wies 1979, p. 225, n. 37). There, a special school reform created parallel classes grouped according to ability (Jaeger and Staueble 1981, p. 71). Hans Lämmermann in Mannheim was to remain the sole school psychologist in Germany for decades. The main reason that the stage of institutionalization was never reached must have been that selection in the schools was based largely on teachers' assessments.[21]

Psychologists were professionally active in these fields only when they were working in the ministerial bureaucracy or in institutes preparing selection procedures for teachers to use. Erich Wohlfahrt, later a Wehrmacht psychologist, worked from November 1933 until June 1935 as consultant on selection matters at the Education Ministry in Saxony (*WPsM*, 1939, H. 3, p. 47). Theodor Valentiner headed the Institute for Youth Studies in Bremen, which was concerned mainly with selection at school and at work. For example, in 1933 the institute was working on a comparison between school entrance examinations and tests of giftedness. In wartime the age at which children began school was to be reduced, so attention shifted to testing if young children were "mature" enough to begin schooling.[22] In Leipzig the Pedagogic-Psychological Institute of the National Socialist Teachers' League was set up under the leadership of Krueger's pupil Heinz Burckhardt following the closure of the Institute of the Leipzig Teachers Association (*Der Thüringer Erzieher*, 1933, 1, p. 59). It developed observation and assessment sheets for teachers selecting pupils for secondary schools (Ottweiler 1979, p. 126). During the Nazi period a number of ministerial decrees altered the regulations for the transition from one level of schooling to the next. From 1941 the transfer to secondary schooling was dependent on "character attitude," "physical suitability," and "mental capacity." The decision lay in the hands of the teacher; the school doctor could be consulted. The test sheets were the link between psychology and the teacher.[23]

---

21 See "Aus dem Tätigkeitsbericht des Mannheimer Schulpsychologen für die Schuljahre 1930/31 und 1931/32," *Z. päd. Ps.*, 33 (1932), pp. 380–83; "Zur Frage des Schulpsychologen an höheren Schulen," *Z. päd. Ps*, 29 (1928), p. 536; W. Stern, "Der Schulpsychologe," *Z. päd. Ps.*, 31 (1930), pp. 380–1; and Kiessling (1931, p. 364).

22 There were yearly reports on the work of the institute in the *Zeitschrift für pädagogische Psychologie*.

23 "Der zweite Hauptschulerlaß..." (1941); Machazek (1941).

A further link between psychology and schooling arose from the introduction of scientific teacher training. The Weimar Constitution laid down that teacher training should meet the general standards of higher education. Smaller states had introduced academic teacher training, and Prussia created pedagogical academies. The "Reform Science" psychology did not stand aside, and was demanded by many teachers or studied by them following their training. Those who did so probably found quite numerous opportunities for professional activity in academic teacher training; but since this was not a practical application, it is not so relevant here.

At the Prussian pedagogical academies, psychology was a compulsory examination subject alongside pedagogy under the 1928 regulations. Kiessling (1931), Hoffmann (1931), and Pfahler (1931) have given accounts of the way psychology was taught at various academies. On 6 March 1933 the pedagogical academies became higher schools for teacher training, as did those of the other states over the following years. The rigorous ideological screening of teaching staff led to losses among the psychologists, such as Hildegard Hetzer (Z, f. 165). On 18 March 1936 new "Regulations for Teaching at the Colleges for Teacher Training" were decreed; they no longer mentioned psychology, referring instead to "character studies and youth studies." The reason given was that in view of the diversification of psychology a new definition of its tasks at teacher training colleges was required. Since the task of the colleges was "to further the students' incidental and planned observation of people, particularly children and young people, and to train their eye for expression and behavior," it would hardly be helpful "to study the psychology practiced currently at the universities." Character studies and youth studies were more appropriate.[24] The new regulations for the first examination for elementary teachers in 1937 listed "character studies and youth studies" as examination topics (*DWEV*, 1937, pp. 461–2). At the same time in the teachers' journals, psychologists were emphasizing the importance of psychology for teacher training and the fact that psychology was not at all estranged from "real life" or practical purposes (Kesselring 1936; Kienzle 1936). We can also find criticism of the reservations and ignorance of some teachers with respect to psychology (Helwig 1936). In 1940 the fact that "character studies and youth studies," and not psychology, were found in the teacher training program was for Simoneit (1940a, p. 114) still one of the "current worries of psychology."

After the introduction in 1938 of a two-semester basic course for high school teachers at colleges for teacher training, which certainly came at the expense of pedagogy at the universities, the training of elementary school teachers was also taken down a peg scientifically and transferred from the higher schools for teacher training to teacher training institutes (Ottweiler 1979, pp. 243ff.). In the "Provisional Regulations for Instruction at Teacher

24 Richtlinien für die Lehrtätigkeit an den Hochschulen für Lehrerbildung, 18.3.1936, pp. 9–10 (Bay HStA, MK 42096).

Training Institutes" from 1942, the fourth-year course included three lessons a week on "basic questions of education," which included "aspects of character studies and an introduction to character assessment" and "growth, maturation, and education."[25] Psychologists were employed at teacher training institutes; for example, Elisabeth Schliebe-Lippert had a position at Vallendar (Z, ff. 132–3). The instruction of trainee teachers was the only professional opportunity open to psychologists in pedagogical psychology until the NSV educational counseling centers were created (see Chapter 8). To provide an exact quantitative judgment, it would be necessary to evaluate the records of the institutions involved, such as the twenty-five colleges for teacher training that existed in 1936–7 (*Kalender . . .* 1936–7, 117, pp. 330–1).

In contrast to developments in the United States, the activities of psychologists in Germany in the clinical field were of only peripheral significance for the professionalization of academic psychology right up until the seventies. The horrors of the First World War had given rise to psychological research and treatment at brain-damage stations. There were at least five such stations at which psychologists or psychologically minded physicians worked: Cologne and Bonn (Walther Poppelreuter), Frankfurt/Main (Adhemar Gelb and Kurt Goldstein), Munich (Max Isserlin), and Halle (Fritz Giese) (Dorsch 1963, p. 81). The station in Bonn went on to become the Institute for Clinical Psychology and the one in Halle the Provincial Institute for Practical Psychology. But a field of clinical-psychological activity in the current sense of the term did not develop. Nor were psychotherapeutic methods developed in psychology. The concepts of psychoanalysis were registered only with reluctance by academic psychology, or else radically rejected (see Brodthage and Hoffmann 1981).

Unnoticed by most of official psychology, a number of psychologists began to turn to psychotherapy, during the Nazi period, of all times. They began to study at the German Institute for Psychological Research and Psychotherapy in Berlin. This institute had begun work after the forced closure of psychoanalytical educational institutions in the winter of 1936–7. During the war the institute was funded by the Reich Research Council as well as the Labor Front and became the Reich Institute (Cocks 1985). Its head, Matthias Heinrich Göring, a cousin of Hermann Göring, hoped to produce therapeutically trained doctors – "doctor-psychologists" (Göring 1938). However, from 1938 onward it was also possible to qualify as an "attending psychologist" either in Berlin or one of the five institute branches that existed in 1940 in Düsseldorf, Munich, Stuttgart, Wuppertal, and Vienna (*Jahresbericht 1940 . . .*, 1942). According to Ministry of the Interior guidelines, this title was for academics who had qualified at the institute as nonmedical psychotherapists. The title was not awarded solely to psychologists, but

25 Vorläufige Bestimmungen für den Unterricht an Lehrerbildungsanstalten (BA:RD 39/8, f. 19).

they constituted a large proportion of "attending psychologists" and training candidates. It had been possible for nonmedical professionals to obtain psychoanalytical training at the Berlin Psychoanalytical Institute since at least 1934, but the Göring Institute was the first to establish such training formally.[26] With the takeover of the institute by the Labor Front in 1939, the first publicly financed institute for psychotherapy was created. It then also became possible for all qualified psychotherapists to become full members of the institute, and not only physicians (Cocks 1985). Thus, the Nazi labor organization enabled psychology to gain ground in its confrontation with the medical profession.

Training as an attending psychologist took two years and included basic medical knowledge, self-analysis, and supervised therapy. In 1939 fifteen candidates were being trained, and thirty-three had qualified. In 1940 the institute had thirty-nine nonmedical academics as members and sixteen as training candidates; according to the yearly report the figures in 1942 were, respectively, forty-two and forty-five.[27] The reports do not show how many of these were psychologists, nor do I know anything of their subsequent professional activity.

In addition to the title "attending psychologist," the institute awarded the title "counseling psychologist" – these were "female welfare personnel, youth leaders, gymnastics teachers, and personnel managers etc." (*Jahresbericht 1941...*, 1942, p. 65) who kept their original occupations but took a short course at the institute. They were not able to carry out major psychotherapy but were trained to provide psychological counseling. After 1941 the title was dropped and they became known simply as "associate" members of the institute.

New legislation on medical treatment, the Health Practitioners Law (RGBL 1, 1939, pp. 251–2), troubled the "attending psychologists." Until then they had carried out psychotherapy under the supervision of doctors, and as members of the Labor Front had had to belong to the "Free Professions" Office. Under the new law, any nondoctor treating the sick was required to obtain a certificate as a health practitioner. Although the law did not specifically include mental disturbances, legal commentators felt that psychotherapy was covered and the "Reichs Health Practitioner Führer" determined that professional treatment of "mental ailments" required approval.[28]

The certificates had to be obtained by April 1939, and the uncertainty

---

26 Personal correspondence from Geoffrey Cocks, 1 April 1982.
27 Sources: Göring (1940), "Jahresbericht 1940..." (1942), and "Jahresbericht 1941..." (1942). Cocks (1985) sees a connection between the training of attending psychologists, the Labor Front campaign to improve general health, and the development of a "*kleine*" (i.e. short-term) psychotherapy at the Berlin institute.
28 "Wer fällt unter das Heilpraktikergesetz?" *Deutsches Ärtzeblatt*, 69 (1939), p. 275. On the law itself: Gütt (1938–9), Kallfelz (1939), Kügele (1974, pp. 37ff.), Schultz (1939), and von Wolff (1941).

146   *The professionalization of psychology*

among the attending psychologists, who did not want to be classified as "health practitioners," was understandable. M. H. Göring also felt that the law did not apply to them because they were an "auxiliary medical profession." This was their classification at the Labor Front. Finally, the Ministry of the Interior informed the institute that the psychologists were regarded as medical auxiliaries under the Law to Regulate the Care of the Sick (28 September 1938). The ministry was preparing the official recognition in accordance with this law, and it would be necessary to write a syllabus and define a professional profile for the nonmedical therapist. To judge from the sources examined, things never got that far.[29] Meanwhile, they were to comply with the professional regulations for German doctors (*Jahresbericht 1941...*, 1942, p. 65).

A new formal regulation for the training of "attending psychologists" accompanied the decree on the Diploma Examination for Psychology (1941). For the admission of nonmedical candidates, a diploma in psychology was now "desirable" – which means it was not a precondition. The legal status of this professional activity was, however, not clear: "After the official recognition of the attending psychologist by the Reich Ministry of the Interior, the newly accepted member will receive a written installation as attending psychologist, on receipt of which he must sign the professional regulations (*Standesordnung*) that accompany it."[30] A footnote says that the regulations were in preparation, but they were probably not completed during the war.

The relatively large number of attending psychologists (and therefore the large number of trainees), together with the various efforts to influence the treatment of attending psychologists under the Health Practitioners Law, indicates that the Berlin Psychotherapy Institute had an interest in the professionalization of these psychologists. This was also probably the reason for their interest in the regulation of a new syllabus for psychology. Gustav R. Heyer, responsible in the institute for the attending psychologists, reported that he and Göring took part in the commission that worked out the Diploma Examination Regulations (*Jahresbericht 1940...*, 1942, p. 11).

This commission also included Arthur Hoffmann, a representative of the colleges for teacher training, as well as representatives of Wehrmacht Psychology and industrial psychology (see Chapter 6). This perhaps indicates the areas of practical psychology that were trying to achieve complete professionalization. Not represented were areas where psychological methods and findings were applied to some extent, but where psychology was not yet a profession. Traffic psychology, for example, was subsumed under general

29 Based on the files of the Deutsches Institut für psychologische Forschung und Psychotherapie (Kl. Erw. 762).
30 Deutsches Institut für psychologische Forschung und Psychotherapie e.V. in Berlin: *Richtlinien für die Ausbildung* (no date; judging by the name and the reference to the DPO, between 1941 and 1944). The document is from a private collection and was kindly provided by Geoffrey Cocks.

psychotechnics. It did not become a professional field for psychologists until the fifties in West Germany when medical-psychological Institutes for Traffic Safety were set up (Dorsch 1963, pp. 148–9). The police had introduced psychological tests for applicants in the twenties; since 1928 senior police officers in Prussia had been given training in pedagogy and psychology (Rupp 1929; Stiebitz 1974). But the police offered no places for full-time psychologists. Forensic psychology did develop as a scientific field. At the twelfth congress of the DGfPs in 1931 under Stern's chairmanship, a study group on forensic psychology decided that it should be constituted as a permanent study group of the society and work together with the legal profession, educationalists, and doctors (Kafka 1932, p. 469). Psychologists were involved in assessing the credibility of witnesses, but were only rarely called to provide expert evidence.[31] Undeutsch (1954, p. 11) reports that the leading representatives of forensic psychology in the twenties, Marbe and Stern, had given evidence in court only a few times. I know of no case before 1945 where a forensic psychologist was professionally active in the penal and judiciary systems. After the disbanding of Wehrmacht Psychology, Wolfgang Hocheimer was asked by the Ministry of Justice whether he wanted a post in the prison in Berlin-Moabit, but after having a look around he refused (Z, f. 57).

War economy and industrial psychologists

The Four-Year Plan of 1936 marked the start of German industry's preparation for war. The increase in arms production caused a temporary upswing in the economy and led to a severe shortage of labor. By 1938 it became clear that these economic policies could no longer be financed. The alternative was either to cut back social improvements (e.g., reduce wages) or to make use of weapons to conquer new markets and raw materials (see Mason 1978). The Nazi leaders opted for the latter, and from 1938 all economic policies were dedicated to the war economy and increasing arms production. A number of measures were introduced to control the labor market: compulsory service, expanding vocational training for skilled workers, introducing compulsory career advice for pupils leaving school, shortening apprenticeships, and lengthening the working weeks. An increase in the involvement of women did not occur before the start of the war. Foreign workers were used to a greater extent, and the deported and prisoners of war were later forced to work in production (Mason 1978, pp. 271–2).

The war economy posed major problems for the industrial sciences. The rationalization of human labor was to be pushed further than in the twenties. The aim was the optimal deployment of labor. A further massive problem in the years after 1938 according to Mason was a stubborn, if indirect, refusal

31 See Sporer's (1982) essay.

to cooperate all along the line (p. 315). He describes how this was countered with terror during the war. The numerous attempts to train the heads of firms in human management as "experts" and "clever psychologists" (Arnhold 1941, 1942) can be interpreted as a further reaction. As foreign workers were brought in during the war, industrial psychologists were faced with yet another problem, the most effective way to use them. On rationalization the industrial psychologist Ernst Bornemann commented:

> This idea, which sometimes caused the human to be forgotten, as in production lines, was discredited at the start of the Nazi regime, and industrial psychology suffered a number of setbacks in the first years after 1933. Later it recovered, in my opinion, and during the war years it had a better standing than it did after the war. (Z, f. 146)

Firms had an economic interest in an industrial psychology oriented toward effectiveness. This had some effect on the demand for psychologists and on their status in firms and in the Labor Front. Contemporary authors wrote that manufacturers were looking out for psychologists during the war, though this may just have been wishful thinking. It would be necessary to examine job advertisements or firm archives to test this hypothesis. Kroh (1941, pp. 2–3) commented on the search for psychologists by private and public employers. According to the personnel manager of I. G. Farben in Ludwigshafen, Albrecht Weiss, there were not enough industrial psychologists in 1944; in a "letter from the industry" published in the *Deutsche Allgemeine Zeitung*, probably from someone in I. G. Farben, it was suggested that other firms should also have courses in leadership. "A precondition for this, however, is that psychological science undertake to train suitable persons to a greater extent, since they are currently available only to a very limited degree."[32] According to Arnhold (1938, p. 38), firms were waiting for psychologists. In retrospect this demand for psychologists has been questioned. Albert Bremhorst, Arnhold's successor at the German Labor Front, feels it was not so large. Psychology was "not a pillar of economic rationalization" (Z, f. 186). This is indirectly confirmed by the industrial psychologist Ludwig Kroeber-Keneth, who did not feel that psychology enjoyed particular recognition in firms (Z, f. 200).

A total of seven full-time appointments for psychologists have been counted between 1937 and 1945. Zeddies's work at the Siemens-Studiengesellschaft has already been mentioned. Rieffert, who had been head of the Reichswehr Psychotechnical Laboratory and then professor at Berlin University until his dismissal in 1937, worked first as a free-lance industrial psychologist and then, from 1 July 1937, full-time for Rheinmetall Borsig (see Chapter 2, note 28). Borsig was one of the most important munitions works in Berlin and suffered from a shortage of labor (Speer 1972, p. 232).

32 "Verachtet mir die Meister nicht . . . ," *Deutsche Allgemeine Zeitung*, No. 82, 23 March 1944.

Rieffert's first task as industrial psychologist was therefore to assign every available employee so that his or her abilities could be used to the full. Aptitude tests were also carried out for promotions. Later, the selection of foreign workers and training for foremen were also focal points (Dunlap and Rieffert 1945). In 1939 Kroeber-Keneth came to the firm of Reemtsma. His two most important duties were to select apprentices and to produce graphological reports on the senior employees. The same points were mentioned in a report on industrial psychology at I. G. Farben in 1939, where Julius Bahle, who had been assistant to Sander at Jena until 1936 and then had gone to Switzerland, was psychological consultant. His exact status is not clear from the available source.[33] In 1939 Erika Hantel began work for Bosch in Stuttgart. She came not from academic psychology but from the German Institute for Psychological Research and Psychotherapy, which took up industrial psychology at this time. Psychotherapy, indeed the whole of the medical profession, had turned to the problem of increasing performance (see Bilz 1941; Grassner 1980). The 1941 annual report of the institute mentions that practical work had commenced in three firms outside Berlin, with members of the institute acting as advisers (Bornemann 1975, p. 24; Cocks 1985).

A new emphasis in industrial psychology became evident here – management training and "psychotherapeutical consulting" (Bornemann 1975). This arose in part to counter increasing problems of disintegration. An additional headache was that the military was continually calling up foremen, for whom replacements had to be found (see Dunlap and Rieffert 1945, p. 10). It was hoped that training foremen in leadership would lead to an increase in output. The productivity of prisoner-of-war workers was also to be increased by such measures (Ansbacher 1950, pp. 44–5). According to Kipp (1980, p. 314), air force industry required more management personnel as a result of re-location and decentralization to avoid bomb attacks. At a psychotherapists' congress in 1940, Arnhold (1941, p. 124) described the selection of leaders as "the key to the future." Foremen should occupy themselves more with individual workers, the hope being that this "in turn motivates the workers for higher output, because everyone works more joyfully and with greater liking for his work if he knows that he is understood and well treated" (Dunlap and Rieffert 1945, p. 10). Reporting on the largest training program for foremen, at I. G. Farben in Ludwigshafen, Viteles and Anderson (1947, p. 2) noted the belief "that the greatest existing handicap to increased production lay in the failure of supervisory personnel to stimulate active cooperation of employees in meeting production demands through the use of appropriate leadership."

In 1939 I. G. Farben began to send foremen to a hotel for recreation and

---

33 Akademische Nachrichten, WPsM, 1939, 1. H. 6, p. 29. Biography: Information from the UA Jena, 6 April 1981.

an exchange of views. Later, courses were set up with psychologists August Vetter and Ludwig Zeise from the Berlin psychotherapy institute acting as advisers. The main aim was to discuss questions of leadership. Afterward a circular, the so-called Kohlhof Brief, was sent out. Unfortunately, this is no longer accessible.[34] The head of personnel, Albrecht Weiss, supported this program; the leadership of the firm not only had to "organize" workers properly, but also had to "handle [them] properly." This required psychological knowledge and training in leadership (Weiss 1944).

The program was also supposed to train foremen in handling foreign workers. In June 1941 there were 1.5 million deported foreign workers and 1.3 million prisoners of war working in industry and agriculture.[35] After Fritz Sauckel became responsible for the deployment of labor on 28 March 1942, foreign workers were recruited in German-occupied areas by mobile squads from the labor offices accompanied by police (Broszat 1975, p. 378). When the workers arrived at the factories, nobody knew what they were qualified to do. There were also language problems, for example, with the Ukrainians. In order to seek out the best workers, the old psychotechnical methods of competitive selection were used. "Psychotechnical methods were applied on a mass scale as they were in the very early days" (Pechhold n.d., p. 310). Another factor favoring the reintroduction of psychotechnics was the success of attempts by Speer's Ministry of Armaments to introduce mass production techniques, which made it possible to use unskilled and foreign workers. "The aircraft industry was the last industry to go over to production line methods in 1944" (Fischer 1961, pp. 84–5). The methods used for selection were provided by the industrial psychology institute of the Labor Front. This was a major part of the industrial psychology work performed, for example, at Borsig by Rieffert or at Hoesch by Bornemann and Galle. While Elisabeth Lucker was active at Krupp in 1942 on behalf of the Labor Front Institute, she was offered a full-time post (Z, f. 49). Obviously industry valued the role of psychology in selecting foreign workers.

During the war there was a further increase in professional industrial psychological activity. In 1942 Hoesch appointed Ernst Bornemann, who had previously been assistant at the Kaiser Wilhelm Institute for Industrial Physiology in Dortmund. A second psychologist, Gerhard Galle, worked there as well (Z, f. 147). They worked on aptitude tests, ways of increasing performance, leadership methods, the internal suggestions system, and the selection of foreigners (Ansbacher 1950; Bornemann 1944, 1944a). At Messerschmitt in Augsburg, where Kroeber-Keneth worked as industrial psychologist from 1943, foreign workers were used in the factory and lived in camps guarded by the SS (Kroeber-Keneth 1976, pp. 200–1; Ludwig 1979, pp. 480,

34 See Comments on Sources. During the war the Amsterdam professor Géza Révész was active at I. G. Farben temporarily. See the obituary in *PsRd*, 7 (1956), p. 57.
35 Mason (1978, p. 310). See also Fried (1945, pp. 54ff.), Homze (1967), and Pfahlmann (1968).

495). The work of Kroeber-Keneth, however, concentrated on the selection of apprentices and on career advice, Because of its military importance, Messerschmitt was to receive the best apprentices (Z, f. 198). From 1942 the industrial psychologist Erika Hantel worked in the aviation industry at the Arado Works in Berlin; from 1943 Walter Jacobsen was head of personnel at Heidenreich and Harbeck in Hamburg, and Siegfried Gerathewohl was at the Bavarian Motor Works.[36] Both Gerathewohl and Jacobsen were involved with the selection of foreign workers, as was the engineer-psychotechnician Hellmuth Schmidt at the Heinkel Works in Rostock (Ansbacher 1950, p. 39). A report on a course at the Charlottenburg Institute for Industrial Psychotechnics (28–9 January 1943) on the contribution of practical psychology to the war makes it plain that the central task of industrial psychology was now the utilization of foreigners, women, and invalids in production. These then were the tasks that helped to establish the first generation of full-time industrial psychologists in Germany. The moral implications of this will be considered in Chapter 9. It is worth noting that just as industry was expected to function effectively, the main demand on psychologists was that they, too, function effectively. The facts that Kroeber-Keneth had been a Communist, that Vetter had not been able to become a university lecturer for political reasons,[37] and that Rieffert had been dismissed from the university all seem to have been relatively unimportant for a post in industry.

The question that now poses itself is why psychologists were able to take over the application of psychological methods in industry in wartime, whereas in the twenties engineers had staked a claim on psychotechnical activities. I suspect that the decisive difference is that engineers were being used for other activities wherever possible. In 1942 experts noted the catastrophic shortage of engineers (Ludwig 1979, pp. 296–7). They were probably glad to be able to use members of other professions for nontechnical activities.[38]

At least as important for the professional development of industrial psychology was the fact that the German Labor Front set up the first central industrial psychology institute in Germany at the end of the thirties. The Office for Vocational Training and Works Management (Amt BuB) had been part of the Labor Front since the takeover of DINTA in 1935. The engineer Josef Mathieu, who had headed a psychological department in DINTA was now charged with setting up the new Institute for Work

36 On Hantel: Cocks (1985); on Jacobsen: letter from Dr. Walter Jacobsen, 10 March 1980; on Gerathewohl: Z, f. 16.

37 At least according to Zeise on Vetter's seventieth birthday in *PsRd*, 8 (1957), p. 156.

38 This did not mean an end to the differences. Parallel to the small "wave of appointments" of psychologists to industrial firms, an engineer writes in an article entitled "Selecting Leaders in Firms" that only the manager could make the selection; psychologists could at the most advise him (Frankenberger 1942, p. 84).

Psychology and Pedagogy in the Amt BuB; the rest of the staff members
were psychologists. The founding was declared openly only in 1941.[39]

The Labor Front also had a Work Science Institute, which is better known
in the literature due to its numerous publications and its annual reports.
This was a central institute responsible for scientific research in all areas
covered by the Labor Front (Marrenbach 1941; *Organisationsbuch*... 1938,
p. 198). The psychologist Walter Jacobsen worked there for a short period
in 1942. The Institute for Work Psychology and Pedagogy was more a
scientific service center for firms. The institute's tasks were described in the
1941 annual report of the Amt BuB as problems of women in industry, basic
and elementary training, selection tests (especially for foreigners), work with
war wounded and invalids, and consulting work for firms. A former staff
member, Carl-Alexander Roos, reported that most of the work up to 1942
actually involved selecting apprentices (Z, f. 162). Later on, the selection of
foreign workers appears to have become the major task.[40]

The institute developed methods based in part on the U.S. Army alpha and
beta tests from the First World War. The instructions and all tests involving
language had to be translated into all the various languages required. The
institute gave instructions on how to use the tests or carried out the work
for the firm. Only a few practical tasks were carried out in the institute itself,
such as reporting on war invalids in the Berlin region. Other work, such as
the training of leading personnel in the firms, sometimes took place in central
courses and sometimes at the firms themselves (see Z, pp. 47–9, 162, 166–7):

> In firms that do not have their own psychologist, but that are large
> enough to provide continued work for a full-time aptitude tester, the
> Institute accepts requests to set up evaluation posts. It provides the
> methods, the equipment and the forms needed, as well as training an
> aptitude investigator.... In small and medium-sized firms ... the Berlin
> Institute only carries out the necessary investigations on site. (Gl, 1942,
> p. 117)

What the psychologists had perhaps wanted, but industry had not provided,
was now made possible by the Nazi labor organization: seven psychologists worked at a central industrial psychological institute that provided its
services to client firms without charge – a "socialization" of industrial psychology research under fascism. According to Ansbacher (1950, p. 40), the test
program for the selection of foreigners involved testing 400,000 deported

39 On DINTA see Hinrichs (1981, pp. 277ff.) and Seubert (1977, pp. 61ff.). I have gone into
the history of the institute and the problem of dating in detail later elsewhere (Geuter 1987a).
Official notice of the founding is in "Jahresbericht des Amtes für Berufserziehung und
Betriebsführung im Kriegsjahr 1941," *Anregungen-Anleitungen*..., 7 (1942), pp. 16–17. Staff
members at the institute were Wilhelm Lejeune, Elisabeth Lucker, Thaddäus Kohlmann,
Martha Moers, Maria Paul-Mengelberg, Carl-Alexander Roos, and Maria Schorn.
40 See the "Jahresbericht 1944" of the work psychology institute in Z, ff. 271–6.

workers in 1,100 factories. This was larger than any previous psychological test action, putting even specialist testing in the Wehrmacht in the shade (see Chapter 5).[41] In 1944, according to the annual report, the institute still gave out 148,360 forms for "the rough selection of alien workers and a short series for German women," as well as holding fifteen introductory courses explaining the methods to 394 employees of various firms (Z, f. 276). The war economy had required the activities of the psychologists in the industrial psychology institute of the Labor Front, and the war had increased their importance. But the role of the psychologist in industry could not increase to the extent that it had an important effect on the professionalization of psychology in other areas. The decisive impulses for this came from the further expansion of Wehrmacht Psychology.

The expansion of the Wehrmacht and Wehrmacht Psychology to 1942

In February 1935 Hitler announced the existence of the Luftwaffe, which had been built up secretly in violation of the Treaty of Versailles (Absolon 1975 III, p. 418). On 21 May 1935 general conscription was introduced. In the legislation the army, navy, and air force were now referred to collectively as the "Wehrmacht." The corresponding term "Wehrmacht Psychology" also dates from this time. The following day saw an "Ordinance on the Mustering System," which detailed methods of registration and mustering (RGBL I, 1935, pp. 609–14). Initially there were nine military regions (*Wehrkreise*), and psychological testing stations were attached to each regional drafting command (H.Dv. 26, p. 11). This meant that as the number of commands grew the number of testing stations grew, as did the demand for psychologists. A tenth *Wehrkreis* was introduced in January 1936; the demilitarized zone was integrated in April; the eleventh and twelfth *Wehrkreise* were set up in October and a thirteenth on 12 October 1937. Following the annexation of Austria, two more were added, followed by a further two in Danzig and Posen in August 1941.

As the Wehrmacht expanded, so the tasks of Wehrmacht Psychology grew. Although Simoneit mentions numerous areas of activity in his standard work (1933), the real practical task was aptitude testing. This was even specified in the "Guidelines for Psychological Testing Stations," which in 1936 stated, "The psychological testing of reinforcements for the Wehrmacht to the extent prescribed is the most important task of Wehrmacht Psychology" (H.Dv. 26, p. 15). The tasks relating to selection were listed in some detail, but everything else was subsumed under the heading "other activities of psychological testing stations" (p. 26). As the number of commissions increased, so did the number

41 Despite this, the selection of foreign workers is barely mentioned in the literature on industrial and labor psychology. It is mentioned by Hofstätter (1967, pp. 325–6), but under the heading "Wehrpsychologie"; it is also mentioned by Misiak and Sexton (1966, p. 114).

Table 7. *Tests in Wehrmacht Psychology, 1936–9*

Tests	Total	Army	Navy	Air force
Tests at army test stations, 1936–7				
Officer applicants and civil servants	6,655	4,820	304	1,531
Other ranks	39,682	12,935	—	26,747
Flying tests for officer applicants	1,026	—	—	1,026
Radio tests for officer applicants	180	157	—	23
Other tests	13,287	8,117	—	5,170
Total	60,830	26,029	304	34,397
Tests at navy test stations, 1936–7				
Officer applicants and civil servants	920			
Other ranks	19,223			
Flying tests for officer applicants	148			
Radio tests for officer applicants	5			
Other tests	514			
Total	20,810			
Tests at army test stations, 1938–9				
Officer applicants and civil servants	10,545	5,268	655	4,622
Other ranks and other tests	116,200	52,071	1	64,128
Total	126,745	57,339	656	68,750
Tests at navy test stations, 1938–9				
Officer applicants	603	499	62	42
Soldiers	9,488	4,035	—	5,453
Applicants	16,896	16,556	—	340
Preselection officer applicants	1,280	1,280	—	—
Total	28,267	22,370	62	5,835

Figures from the annual reports (RH 19 III/494) for the financial year 1 April to 31 March.

of psychological tests. The declared aim of "building up a peacetime army of (initially) 36 divisions made it necessary to increase fivefold the 3,858 officers for the 100,000 strong army" (Absolon 1980, p. 247). By the autumn of 1938 the number of active officers had increased to 20,812; including reserve officers it was 89,075 by 1 September 1939. Four years later 246,453 officers were enlisted, not including the air force and navy (pp. 250ff.). In addition, the groups of servicemen subjected to tests also increased. In 1936–7 these were officer and corporal applicants, reserve officers in the Luftwaffe, and specialists in the army, air force, and navy. In 1938–9 the list was extended to include prospective noncommissioned officers, applicants for noncommissioned officer schools, reserve officer candidates, balloonists, and army accountant candidates.[42] Above all, however, demands grew with the

42 Account based on the annual report. See also the regulations for the test stations, H. Dv. 26, pp. 21–2, 24–5, 67–8. Later, however, these limits were not adhered to.

Table 8. *Number of specialists' tests for the army, 1928–42*

1928	1,487	1937	15,525
1929	2,489	1938	41,551
1930	2,506	1939	66,633
1931	2,697	1940	136,691
1932	2,980	1941	199,743
1933	1,748	1942	81,729
1934	6,565	(Jan.–Mar.)	
1935	6,790	Total	581,614
1936	12,750		

Intelligence personnel	329,565
Gas tracers	121,836
Motorized personnel	101,778
Range finders	13,998
Others	14,437

Source: Flik (1942, p. 1).

rapid and enormous expansion of the air force. In 1936–7 the majority of tests were already being carried out for the Luftwaffe. A comparison of tests for army members in 1936–7 and 1938–9 demonstrates the growth (Table 7). No such exact figures exist for other years. The statistics clearly show three things: the growth of the overall number of tests, the high proportion of tests of other ranks, and the large number of tests for the Luftwaffe.

The number of tests on other ranks were more numerous than those for potential officers, but qualitatively less important. The latter lasted two or three days at first and later a day. They employed more character diagnostic methods (which must have made them more interesting for the psychologists) and were under the scientific control of the test station's chief psychologist. The provisions governing them were the most extensive in the "Guidelines for Psychological Testing Stations" (H.Dv. 26). They were, after all, matters concerning future military leaders. The conduct of officer candidate selection examinations by psychologists also gave rise to the greatest controversy about psychological tests in the military (see the section entitled "The Occupation of Army Psychologist before the Reintroduction of Conscription in 1935"). The tests for other ranks were all specialist examinations. At first drivers and radio operators were tested; after 1932 range finders, and from 1936 onward specialists in the tank corps, and from 1940 gas tracers, were added. The number of such tests increased markedly with the beginning of the war. Figures are available for the years 1928–42 (Table 8).

The increased workload led to reinforcement of the psychological personnel in the test stations. From July 1935 a fifth psychologist post was created at the stations, from August 1936 a sixth and seventh, and a year later an

Table 9. *Posts for psychologists in the army and navy,*
1935–8

	Test stations: army + navy	Posts/ stations	Posts at inspectorate	Total
July 1935	10 + 2	5	9	69
April 1937	14 + 2	7	15	127
August 1937	14 + 2	8	15	143
July 1938	17 + 2	8	18	170

Collated from material in RH 12-2/101; RH 19 III/686 and 494.

eighth. In 1938–9 two aviation psychologists were added at each station. For the navy each of the two test stations employed five psychologists in July 1935; on 1 April 1937 it became seven, 1 August eight, and in 1938–9 it was increased to nine, in addition to two psychologists at the navy Inspectorate of Education. Toward the end of the war, there were ten civil service posts for marine psychology in both Wilhelmshaven and Kiel. During the war numerous conscripted psychologists were also active as auxiliary psychologists. Bearing in mind the increase in the number of military commands and the beefing-up of the Wehrmacht Psychology headquarters in Berlin, this all adds up to a jump in the number of permanently budgeted posts for psychologists in the Wehrmacht, which must have led to a great demand on the academic labor market. This demand was further increased when Luftwaffe Psychology became independent. The total numbers of budgeted posts in the army and navy between 1935 and 1938 are shown in Table 9.

From 1935 onward the Luftwaffe enlisted its own recruits (Völker 1967, p. 118). At first many recruits were still released from the army and the navy, which is one reason why the statistics in Table 7 show army and navy tests for the Luftwaffe. Starting in 1938 flyer schools were set up. The selection of airmen became, according to Gerathewohl (Z, f. 17), the "heart of Luftwaffe psychology." There had been a number of losses during training, especially involving technically complicated planes such as the Do 17, Ju 88, and Me 109. These became more frequent during the war; the head of training for the Luftwaffe was understandably most interested in the selection of the flying crews (see Gerathewohl 1950, p. 1037).

When the existence of the Luftwaffe was made public, it had only about 900 officers (Völker 1967, pp. 55–6). A top priority was therefore the rapid recruitment of qualified personnel. On 1 April 1936 the first officer cadets to be recruited and trained independently received commissions. By October the officer corps had grown to more than 5,500; by 1939 it had reached about 15,000 (pp. 120ff.). Psychological tests were obligatory for all Luftwaffe officer candidates.[43] Since the applicants for the flying corps also received training as pilots, they were also tested for their suitability as flyers. In January 1935 the Luftwaffe set up its first enlistment center for flight officer candidates; the psychological tests were still carried out at this time by the army centers (p. 54). At the Psychological Laboratory of the War Ministry two psychologists were specialists in pilot psychology at that time. These were scientifically integrated into the laboratory, but were formally assigned to it from the Aviation Ministry (RH 12–2/101, f. 372). Soon afterward the laboratory was divided into departments, including an aviation psychology group (RH 19 III/494, ff. 196ff.). According to the annual report for 1936–7, flying tests were not handled very rigidly. The individual testing stations also ordered psychologists to carry out tests at Luftwaffe selection courses or at its enlisting centers for officer candidates.

The increased activity made Luftwaffe Psychology more and more independent between 1938 and 1940. Political decisions also played an important role, increasing in general the autonomy of the services in the Wehrmacht. After the reorganization at the top of the Wehrmacht early in 1938, a directive transferred the authority for decisions on personnel matters concerning officers and civil servants from the War Ministry to the commanders in chief (Absolon 1979iv, p. 157). In the late fall of 1938 the commander in chief of the Luftwaffe demanded that its officer candidates be separated from the general group of officer candidates (RH 19 III/494, f. 24). In November 1938 the Luftwaffe published a new leaflet for prospective officers outlining the application procedure. The leaflet mentioned that a psychological test would not be included, but that there was an aptitude test (*DWEV* 1939, pp. 52ff.).

In 1938–9 three officer candidate enlistment centers were set up for the Luftwaffe in Berlin, Hannover, and Munich; in mid-1939 a fourth was set up in Vienna (LVBL 1939 A, p. 149). Each test center employed three psychologists.[44] A group solely for Luftwaffe Psychology was created at the Personnel Office of the Aviation Ministry (RLM) in May 1939 with four psychologists (RH 19 III/494, f. 243; Metz 1939). In July 1939 a section of the Luftwaffe Personnel Office was absorbed by the newly created Luftwaffenwehramt, including a group for aptitude testing with five posts for psychologists. Officer selection remained with the Personnel Office (RH 2 III/14, 21).

The psychological testing of volunteers was now also treated as an independent task of the Luftwaffe. On 11 May 1940 independent test centers at the enlisting and demobilization offices of the district air commands were announced (BLB 1940, p. 264). All aptitude testing for the air force had to be conducted by the Luftwaffe itself from July onward (LVBL 1940, p. 451).

43 Memorandum, "Der Offiziernachwuchs der Luftwaffe," August 1935, RL 5/920, f. 159.
44 With the start of the war the psychological testing at these posts was temporarily stopped and the Luftwaffe psychologists were loaned out to the army stations. At the same time a start was made at setting up selection posts for the other ranks (Metz 1939a).

The number of psychological test centers must therefore have corresponded to the number of district air commands, of which there were ten in 1938 (see Völker 1967, pp. 84ff.). By 1940 there were fourteen. Since each center initially had two psychologists, this meant that in 1940 some twenty-eight psychologists were working at these test centers, twelve at the enlistment offices for officer candidates and five at the Luftwaffenwehramt. The test centers may already have had more posts at this time, but exact figures are not available. Whether or not this is so, before long at least forty-five posts had been created. A large proportion of these were probably held by the aviation psychologists who had been appointed by army test centers in 1938–9, but who are not included in the statistics on army posts (Table 9). The total number of posts for psychologists is certainly higher. Psychological facilities were also being created elsewhere in the Luftwaffe. Pilot training regiments and training units for antiaircraft artillery had their own aptitude testing centers in 1940 (I C, Ge 1, f. 2). Each Training regiment had one or two psychologists (Gerathewohl 1950, p. 1048). Metz (1942) wrote that in 1941 each antiaircraft auxiliary unit, each district air intelligence regiment, and some air intelligence regiments had their own aptitude test facilities.

Metz also listed who was tested: officer candidates, candidates for higher administrative posts, noncommissioned officer candidates, volunteers and soldiers in the flying corps, antiaircraft artillery personnel, and air intelligence personnel for a number of specialist tasks. Thus, no end of tests were conducted by Luftwaffe Psychology (see Gerathewohl 1950, p. 1028). The enormous expansion of the Luftwaffe presented a gigantic task of personnel selection. This in turn initiated an unprecedented job-creation program for psychologists. According to Gerathewohl (I B) about 150 psychologists were working for the Luftwaffe in 1942, a figure that might, however, include conscripted psychologists.

Where the psychologists in any of the armed forces had budgeted posts, they were Wehrmacht civil servants (see the section entitled "The Occupation of Army Psychologist before the Reintroduction of Conscription in 1935"). They were thus subjected to the civil service laws and the career requirements these involved. One important point was that those appointed should, as a rule, meet defined qualifications. Since 1934 the army had been trying to draw up career requirements for psychologists, but these were not published until April 1937, two months after the new general laws on civil servants had been proclaimed; they came into force three months after the new law on 1 October 1937. Since 1934 the Wehrmacht had been endeavoring to set up uniform civil service regulations (Absolon 1979, III, p. 251). The general career regulations of 1933 required that "a civil servant has the necessary or usual previous education" for his intended career (p. 255). Paragraph 4 of the Reich civil servant regulations of 1936 required that for higher posts, which included Wehrmacht psychologists, appointments could be made only if the obligatory state examinations for the career had been passed (RGBL

1936 I, pp. 893ff.). But there was no state examination for psychologists, nor were there career regulations. Thus, the only way they could become civil servants was under an exception in Paragraph 5 after having worked for three years with a private contract. According to the new law of 1937 it was possible to be accepted as tenured civil servants only if one had passed the appropriate examinations and served for a probationary period, or had been in office for five years (Para 28; RGBL I 1937, p. 45).

The policy of the Wehrmacht Center for Psychology, according to its deputy military head, was to press for lifelong appointments as civil servants and not revocable or temporary posts (Baumbach 1939). The career regulations for Wehrmacht psychologists (1937), formally based on those for lawyers and the higher administration of the armed forces, provided the necessary precondition.[45] Previously, psychologists had been installed as psychologist aides (*Hilfspsychologen*), if without a doctorate, or as assisting army psychologists (*Heerespsychologen*) with a doctorate, with the prospect of later becoming a full army psychologist or even a senior army psychologist. With the new career regulations of 1937 a three-year probationary period was introduced, at the end of which there was an examination to assess the suitability of the prospective civil servant. This consisted of an oral examination and a dissertation to be completed in six weeks. Successful examinees were awarded a diploma as "assessor of army/navy/air force psychology." An appointment as assessor could then follow, and where the budget allowed, also a post with tenure.

The regulations required prior qualification at a university, with a major in psychology. This was understood to mean a doctorate with a psychological dissertation. Since the doctoral degree was not a state examination, and was not yet available even at some universities, the Wehrmacht Center for Psychology sought an alteration in psychology teaching. I will consider this in detail in Chapter 6. In 1937 the first three-year probationary period began, and in 1940 the first assessor examinations of Wehrmacht Psychology were held (Simoneit 1940b). Since the Diploma Examination Regulation (DPO) for psychology was in the offing (it was decreed on 16 June 1941), the Army High Command determined on 15 May 1941 that in the future senior civil servants in personnel testing should be recruited from among psychologists with the diploma, if their studies had been completed after 1 April 1941, the date when the new DPO came into force (UAT 148). Luftwaffe Psychology planned the same arrangements (Metz 1942, p. 18), as did the navy (Mierke 1942), but matters appear to have gone no further.

The career regulations were a decisive advance for the professionalization of psychology in the Wehrmacht. They meant the recognition of the qualification obtained after studying psychology for a defined activity in the

45 The complete regulations are in UAT 148; extracts were published in *Z. ang. Ps.*, 53 (1937), pp. 261–4. The revised regulations for the army were published in *Heeresverwaltungsver-fügungen*, 15 (1941), pp. 182–7.

Wehrmacht. The Luftwaffe psychologist Kreipe (1941), for example, saw the career regulations as protecting officers against charlatans. Newly recruited psychologists now had to be formally qualified, and the field of aptitude testing in the Wehrmacht was thus a professional monopoly of the psychologists. However, the stricter initial requirements caused some problems, since they reduced the number of eligible applicants for appointments. Temporary alleviation was provided by employing some psychologists who did not meet the career requirements (Simoneit 1940, p. 41). This was why in 1938 only 92 of 170 budgeted posts (see Table 9) were tenured (81 lower senior civil servants and 11 upper; Simoneit 1940, p. 40).

The jump in the number of posts for psychologists posed problems in any case. When new psychologists were sought to start on 1 April 1939, the younger ones were to be employed as probationaries and the older ones as "civil service auxiliaries" (if they were already civil servants) or otherwise as "scientific auxiliaries."[46] The war only worsened the shortage. In 1940 twenty-five posts were unoccupied in Army Psychology alone. "In addition to this demand will come the extraordinary demand of Luftwaffe Psychology in the near future" (Simoneit 1940b, p. 68). Both army and navy asked academic psychological institutes for the names of younger psychologists. (I have no documents about the Luftwaffe.) Late 1941 saw the military and scientific heads of Wehrmacht Psychology, von Voss and Simoneit, conduct an inspection tour of the academic psychological institutes.[47]

When the war began, large numbers of teachers, philosophy and psychology professors, lecturers, and psychologists active in other fields were drafted as additional psychologists. Among the first were the psychology professors Gottschaldt, Keller, Straub, von Allesch, Heiss, Kroh, Pauli, Metzger, Herwig, and Jesinghaus; the philosophy professors Julius Ebbinghaus, Rothacker, Faust, and Ziegler; the lecturer and later professor of psychology and pedagogy Hans Wenke; and the psychologists Friedrich Dorsch and Walter Schulz (*WPsM* 1939, H. 10, p. 87). Virtually all representatives of psychology at the university institutes listed in Table 5 for 1939 did service as Wehrmacht psychologists.[48] Associate professors such as Siegfried Behn, Adolf Busemann, and Otto Tumlirz and assistants such as Rudolf Hippius, Richard Kienzle, and Hubert Rohracher were also called up to Wehrmacht Psychology (*WPsM*, 1940, H. 2, pp. 113–14). Anyone who studied psychology at that time did so primarily with the prospect of becoming an official in the Wehrmacht.

46 Von Voss to Bauemler, 9 August 1938, NS 15 alt/22 f. 58844.
47 Inquiry of the Psychological Laboratory, 21 June 1935; inquiry of the navy test station in Wilhelmshaven, 4 November 1939 (UAT 148); report by Pfahler of the visit of Simoneit and von Voss (UAT 131/130, enclosure to 70); observation of this trip in RMWEV: REM/821, ff. 212ff.
48 Based on an evaluation of extensive collected data. No information was obtained for Anschütz, Moede, or Wilde. Lorenz worked as a physician in the Wehrmacht.

According to Simoneit (1972, p. 105) 250 psychologists were employed in the Wehrmacht at its peak. After the war von Voss (1949, p. 10) mentioned a figure of 150 fully employed psychologists and 40 psychologists from other fields. The sum of the figures for the individual services is larger, but this also includes psychologists who had been drafted. According to Gerathewohl (I, B) the Luftwaffe had about 150 psychologists, which seems to me to be on the high side. In 1938 Army Psychology had employed 170 psychologists. For the subsequent period I have exact figures only for the Inspectorate for Aptitude Tests, which shrank when Luftwaffe Psychology and the Ethnic Psychology Group[49] were separated out, but which expanded again during the war without ever reaching the strength it had in 1938 (H6/840). In addition, there were the auxiliary psychologists at the test centers, of whom there were 95 in the army alone in 1940 (Simoneit 1940c, p. 1). The navy had 42 to 45 psychologists and assistant psychologists toward the end of the war (Feder et al. 1948, pp. 16–7). All in all this gives a total of about 450 psychologists in the Wehrmacht. Perhaps Kroh was not exaggerating when he said in 1941 that "in Wehrmacht Psychology at present some 500 individuals are employed, capable of carrying out psychological selection of a high standard" (REM 3147, f. 21 Rs). Thus, in the eight years since 1933, when 33 psychologists had been employed, their numbers had increased fourteen-fold.

These could not all have been fully qualified psychologists. Whereas in the past psychologists had worked in other areas – for instance, in career advice – the situation in the Wehrmacht was now reversed: people from other fields were occupying posts or carrying out the duties of psychologists. This must have encouraged efforts to find new psychology students, who after a short university course and a state exam would be able to start careers as Wehrmacht psychologists.

Thus, increased demand for psychological testing as a result of Nazi militarization proved to be the decisive factor for the great demand for psychologists in the late thirties, providing the main impulse for the professionalization of psychology in this period. The years 1931, 1933, and 1935 saw small increases in demand, followed at the end of the decade by a great leap. The disproportionately large problems that the Wehrmacht faced with selection, the only area in which psychology had really proved itself, led to the first expressed need for psychologists from an institution providing posts, the first career regulations, and then – eminently important in Germany – the first official civil service positions for psychologists. This development was

49 This group had existed in the Inspectorate, studying questions concerning southern Europe, Czechoslovakia, France's African army, and the Red Army (MA:RH 2/v. 981, ff. 116ff., 179ff; RH 2/v. 2981, ff. 40ff.; RW 6/v. 104); the Ethnic Psychology Group was transferred in December 1938 to the Wehrmacht Supreme Command (RH 19 III/494 f. 239). The former army psychologists Block, Stupperich, and Wünsche were then probably active in Wehrmacht propaganda.

certainly helped by the fact that selection in the Wehrmacht was in the hands of psychologists right from the start, in contrast, for example, to the situation of career advisers at the labor offices. In other areas the profession expanded more modestly, although progress was made, especially as a result of the war economy. Certainly the powerful position occupied by the Wehrmacht and the resources at its disposal also help to explain why it was here, rather than in social services or other areas, that the most posts were created for psychologists. Somewhere behind came industry and the Labor Front, followed by employment offices. This dominance of the Wehrmacht led to careers in psychology becoming very much a matter for men right from the start, in contrast to the situation in the United States, where many women were working in applied psychology at this time (Napoli 1981).[50]

50 For Kroh an indication of the inferiority of Italian military psychology was the fact that persons "of the female gender" worked there (REM/3147, f. 21 Rs.).

5

Legitimation strategies and professional policy

If a professional group is to become fully professional, it is advantageous if it is able to give a convincing account of the usefulness of its activities and to present a united front. This chapter will look first at how psychology presented itself and what claims it made for itself. Second, it will consider to whom these claims were addressed and the expectations they aroused. Finally, it will examine which steps were undertaken by the profession to achieve a uniform representation of group interests.

It is not usual to raise the question of legitimation strategies in the literature on the history of psychology. The term "legitimation" here refers to attempts to use specific arguments to prove the necessity or usefulness of psychology to those important for the recognition of the subject. In Chapter 2 legitimation strategies were considered as an aspect of attempts to institutionalize academic psychology. There the aim was to show the usefulness of psychology to related subjects and to the science administration in the restricted context of university appointments. Here strategies will be considered in the wider framework of professional politics. This will involve examining general and programmatic scientific texts to determine which legitimation strategies they express, irrespective of their methodological and theoretical tendencies.

Individual scientists are also engaged in self-legitimation vis-à-vis the scientific community or the state administration. Such attempts at legitimation increased as infighting flourished under the Nazis, and political careerists abounded. Scientists distanced themselves from others or claimed that their theories were the ones closest to Nazi ideology. They sought to dedicate works to the Führer, as Poppelreuter (1934) had done in his book on Hitler, or sought to win favor in the Nazi Party with their publications. Scientific papers were also weapons in university politics. Here I do not want to consider the extent to which individual suggestions about the possible role of psychology under the Nazis were designed to further individual careers or to strengthen the position of schools. They will be interpreted only in terms of the general legitimation strategies for the subject that find expression in them.

Psychological knolwdge can have two functions (see Chapter 3): it can be "social-technical" in a wider and not pejorative sense, that is, knowledge for the judgment and alteration of abilities, patterns of behavior, and forms of experience, or it can serve to interpret or explain phenomena of human experience and behavior, without drawing directly practical conclusions; as such it can be both critical and apologetic.[1] Thus, psychology can emphasize its relevance in one of two ways, by concentrating on either its practical or its scientific-theoretical usefulness. In terms of professional roles these correspond to the expert or the scholar. These two strategies can be found throughout the history of psychology in this century. In phases of active professionalization the predominant strategy has emphasized practical usefulness. My hypothesis is that this strategy was more important for success in professional politics during the Nazi period than that of theoretical legitimation.

For a profession looking for support, the problem of legitimation under the Nazis was particularly acute because of the enormous increase in state pressure on the sciences to prove their worth. We can find explicit efforts in psychology to meet these ends, in which legitimation in terms of theoretical value was displaced more and more by legitimation in terms of practical usefulness. The offer to provide backing to the Nazi ideology gave way to a practical instrumentalization of psychology for war preparations and war itself. Whereas some psychologists at first tried to gather the profession to a common weltanschauung, they later united around practical tasks.

When considering the establishment of new chairs and the appointments made, we saw that in the thirties the practical importance of the subject and the ability of a candidate to provide practically relevant training became increasingly common arguments. This shift, corresponding both to the professionalization of psychology and to the change in Nazi science policy, was registered by Gert Heinz Fischer in 1942:

> The struggle of German psychology now [following the new examination regulations] enters a new stage. The focus of attention is no longer on worldview or the refutation of Jewish-liberalist ideas and the defense against numerous misunderstandings of the tasks and goals of the subject, but rather the internal formation of the professional group,

---

1 As soon as a socially critical psychology confronts reality, it too gets tied up in the first type of function. Marcuse (in the epilogue to *Eros and Civilization*) notes that such a psychoanalysis can be critical as a theory, but that as a therapy it will always be confronted by the limitations of the reality principle. I do not want to say that theoretical psychology has only an ideological function, nor that applied psychology has only a technological function. Applied psychology, which can create the appearance of objectivity with its methods, while its criteria in fact remain concealed from those being examined, also serves as an apologia for a certain type of reality even if it functions as social engineering.

the development of close links between theoretical research and practical tasks, in which the mental reality of human beings is vividly reflected as the natural basis of psychological research, and the continual direction of knowledge to the achievement of an image of man, the progressive illumination of which has been indicated from various perspectives by previous approaches and schools. (1942a, p. 12)

Ideological or practical usefulness: part I

In the early stages, psychologists argued that the academic institutionalization of their subject was important because of its relevance to other disciplines. After the use of psychology in the First World War, however, practical relevance became an increasingly prominent argument. Psychotechnics was advanced in nonacademic speheres as a useful instrument with which to solve some of the problems arising from the difficulties of the postwar economy, an argument addressed to those who either employed psychologists for practical tasks or used psychological methods in their own work. Two factions formed among psychotechnicians, one around Stern and Lipmann, the other around Schlesinger and Moede. They did not agree "in whose interests, from whom and under which conditions psychotechnics should be carried out" (Jaeger and Staeuble 1981, p. 84). The first group favored cooperation with state institutions and wanted to establish psychotechnics in vocational guidance and schools; the second group sought cooperation with industry. The latter won the argument.

This decision was influenced to no little extent by financial considerations. The fact was that, especially during the inflation, industrialists were the only source of funding for research ("Die 1. Tagung...," 1923. p. 399). When psychologists (e.g., W. Stern at Osram) were criticized for doing research work that was subsequently used to the disadvantage of blue- and white-collar workers, Moede replied that private contracts from industrial firms were the only way he could keep his institute going (REM/2289, ff. 25ff.). Industrial firms were not the only ones looking for profitability. Giese (1925, p. 769) complained that bureaucratic authorities, too, put up funds only when they promised to show a return. It is therefore not surprising that psychotechnicians tried to emphasize the economic effectiveness of their activities.

When administrative power was put in the hands of the Nazis, legitimation had to be sought in new ways. At first the new administrative apparatus had less "an open hand" for practical matters, as Marbe had assumed governments would have in 1921 (see Chapter 2), than "an open ear" for the ideological line. The strategy of demonstrating the scientific and theoretical relevance of psychology, corresponding to its traditional representation as an auxiliary to philosophy, thus became under the Nazis a strategy of supporting the Nazi weltanschauung. However, the Nazis did not need to invent a global political understanding of the role of psychology in society. Krueger (1932,

p. 71) had already declared at the twelfth congress of the German Society for Psychology in 1931 that psychology must make its contribution toward solving the "cultural crisis." Once the Nazis had come up with their "solution," the psychologists could then help to place it on solid scientific foundations. This was characteristic of the ideological legitimation strategy under National Socialism.

In 1933–4 there was a controversy – perhaps the last one to be conducted openly until 1945 – about the strategy to be adopted. The attitudes of the authors were influenced by their political views. In 1933 Adolf Busemann published an article in the *Zeitschrift für pädagogische Psychologie* entitled "Psychology in the Midst of the New Movement," to which Otto Bobertag replied in the *Zeitschrift für Kinderforschung* in 1934 under the title "On the Struggle for and against Psychology." Both articles concerned themselves with the position psychology should occupy in the new political situation. Busemann (1933) argued that in the past psychology may have concerned itself too much with the abnormal, but that it now had adequate concepts for all aspects of the mind. At the time of a "German movement" for the "formation of the life of the *Volk* according to the needs of the German soul," there was therefore no longer any reason "to restrict the living space (*Lebensraum*) of the science of the soul" (p. 199). With that, a new argument was advanced to increase the support for psychology: since it considered the specifics of the German soul it must be furthered. For Busemann the struggle against psychology had become an anachronism.

Bobertag replied that it was more important to demonstrate the practical usefulness of psychology. Psychology should not be restricted, because it represented an invaluable tool with which to solve many social problems, especially in education (e.g., selecting teachers or pupils). This fact had "virtually nothing to do with the special aims or content of some movement or other" (1934, p. 190). Bobertag's strategy of demonstrating the practical value of psychology to the "new movement" by telling them "the way things really were, and not the way they can be ideologically fixed up" (p. 199), must be related to his anti-Nazi viewpoint. This is apparent in the article and prevented him from currying favor in the way Busemann did. According to Baumgarten (1948), Bobertag was a lone voice crying in the wilderness. He was at that time in charge of the test psychology section at the Central Institute for Education and Teaching in Berlin; he committed suicide on 25 April 1934 (Geuter 1986, p. 147). The political position of Busemann is not clear from the available sources. At any rate, in this context the important thing is that the strategy he proposed was to have a major influence on the public presentation of psychology in the first years of Nazi rule.

The presentation of ideological-political usefulness

The dominance of this strategy was clearly demonstrated at the first large collective presentation of the subject at the thirteenth congress of the DGfPs in October 1933 in Leipzig (see Geuter 1979). The new committee of the society (in office since March) had relocated the conference from Dresden and attempted to ensure that "the practical significance of psychology for central aspects of current German life is [treated] adequately."[2] At the annual meeting of physicists a month earlier, their president, Max von Laue, had compared the Nazi attitude toward Einstein and relativity to the Inquisition's treatment of Galileo (Beyerchen 1977, pp. 64–5). But the psychologists who did the talking at their congress seemed to want to demonstrate to the new state how useful and important they could be. They were presumably seeking the recognition they had not been granted during the Weimar Republic. The new chairman, Felix Krueger, defined the task of psychology as assisting the mental renewal of the German *Volk*, the renewal represented politically by the "movement." The Führer of the new state knew that the *Volk* did not live by bread alone – it needed a worldview and this was where psychology could make itself useful. Various speakers attempted to show the merits of psychology by demonstrating its congruity or compatibility with the Nazi weltanschauung. Some even claimed that psychology could be of service for the impending ideological-political tasks. This strategy was aimed at the new state apparatus, which was already largely in the hands of the Nazis.

There were various ways in which the ideological value of psychology could be demonstrated. The attempt was made to distance psychology, a German science, from "Jewish contents" and Jewish scientists; the agreement of its development with Nazi ideology was demonstrated, or the claim was made that the theories of psychology provided a foundation for Nazi ideology. Finally, new areas of research such as genetic and racial psychology were put forward, and psychology was proclaimed to be responsible for answering questions of importance to the Nazis. I can only give some examples of these various strategies here, but it would be possible to write a whole book about the way scientists were willing to employ psychology to sing the praises of National Socialism.

To start with, however, it seems appropriate to inquire whether there was a "party program" for psychology on which these apologias could be based. I know of no programmatic statements from the NSDAP that expressed such clear ideological expectations. Probably psychology was too insignificant socially, and less important as a subject than eugenics, racial theory, and others. In the theoretical organ of the Amt Rosenberg, the *Nationalsozialistische Monatshefte*, the first article appeared on psychology in 1943. In it, psychology was called on to turn its attention to the "great questions of the essence and nature of the mental stirrings and processes of the community" (von Werder 1943, p. 244) and to penetrate to the concrete mental reality of the workers and peasants. At the thirteenth congress of the DGfPs in 1933 the Saxon state minister Wilhelm Hartnacke demanded that psychology tackle questions of

2 *Arch. ges. Psych.*, 88 (1933), p. 420.

the formation of the person, such as the problem of heredity and environment. But this was oriented toward the area in which Hartnacke himself published rather than being a general state or party program for psychology, even though that was the way Hartnacke presented it.

In order to prove that psychology was a "German" science, individual psychologists repeatedly dissociated themselves from psychoanalysis. Their tactics went as far as verbal abuse and calls for persecution. In 1933 Sander commented in an educational magazine that German psychologists were conducting a "struggle against the subversive theories of the Jewish 'dissolution' of the soul, of psychoanalysis" (p. 12). Kroh avowed that as far as psychoanalysis was concerned "no German psychologist of standing . . . had given up his critical attitude at any time" (1933, p. 321). If he meant academic standing, then this was not said without some justification. Dissociation gave way to denunciation when Volkelt (1939, p. 35) spoke of eliminating the "alien race instinct psychology," and later voices joined in Julius Streicher's frenetic chorus against "perverse Jewish psychoanalysis." Otto Tumlirz, for example, an Austrian development psychologist, wrote in a review of the book *Psychoanalyse des Kindes* by Melanie Klein that he wished "to spare [himself] from following the author's train of thought, spiraling around perverse sexuality," and he expressed the hope "that such 'psychological' literature would soon disappear completely from the German-speaking region."[3] I will spare myself and the reader further examples of this train of thought, but it should be mentioned that the dismissal of Jewish professors was cited as proof of the "German" character of psychology (Anschütz, 1941, p. 254; Schliebe 1937, p. 195).

The usual method of demonstrating the true German *Volk* or ideological nature of German psychology involved reinterpreting the content of an individual theoretical system by its main representative or one of his disciples, sometimes with an added tirade against "Jewish science," especially in Erich Jaensch's work (e.g., 1934, p. 12). Jaensch (1938b) and Friedrich Becker (1938a) of his school, for example, were of the opinion that the existing intelligence tests corresponded to the "Jewish way of thinking," but not to the German type of intelligence. Without launching such attacks, others, like Metzger (1938, 1942) and Pfahler (1935), developed systems demonstrating the congruity of their scientific views with Nazi ideology. According to Metzger the conflict of Gestalt and holistic (*Ganzheit*) teaching with associationism in psychology corresponded to the political conflict between the "folkish worldview and the influence of the West" (see Chapter 1, note 17, and Geuter 1987, pp. 91ff.). Sander also drew parallels between the ideology of the "folk community" (*Volksgemeinschaft*) and *holistic* psychology (*Ganzheitspsychologie*), and went on to justify the "elimination of parasitically proliferating Jewry" and "making infertile all those of inferior genetic

---

3 *Zeitschrift für Jugendkunde*, 4 (1934), p. 142.

makeup" on the basis of a Gestalt law he had devised about the expulsion of that which is "alien to the Gestalt" (1937, p. 642; see also 1933, p. 12)

This demonstrates the transition from a defensive strategy, concerned with proving consistency and loyalty, to an offensive attempt to legitimate Nazi policies. For Sander the function of psychology was "to become a helpful tool for the aims of National Socialism" (1937, p. 649). Jaensch was at the forefront when it came to assigning psychology a role as a buttress to Nazi policies. Right after the Nazis came to power, he made a number of programmatic contributions about the "struggle" or the "situation" or the "tasks" of psychology. He saw psychology, or as he preferred to say, "psychological anthropology," as occupying a key position in the realm of science and ideology. He interpreted the political volte-face as a "cultural turn" (*Kulturwende*) in which certain basic mental forms came to prominence. Since psychological anthropology dealt with these basic forms determining weltanschauung, it was superior to other sciences. Even racial studies would form only one of its branches. Psychology thus became fundamental to all Nazi ideological endeavors.

Jaensch then set to work fleshing out his programmatic outline. Earlier he had combined folk-psychological ideas with his typology of personality, which classified individuals according to the extent to which they integrated certain mental functions, such as emotion and imagination. He now developed this further into a theory of the struggle of the "Northern integration type" against the "Jewish-liberal dissolution type." He was the only psychologist to integrate anti-Semitism systematically into his theory. In his book *Der Gegentypus* (The antitype, 1938) he aimed to use psychological anthropology to provide scientific support for the "emotional and instinctive" contributions of the "movement," as he wrote to the ministry (REM/2606, f. 19). Thus, psychology was to show the scientific necessity of Nazi actions. When Jaensch took over the chairmanship of the DGfPs in 1936, his first circular to members disclosed the "top priority" of psychology: "To do the scientific spade work for the political and cultural institutions charged with the new formation of all German life, and to place our scientific work at their disposal."[4]

Legitimation strategies become fairly obvious when it was a matter of getting public funds for research projects. I have not been able to study the lines of argument used in actual applications, but the proposed topics of some projects clearly indicate that not a few psychologists felt they could make themselves useful with research on racial psychology. Rieffert planned a project at the Berlin Psychological Institute entitled "The Psychology of Jewry," using the methods of Wehrmacht Psychology. The project was approved by the Nazi Race Policy Office in a letter to the ministry (REM/2606, ff. 1–16; see Geuter 1984). In Tübingen, Gerhard Pfahler's

4 NS 15 alt/17, f. 56942. For more details on Jaensch, see Geuter (1985a).

research (1941), "Racial Cores of the German *Volk*," was financed by the German Research Society (UAT 148). In Leipzig, Hans Volkelt also applied for funds in 1939 for "purposes of racial psychology, type psychology and expression research" (UAL Phil B I/14[37], V, f. 12); in Halle, Kurt Wilde received support for hereditary psychology research from the Alfred Rosenberg Foundation (NS 15/242, f. 24). Research along these lines was also carried out in Hamburg, Jena, Marburg, and Rostock, among others. We can interpret the establishment of these types of research, and the numerous psychological studies linked to racial views, as attempts to improve the reputation of psychology by choosing politically desirable approaches and research topics.

I cannot resist giving one example of the stupidity and bold-faced arrogance with which psychology was degraded by apologists for expansionist Aryan supremacy. Two of Jaensch's pupils, Heinrich Ermisch (1936) and Siegfried Arnhold (1938), wrote about the relationship between integration type and breeds of chicken. A comparison of the pecking behavior of "Nordic" and "Mediterranean" chickens showed that the "Nordic" chickens had a firmer "integration core." They pecked a little slower, but more accurately and regularly, and were better at taking their place in the group. The southern chickens were less balanced, unsettled, and – as was to be expected of such bad types – sexually mature at an earlier age. All this in the *Zeitschrift für Psychologie*, proving empirically Jaensch's methodological hypothesis: that the chicken yard was an area for "research and demonstration concerning human racial questions" (Jaensch 1939).

In fact, the ideological usefulness of psychology for the Nazis seems to have been so small that the strategy of ideological legitimation did not benefit psychology all that much. First, the Nazis did not want to have scientists going around providing support where they saw fit. Heinrich Härtle (1941) from the Amt Rosenberg was one who wanted to see a clear separation of science and weltanschauung. The Amt Rosenberg presumably felt that establishing the day-to-day nuances of Nazi ideology was its turf, and may well have found it inconvenient to have positions supported in a scientific paper that at the time of publication might no longer be current.[5] Second,

---

5 The Amt Rosenberg distanced itself from Jaensch in 1941, the year after his death. When his successor in Marburg, Gert Heinz Fischer, wanted to publish a selection of Jaensch's essays on youth leadership and Nazi education together with the Press Office, the local branch of the Hitler Youth, and the writer Dr. H. Böhme, Alfred Baeumler commented: "The publication of essays of the late Prof. Erich Jaensch ... seems to me to be impossible. The case of Jaensch is similar to the case of Krieck. Erich Jaensch claimed emphatically to represent the Nazi psychology, and this since the twenties. He belongs to those otherwise admirable professors with a nationalist approach who never had an inkling of the revolutionary in National Socialism, and who were always making comparisons between the past and the present. The publication of the essays of Erich Jaensch ... would give the publishing house a German-nationalist tone that would meet with the disapproval of many" (NS 15/201, ff. 68ff.).

psychology, in contrast to some other disciplines, was not directly involved with Nazi scientific projects, which were generally run by the SS (see Kater 1984) and the Amt Rosenberg. Third, where it did attempt to provide ideological support, psychology usually ended up borrowing racist concepts from the natural sciences.[6] And finally, not all psychologists were prepared to go along with this strategy.

The presentation of practical usefulness

While the Nazis were concerned with the ideological mobilization of the people and with winning over those intellectuals who were not being dismissed from the universities as Jews or radicals, many psychologists felt that there was more to be gained by emphasizing their own ideological and political values. The topics of conferences and programmatic speeches show, however, that in the thirties there was a gradual shift to the presentation of practical usefulness. This was a strategy that psychotechnicians and military psychologists had been following anyway. At the congresses of the DGfPs in 1933 and 1934, the opening speeches by chairman Felix Krueger concentrated on the ideological and theoretical reorientation of psychology. In 1936 in Jena there were specific references to the practical opportunities for psychology in the Wehrmacht, the employment offices, and the Labor Front. Krueger praised the chances for new careers in the Wehrmacht and announced the presentation of honorary membership in the society to Colonel von Voss, the military head of Wehrmacht Psychology. At the society's thirteenth congress in 1933 the main lectures included "The Germanic Soul" by Ludwig Ferdinand Clauss and "The Anti-Type of the German Folk Movement" by Jaensch. At the fourteenth congress in 1934, held under the motto "Psychology of Community Life," a central contribution was "Race and Heredity." The fifteenth congress in 1936, entitled "Emotion and Will," presented mostly research on volition from Wehrmacht Psychology. At the sixteenth congress in 1938 all Wehrmacht psychologists were ordered to appear in uniform (Z, ff. 13, 33), providing a setting suitable for the contents of the proceedings. The congress was held in Bayreuth, the seat of the Nazi Teachers' League, and it was also planned to show the role of psychology in educational matters. Questions of "psychology and race theory" were to be considered at the seventeenth congress planned for 1940 (REM/3147, f, 10). As the racist program of the Nazi Party became a visible reality, race psychology no longer focused on showing ideological usefulness, but on

---

6 Jaensch based his mixture of typology and racism on the natural sciences, not least of which were the physiological theories of his brother Walther. This also casts doubt on the theory of Wyatt and Teuber (1944), discussed in Chapter 1, that the psychological theories melded more easily with Nazi ideology the more philosophical they were. Perhaps it would be more accurate to say that the Nazification of a theory was easier when its speculative component was large, so that it offered more scope for arbitrary change.

assisting "empirically" in racist policies (see Chapter 8). In fact, the congress did not take place due to the war (see Geuter 1984a).

Right from the start, the aim was to demonstrate how practically useful psychology could be to those interested in having psychologists solve their problems. We have already seen this in the relationship between psychotechnics and industry. The strategy was also adopted in Wehrmacht Psychology. Naturally the interests of the Wehrmacht were different from those of the party or a government administration dealing with university psychology. Effective selection was required, and the legitimation for psychology was that it could provide it.

In 1934 Max Simoneit, scientific head of Military Psychology, writing in the first book of a series entitled the "Theory of Practical Human Knowledge," edited by the Psychological Laboratory of the Reichswehr Ministry, wrote:

> The series begun by the reflections presented here will have to show the scientific foundations from which the practical selection of persons for leadership purposes can be carried out. The proof of scientific possibilities also provides a justification for the offer made to the State by psychology ... If psychology proves that it can solve these tasks more correctly than others, then it would be a neglect of duty if one did not wish to involve it in this task. (Simoneit 1934a, pp. 13–4)

The series thus substantiated the claim of psychology to be in charge of selection. Simoneit thought that psychology must show the officers "that it was better able to carry out selection" (1939b, p. 4). This was done practically, by means of selection controls, but also programmatically. For example, Simoneit wanted to tackle the problem of diagnosing will, because psychology would lose "significantly in importance to the military" if it failed: "Therefore we place the practical-diagnostic problem before the theoretical" (1937, p. 5). Theoretical work might be an essential aspect of Wehrmacht Psychology, but "only when it is responsible for providing psychological diagnoses for state purposes does psychology really become important" (1938a, p. 1).

Wehrmacht Psychology was to demonstrate just what psychology as a whole could do. Simoneit (1940) ended a historical review of Wehrmacht Psychology with some thoughts on the "political significance of applied psychology at the national level." Wehrmacht Psychology had fulfilled a "scientifically historical role." It had

> clearly demonstrated the possibility and usefulness of the application of a science for the selection of persons by the state.... Other state organs with problems of selection have been shown an example.... Scientific psychology is to be given an advisory role in the procedures of selection of persons by the state. A special merit of the Wehrmacht has been to show that this is possible and useful. (1940, pp. 56–7)

A satisfied Simoneit noted that "particularly due to the performance of military psychology public opinion about psychology is becoming more positive" (1940a, p. 119).

For Simoneit the presentation of the practical value of psychology in the Wehrmacht also provided the best argument for its use in other areas. This raises a general problem of this legitimation strategy. If psychology's practical relevance is determined solely in terms of existing external demands, coming in this case mainly from the Wehrmacht, then it faces the danger of adopting a purely affirmative stance toward the status quo.

In the universities, we find not only research projects in support of Nazi ideology, but more and more projects involving applied research, and we even find affiliation with practical facilities such as educational counseling centers (see Chapter 8). This demonstrated the practical importance of psychology, especially during the war. At the Leipzig Institute, Wilhelm Wirth investigated shooting psychology, a project that interested Wehrmacht Psychology (RH 12-2/101, f. 369). In Tübingen work was carried out on the system and structure of the senses of positional orientation for Air Force Psychology (RH 19 III/494, f. 196). A series of projects was conducted for the Amt BuB of the German Labor Front, on such topics as maintaining work performance in old age, the practical importance of the results of apprenticeship examinations, the link between "genetic makeup" and vocational competence, vocational training, and the use of basic courses to identify technical abilities in young people.[7] The institute in Jena was charged by the Medical Corps in Weimar in 1944 with the examination of the brain-damaged in the Wehrmacht, as well as the so-called psychopaths and problem cases, neurotics, and forensic cases (H 20/495). In Halle they cooperated with Siemens on lighting for workplaces, as well as carrying out secret work on dogs for the intelligence section of the Army High Command.[8]

Such projects were not only of interest to the clients. They promised opportunities for research and recognition for psychology. The dog study in Halle, for instance, was quoted in applications for funds. Such research also made it possible to exempt staff members from military service. Practical achievements were considered to have a beneficial effect on public opinion, as can be seen from the exhibits of the Psychological Institute at the Berlin University Week in 1938. According to a press report, the labor psychological work of Hans Rupp was given prominence, although he was an outsider at the institute. There were only a few examples of experimental research.[9]

The practical usefulness of psychology was particularly emphasized when

7 See correspondence between the institute and the Reichswirtschaftskammer, November 1940, in UAT 148 and REM/2162, f. 5.
8 UAH PA Wilde, application dated 17 February 1940. The research is documented for 1939–41; see ibid. and UAH Rep 4/282.
9 *Nationalzeitung.* Gelsenkirchen, 26 November 1938. On Rupp's role at the Berlin Institute see, e.g., Z, f. 79.

it came to developing university courses and new examination regulations, because it was necessary to demonstrate a need to train professional psychologists. We shall consider this in the next chapter. In a commentary on the introduction to the new DPO, Kroh expressed quite clearly that psychology should become an applied science aimed at allocating and preserving "human material":

> It is becoming... plain that psychology has ceased to be a science for connoisseurs. With activities such as selection, evaluation, control, guidance, and care for the mental hygiene of the healthy members of our *Volk*, with aid and advice for the susceptible, the endangered, and the inefficiently functioning, it is becoming deeply involved in the necessary tasks of regulating, maintaining and strengthening the *Volkskraft* as a whole. (1941, p. 8)

Thus, at a time when the war was placing demands on it and new examination regulations were being drawn up, university psychology resorted more and more to legitimation based on practical usefulness. The decisive criteria for any science was its importance for the war effort. In Luftwaffe Psychology, Siegfried Gerathewohl therefore proposed abandoning the diagnosis of "undeveloped and latent talents," since wartime psychology was not concerned with questions of research but with arriving at practical evaluations and rapidly assigning individuals to posts where they could give of their best (I C Ge 5, f. 12). When delegations of German and Italian psychologists met in Rome in 1941, the German delegation listed as successes only work in the Wehrmacht, labor and industrial psychology, and psychotherapy. In a report to the ministry on the meeting, Kroh wrote that these were the areas "with a high degree of importance for the war" (REM/3147, f. 19). In contrast to the case with the natural sciences, less research and more practical application and training functions were regarded as psychology's important contribution to the war. Research funds from the Reich Research Council or the Wehrmacht went to other, more important disciplines. But whereas natural scientists had to prove that their research projects were essential to the war effort, academic psychologists needed only to point to the importance of training Wehrmacht psychologists to obtain exemptions from active service (see Ludwig 1979, pp. 216ff., esp. 238–9, 252).

Ideological or practical usefulness: part II

Various centers of power funding research, teaching, or other activities expected different things from psychology. The Wehrmacht did not have the same interests as the bureaucracy or the Nazi Party. The universities, as an important part of society's ideological apparatus, did not expect academic psychology to concentrate exclusively on demonstrating its practical value. Of the four centers of power in Neumann's analysis (see Chapter 1), one would expect the party to be most interested in the ideological usefulness, and industry and the Wehrmacht primarily in the practical applications, of psychology. But in addition to these main centers, we must also consider the diverse party institutions, each with its own interests. In the case of the Amt Rosenberg, which tried to influence the development of Wehrmacht Psychology, these have been documented in detail by Bollmus (1970). Psychology had to attempt to justify itself on various fronts. For Wehrmacht Psychology it can be shown that under pressure it adopted a double strategy. The conflict between attempts to present an ideologically pure psychology and also to demonstrate its practical effectiveness can also be observed in labor psychology, but the case of Wehrmacht Psychology is more interesting, not only because it was the largest psychological organization, but also because internal material is available that makes it possible to interpret scientific texts as documents of attempted legitimation.

The Amt Rosenberg attacked Wehrmacht Psychology at a time when Hitler had replaced the War Ministry by the Supreme Command of the Wehrmacht (1938) and had appointed himself supreme commander of the Wehrmacht. The Psychological Laboratory that had been attached to the ministry reemerged in June 1938 as the Wehrmacht Center for Psychology and Race Studies (HVBl, 1938, p. 160). The inclusion of race studies in the title indicates a desire to emphasize conformity with Nazi ideology. In the Amt Rosenberg the new name was seen as a deception (NS 15 alt/22, f. 58817).

Perhaps aroused by the new name, or as part of general measures for "coordination" (*Gleichschaltung*) in the army, Martin Bormann, then chief of staff to deputy führer Rudolf Hess, wrote to the Amt Rosenberg on 6 July 1938 asking its opinion on the psychological tests in the Wehrmacht.[10] The reason may also have been that at the sixteenth congress of the DGfPs (1–3 July 1938) the military head of Wehrmacht Psychology, Colonel von Voss, had presented impressive figures on the increased numbers of assessments and posts and had emphasized the fact that there was "only one National Socialist Wehrmacht Psychology" (Klemm 1939, p. 3). As a result of Bormann's inquiry, Alfred Baeumler drew up a lengthy statement directed against Simoneit. Baeumler had already produced a negative report on Simoneit for the Nazi Lecturers' League in Berlin in 1935. At that time the Wehrmacht was trying to get him an appointment as professor. Baeumler, professor of philosophy in Berlin, had drawn on Simoneit's publications in educational psychology before 1933 and had criticized Simoneit's pursuit of an individualistic psychology based on the search of the individual for happiness and defining the state and the community in terms of the interests of the individual, in agreement with the ideas of the Social Democrats.

10 NS 15 alt/22, f. 58818, for a reference to the letter. From f. 58793 it seems that the new name caused misgivings toward the deputy of the Führer.

Baeumler's arguments can be studied in a politically biting review of a book by Simoneit in the journal of the Nazi Teachers' Organization in Berlin.[11]

Replying to Bormann in August 1938, Baeumler again attacked Simoneit's ideological tenets, but not the methods of aptitude testing used by Wehrmacht Psychology. Simoneit was once again accused of giving the individual priority over the state and the community. Furthermore, he also applied this idea to "soldiery," which he defined in terms of the personality of the soldier instead of in political terms. In an internal draft of the reply, Baeumler had gone further and accused Simoneit of having conformed outwardly without having a proper inner relationship to race thought. He also criticized Wehrmacht Psychology, which was an inflated affair, in which psychological assessments practically had a much greater weight than was claimed. The first draft of Baeumler was much sharper in tone than the official reply that was finally sent to Bormann. It is interesting that the document avoided rejecting the methods of selection, only criticizing Simoneit's weltanschauung despite the fact that Bormann had registered his "objection to the psychological examinations in the Wehrmacht."[12] According to a study by Bollmus (1970), the Amt Rosenberg was a relatively weak section in the Nazi apparatus that seems to have been able to secure its influence only by playing the role of ideological censor of Wehrmacht Psychology, or of its leaders. Baeumler's aim was to put an end to it completely. I do not know whether or how Bormann pursued the matter. In the files that I was able to inspect there was no further reference to action by party sections against Wehrmacht Psychology.[13] This, therefore, does nothing to support the hypothesis that the party was uniformly opposed to Wehrmacht Psychology (see Chapter 8).

The Army High Command does not seem to have adopted the line of the Amt Rosenberg. In his "Guidelines for Training Officer Cadets" (20 October 1938), the new commander in chief, Walther von Brauchitsch, did include a comment that "racial selection... left something to be desired," but this was not linked to the preceding argument that officer cadets lacked the necessary enthusiasm (MA RH 37/2378). Nevertheless, the situation was obviously precarious enough for Wehrmacht Psychology. In 1938 the legitimation strategy was altered to make the race-psychology orientation plain, both within the Wehrmacht and externally. The best documents for the investigation of the changing tactics adopted in the Wehrmacht are the annual reports of Wehrmacht Psychology, presenting an account of its usefulness and its successes. The reports for 1934–5 and 1936–7 sound optimistic and present a great many figures to demonstrate the success of

---

11 For details on the whole matter see the file "Simoneit 1935–42" of the Amt Rosenberg, NS 15 alt/22. For the review see: *Nationalsozialistische Erziehung* 5, No. 2, 11 January, 1936, pp. 27–8.
12 See note 10.
13 The files of the party chancellery, which are in a much worse state than those of the Amt Rosenberg, were not reviewed.

psychological tests in selection work. In 1936–7 there is a comment on the immense expansion of Luftwaffe Psychology. On race psychology we can read the reserved comment in 1934–5 that scientific criteria for the evaluation of anthropometric data were lacking, and in the 1936–7 report there is a one and a half page description of the work of Group B of the laboratory that dealt mostly with twin studies, described in the report as providing a basis for determining the "racially based" psychophysical constitution. Unfortunately, the 1937–8 report could not be traced. But the report for 1938–9 sounds completely different. With the exception of the section on naval psychology, the texts seem to be very defensive. In contrast to earlier years, it is emphasized that military psychology was only advisory and never took any decisions. In the section entitled "General Military Psychology Work" pride of place was given to the "intensification of work in the field of racial psychology," followed by Simoneit's war psychology studies as a step toward "further... reorientation of military psychology." For the first time in the available reports there is a reference to a "strengthening of the inner bond with the National Socialist weltanschauung." Training of all Wehrmacht psychologists in race psychology was envisaged, and a group was set up at the headquarters especially for race psychology. Questions on kinship were included in the test procedure, and kinship research was included as an area of research in military psychology.[14]

This trend toward defensive arguments suggests that there had been a confrontation behind the scenes probably as a result of criticism from the Office of the Deputy Führer and the Amt Rosenberg. Working on this assumption, we find that some of the publications of Wehrmacht Psychology are easier to understand. A close reading of the "Guidelines on the Psychological Examination of Officer Recruits in the Wehrmacht" by Simoneit (1938) shows discrepancies that are otherwise difficult to explain. In contrast to earlier publications Simoneit describes the situation as if Wehrmacht Psychology had always thought in terms of race and constitution typology: "This fundamental of race science [the inheritance of essential properties of the race] has been known and common currency in Army Psychology since its introduction into practical life. All accusations that racial aspects have been neglected draw the counteraccusation of ignorance upon them" (1938, p. 20). But in fact this aspect was not to be found in Simoneit's earlier writings, nor more remarkably is there any mention of it actually being applied in this text. A few pages later, when he becomes more specific, he only states that the psychologist knows the periods of highest productivity of a Mecklenburger, Berliner, and Saxon.

The attacks by the Amt Rosenberg led to new efforts at self-justification by Wehrmacht Psychology, as seen in a flood of new publications. In 1938

14 See the annual reports under RH 12-2/101, ff. 367ff. and RH 19 III/494, ff. 187ff., 237ff.

the Wehrmacht's psychological center began issuing a journal (*Soldatentum*) focusing on the relevance of race science to military psychology. The new spokesman for race psychology, Erich Zilian, published numerous essays on the subject. In October 1938 Zilian spoke to a study group of the *Deutsche Gesellschaft für Wehrpolitik und Wehrwissenschaften* on race studies in Wehrmacht Psychology (1939a, b). In 1939 the Wehrmacht's psychological center issued a "Race Diagnostic Atlas." It contained picture sequences of "race types" in accordance with the quasi-official doctrine of H. F. K. Günther (1926), with the recommendation to use this classification in all cases.[15] Research into kinship and genealogy now also received support (see Dirks 1941, 1942; H. R. G. Günther 1939, 1940). But the part of psychology that established itself in the Wehrmacht was not race psychology or hereditary psychology. Simoneit was certainly right when he commented in 1940 (p. 17): "The psychology of soldier selection has met with the most recognition in military practice and has received the most encouragement as a science." A further alteration of the name of the center to Inspectorate for Aptitude Testing on 23 August 1939 is an expression of this fact (HVBl, 1939 B, p. 266).

This example shows that psychology quickly adapted its legitimation strategies to the diverse interests registered in various quarters. As will be seen again when we consider the preparation of the DPO, the decisive criterion for the acceptance of psychology was its legitimation as a practically useful discipline. This was true for the military services, the bureaucracy, and industry.

We should also take a look at the conflict between ideological and practical legitimation in industrial psychology. Its standing had fallen, and at the end of the twenties it was in as much difficulty as academic psychology. In the new situation in 1933 the psychotechnicians reacted by using both legitimation strategies. They distanced themselves from a "psychologistic blind to reality," as the appeal of the Society for Psychotechnics put it in June 1933.[16] The old psychotechnical methods and intelligence tests, which were under fire from characterologists anyway, were now attacked with new political arguments. Friedrich Becker, from Jaensch's school, criticized certain intelligence tests for favoring the "Jewish form of intelligence." These "cerebral-intellectual meddlings" were especially to the taste of the so-called dissolution type of cultural decline according to Jaensch (Becker 1938a, pp. 33, 94). Schliebe (1937, p. 198) took pains to emphasize that the aptitude tests then employed by the army, the railways, and others had nothing to do with the old "Jewish" psychotechnics. At the same time there were claims that psychotechnics (although this name was hardly used any more) could

15 The atlas itself has not been preserved; Zilian refers to its introduction (1939c) and describes it (1939d); see also Zilian (1939i).
16 *Ind. Pst.*, 10 (1933), p. 11.

provide empirical foundations for the new ideological doctrines. Giese expressed this opinion at the fourteenth congress of the DGfPs in 1934; by comparing the typical performance and function ideals of the German "tribes" and their value systems, it would be possible to develop an empirically based mental study of tribes and then "to publicly propagate the tribal-psychological evaluation of the German people" (1935, p. 202). Similar studies were considered in Chapter 3.

   There were also those who, without necessarily having different political views, adopted the other strategy of demonstrating the effectiveness of psychotechnics in industry. This different type of justification was linked to different scientific concepts, of which Moede was a leading proponent. In the twenties he had come down on the side of industry in the controversy over whether psychology should relate to industrial needs or should seek cooperation with state and public institutions. Moede criticized typology, which might perhaps be useful for other purposes, but not for determining vocational aptitudes (1935, p. 135). Neither could actual abilities be recognized from genetic history. The thing to do was to establish performance potential in terms of vocationally relevant criteria (see Moede 1936). Even Jaensch felt the need to remark that by taking into consideration typological and racial factors in practical psychology the methods for testing specific vocational performances, proposed by Moede, Valentiner, Herwig, Hische, and others, did not become less significant but received valuable augmentation (Jaensch 1938d, p. 18). Whether he wrote this to guard against methodological schisms or because he was concerned with both the ideological and the practical usefulness of psychology is an open question. In any event both industry and the German Labor Front encouraged and respected the effectiveness of psychology in industry. The head of Amt Bub of the Labor Front, Karl Arnhold, described the evaluation of performance as the most reliable way of establishing work abilities and the basic structures of work behavior (1938, p. 37).

   In practice this attitude was seen at its clearest in the use of classical methods of aptitude diagnosis for the selection of foreign workers during the war (cf. Chapter 4). This involved intelligence tests (involving simple tests of logic, combinatorial ability, and calculations), spatial perception (e.g., with Rybakov figures), cog wheel tests for technical understanding, and wire bending to establish manual dexterity (Schorn 1942). The ten simple intelligence questions initially used were based on the U.S. Army mental tests used in the First World War. In Becker's work just quoted these tests did not come off quite as badly as the "Jewish" methods, even though they did give an advantage to the dissolution type. However, Becker (1938a, p. 85) had picked out the Rybakov figures as tasks typifying the "frivolous," "unrealistic" intelligence of the dissolution type. Nevertheless, the test was used for selecting out deported and prisoner-of-war workers. This was because "they can be carried out quickly...and protect the firms from making severe mistakes," as the newsletter of the Amt Bub put it.[17] In a pamphlet written for use in factories entitled "Selection and Deployment of Eastern Workers," a district chief for technology praised the advantages of psychotechnical selection (using this term) in the stark language of the technologist. The psychotechnical selection of deportees was recommended "when these people are to be used for more difficult tasks, which require a certain degree of technical ability, speed of reaction etc." (Haas 1944, p. 24). They met the need for tests that were "flexible, cheap and short" (p. 24). When testing people who were being compelled to do work related to the war effort, the assessment of the personality as a whole was not important.[18] A team of five could conduct 1,000 to 1,500 tests in two days, and up to 50 tests simultaneously (Schorn 1942, p. 212). In all, some 400,000 workers were tested. Thus, when a Nazi organization such as the Labor Front was interested in psychotechnical methods, which were employed and obviously appreciated in industry, then these were used, ideological attitudes notwithstanding. The psychologists involved seemed to be aware that for their work the criterion of usefulness was what counted.

Inward and outward unity

The examples given show that leading representatives of the discipline favored different strategies. Whereas some academic psychologists initially tended toward one strategy, those who were practically active were bound to go for the other. However, the attitudes of academic psychologists shifted increasingly, so that during the Third Reich a tendency toward a unity of theoreticians and practitioners can be observed. As the German Society for Psychology became more and more representative of all psychologists, it made efforts to achieve a unified presentation of psychology and to bridge the organizational gap between academics and practitioners. Its executive committee was a decisive factor in psychology's professional politics.

In the early twenties academic psychology and practical psychology had largely presented themselves as separate entities. The practically oriented psychologists created their own organizations, which I will consider later, and their own platforms for self-presentation. In 1922 the first meeting of the Group for Applied Psychology in the Society for Experimental Psychology was held, and this was commented on later as follows: "For the first time applied psychology proclaimed its right to independent recognition."[19] The

---

17 "Die deutsche Arbeitsfront fördert rationellen Einsatz ausländischer Arbeitskräfte," *Anregungen, Anleitungen für Berufserziehung und Betriebsführung*, 7 (1942), p. 52; cf. Geuter (1987a).

18 A collective review in the *Industrielle Psychotechnik*, 20 (1943), p. 150, states, "While we find it important for the training of German hands to assess the whole person ... for the employment of foreign hands it is necessary to assign them to work where they are able to be most productive and have the greatest practical use."

19 *"Die erste Tagung...,"* 1923, p. 392.

Association of Practical Psychologists (Verband der praktischen Psychologen) was then founded, and at subsequent meetings the results of applied psychology were presented, often in the presence of representatives from industry and bureaucracy (Wies 1979, pp. 52–3). At the academic congresses of the Society for Experimental Psychology, a forum for various tendencies, schools, and fields, applied psychology had become increasingly prominent since the early twenties. At the Vienna congress in 1929, where the name of the society was changed, the major survey paper was dedicated to the topic of psychotechnics.

Congresses generally reflect the lines of development of the subject matter of a discipline, as well as the internal balance of power, factional strengths, and scientific orientation of the Executive Committee, but when a science is to be brought into line their function changes, and they become a means of imposing certain ideas from above. From 1933 the congresses of the DGfPs were focused on topics that had been checked out with the political powers, as Krueger reported to the members of the Executive Committee in a circular before the fourteenth congress in 1934 (NS alt/17, f. 56949). This certainly corresponded to psychologists' wish to create an impression of unity. Any discussions would have made it seem less likely that they would be able to comply with the requirements of the time or to provide clear, practical guidelines for action. By choosing focal topics, the committee was also able to enforce its ideas about the subject. This is indicated by the words Kroh later chose to praise the new organizational form: since 1934 the congresses had "not only become more unified themselves, they had also exercised the function of alignment in the development of the subject" (1944, p. 187). In view of the numerous research results it had become necessary, Kroh felt, to concentrate on specific areas. But as we have already seen in this chapter, the topics actually chosen provided a tool to push through certain concepts. In the course of preparing the fourteenth congress in 1934, a circular naming race, the psychology of the community, and education for community as the three main topics, and lumping everything else together in a fourth group with the comment that other contributions should not be completely excluded (Hartshorne 1937, pp. 116–17, n. 3), was clearly an attempt to bring the members "into line."

At the same time the editorial boards of the journals were also trying to get themselves into line. The editor of *Psychotechnische Zeitschrift*, Hans Rupp, wrote in February 1933 that it was an obligation to examine whether to continue on the path hitherto pursued or to pay more attention to particular tasks. The first new tasks he named were studies about character, ability to lead, and selection and advice for higher vocations (Rupp 1933). *Industrielle Psychotechnik* published an appeal by the Society for Psychotechnics, signed by Moede, Couvé, and Tramm, that began with the words: "All practitioners and scientists in the fields of applied psychology and psychotechnics who are for the new state must at last come together.

Innumerable new tasks are awaiting realization."[20] In the *Zeitschrift für Jugendkunde* Sander presented a program to "draw a scientifically sound" picture of German youth, which was seized by the holy desire for completeness and the wish to "cut off everything alien to... the German *Volkheit* and everything that overgrows it parasitically." The youth had found in Adolf Hitler the man "chosen to lead this desire and this will to its goal" (Sander 1934, pp. 1–3). In the *Zeitschrift für angewandte Psychologie* new editors Klemm and Lersch emphasized the special importance of the current "inner reorientation" of applied psychology. In view of the "organization of our state life in terms of leaders and followers" they declared a particular interest in "questions of a characterological anthropology, especially in areas of education and occupations."[21]

Despite all the proclamations, the unity of psychology could not be achieved. There was, however, a sort of unity of psychologists that arose from the practical activities of virtually all academic psychologists in the Wehrmacht and from the advances of the profession. At least Wehrmacht Psychology helped to create the type of university teachers, involved in research and application, that Kroh (1938) felt would guarantee the connection between theory and practice.

The politics of the professional organizations

The unity of psychologists was reinforced by the bridging of divisions between various professional organizations within the DGfPs, which embodied the interests of professional activity during the preparation of the DPO. This unification represented an element that is generally decisive for the professionalization of an academic discipline. It also conformed to the Nazi principle of corporate professional organizations representing entire occupations. In accordance with the "Führer principle," it was the chairman who represented the organization.[22] The material does not support the conclusion that this unification was compulsory.

Among other professional psychological associations before 1945, the most important was the Association of Practical Psychologists (Verband der praktischen Psychologen), which was initiated and supported by university teachers in applied work. In 1920 Giese had demanded that after establishing

20 *Ind. Pst.*, 10 (1933), p. 11.
21 "Vorbemerkung der Herausgeber," *Z. ang. Ps.*, 46 (1934), p. 2.
22 The unification probably owed more to the special phase of professionalization than to Nazi principles. In the United States there was a similar development during the war. The American Association for Applied Psychology, founded in 1937, was absorbed by the American Psychological Association in 1945 after their joint efforts during the war. Napoli writes of the unity of psychologists in the United States, "During World War II professionalism, unity, and patriotism were more thoroughly intertwined than at any other time" (1975, p. 273).

an appropriate syllabus an association of diploma psychologists should be set up, like similar associations for doctors, technologists, and lawyers.[23] In 1917 Stern had suggested founding a society for aptitude testing (Wies 1979, p. 38). At its seventh congress in Marburg (1921) the Society for Experimental Psychology instructed a subcommittee to organize a congress for applied psychology in Berlin, at Easter of 1922. The purposes of the subcommittee were more scientific. The announcement that it had been set read: "Outside the society the Association for Practical Psychologists has been founded to protect the corporate interests of practically active psychologists. Marbe was elected as chairman, Moede as vice-chairman, and Lipmann as secretary."[24] Stern and Lipmann organized the subcommittee, which then met together with the association in 1922.

Fritz Giese (1922, p. 6) felt that the Association for Practical Psychologists should further the interests of practitioners, who earned their daily bread by means of psychology. But it was dominated by university teachers who also engaged in applied work. Karl Marbe remained chairman until 1927, when he took over the chairmanship of the Society for Experimental Psychology. From 1922 the Executive Committee also included the university teacher Walter Moede and Curt Piorkowski, who worked at an employment office (Dorsch 1963, p. 92). In 1927 they were joined by Professor Walther Poppelreuter and in 1930 by the military psychologist and university professor Johann Baptist Rieffert. By 1930 the Executive Committee included Moede, Rieffert, Narziss Ach, Gustaf Deuchler, Poppelreuter, and Giese – a gathering of professors.[25] It was not until 1933 that a psychologist from the Labor Administration, Johannes Handrick, became chairman. The association added the term "German" to its name in the mid-twenties to become the Verband deutscher praktischer Psychologen (VdpP); criteria for membership were laid down at a general meeting in 1926. Prior academic training was required, with a scientific dissertation as well as proof of practical activity.[26]

Between 1933 and 1945 few activities of the association can be traced. It is possible that in 1933 Otto Lipmann was still secretary. He was Jewish, and after his Institute for Applied Psychology had been destroyed, he died on 7 October 1933, probably by suicide. In 1935 Handrick attempted to affiliate the VdpP with the Labor Front or the Nazi Teachers' League in order to have the protection of either name for applied psychologists. The Science Ministry rejected the request due to insufficient public interest (21 March 1935).[27] The VdpP was nowhere to be seen at this time. There were efforts

23 *Z. ang. Ps.*, 16 (1920), p. 392.
24 Ibid., 18 (1921), p. 395.
25 *Ind. Pst.*, 7 (1930), p. 186; *Prakt. Psych.*, 4 (1922), p. 32; "*Verband* . . .," 1927, p. 159.
26 *Ind. Pst.*, 3 (1926), p. 245. On Handrick, NS 15 alt/17, f. 56949.
27 Communication from Helmuth Schuster, Aachen, who inspected the relevant documents in Merseburg, East Germany.

to reactivate it during the thirties.[28] At the end of this section we will see that it reappeared in 1942 under the name Deutsche Vereinigung für praktische Psychologie (German Association for Practical Psychology).

In 1931 the Reich Association to Further Practical Psychology was founded, with Poppelreuter as chairman and Lipmann as secretary. At its twelfth congress in 1931 the DGfPs joined the association. Its aim was to bring scientists and practitioners closer together.[29] I know nothing about its subsequent development. In 1932, when Stern, a founding member of the Reich Association, was chairman of the DGfPs, cooperation developed between the Reich Association, the VdpP, and the DGfPs. The three organizations formed a "group of an informal character."

> Two questions of common interest were immediately taken up: The work group noted with approval that the German society [DGfPs] and the association [VdpP] had set up a joint committee to develop a draft for a future examination for practical psychologists (chairman: Ach), and the congresses of the various organizations are to be related to one another.[30]

After the thirteenth congress in 1933 a joint meeting of the Reich Association and the VdpP was to be held in Halle on questions of practical psychology. It obviously never took place. The tentative approaches of the practically oriented associations and the academic society, motivated by common professional interests, were replaced under the Nazis by a rapid unification under the DGfPs. The oldest and largest organization of psychologists then concentrated all professional policy activities in its hands.

Meanwhile, in the DGfPs those active in applied psychology had grown in strength, and they were arguing that their work should be given more prominence. Perhaps an indication of the improved standing of such researchers among academics was the election in 1927 of the chairman of the VdpP, Marbe, as chairman of the Society for Experimental Psychology to succeed Georg Elias Müller, who had held that post since the foundation of the society in 1904. Other practically inclined university psychologists such as Ach, Poppelreuter, and Stern took on Executive Committee duties. An even clearer sign of the integration of practitioners was the election of the first nonprofessor to the Executive Committee in 1933. This was Johannes Handrick, a senior administrator in the Labor Ministry and at that time chairman of the VdpP. In 1935 Max Simoneit, head of Wehrmacht Psychology, followed and was even in the running for the chairmanship a year later. Also in 1935 the first nonpsychologist became an honorary member of the society – the military head of Wehrmacht Psychology, Colonel Hans von

28 After a conversation with Ach on 29 October 1938, Hische notes that he had heard "of the intention to bring the Association of Practical Psychologists back to life" (UAT 148).
29 See Kafka (1932, p. 470) and Wies (1979, p. 54).
30 *Z. päd. Ps.*, 33 (1932), p. 303.

Voss.[31] In the DGfPs the professional interests of practical psychologists now began to coincide with the academic interests of university psychologists. Both were concerned to see university qualifications introduced for professional psychologists. In addition to the organization of congresses, this aim was at the center of the activities of the DGfPs. The complicated coordination of two or three different associations was no longer required to formulate the psychologists' position. The Executive Committee could now claim to act as the representative of the entire professional group, unified in its membership of the society.

The old conflicts did not break out again until the examination regulations were being prepared and after they had been passed. Moede presented the ministry with a draft that differed from the ideas of the Executive Committee of the DGfPs (see Chapter 6). In 1942 he appeared to be taking advantage of the disbanding of Army Psychology and Luftwaffe Psychology to strengthen the position of psychology in industry. On 27 June 1942 he sent a report on the German–Italian Psychologists' Meeting (which had been held a year before) to Heinrich Dahnke from the Academic Exchange Service, who was also responsible for international congresses at the ministry. In his accompanying letter, written three days before the official end of Army and Luftwaffe Psychology, he remarked laconically that the meeting was without practical success because "Wehrmacht Psychology, which took up a great deal of the discussions . . . is to be closed down in the army and the Luftwaffe." But the testing used by the railways was not presented at the meeting. The DGfPs, wrote Moede, represented only a section of German psychologists, "and a faction of German psychology. The collapsed Wehrmacht Psychology had put its stamp on it" (REM/3147, f. 38). On 22 August 1942 Moede reactivated the VdpP as the Deutsche Vereinigung für praktische Psychologie. A "general meeting" attended by eleven people chose Moede as chairman. He told the ministry on 1 September that this was the "organization of German practical psychologists with scientific training" (f. 40). The Executive Committee of the DGfPs was bitterly opposed to Moede's intentions. Kroh appealed to the unity of psychologists, which was to be protected against all division. In the ministry the view was that there should be a unified organization of psychologists under a unified leadership.[32] We will not pursue the internal squabbles surrounding Moede's actions here, but in the next chapter we will be able to look at his ideas for a psychology syllabus, which he hoped would be given more weight by the reestablished association.

31 NS 15 alt/17, ff. 56949, 56951, 56954.
32 On the whole matter see Moede to Pfahler, 28 August 1942; Moede to the committee of the DGfPs, 11 November 1942; Kroh to Moede et al., 30 December 1942. All in UAT 148. Further, REM/3147, ff. 38–43; REM/821, ff. 163–71, 322–6, 340.

6

University courses in psychology and the development of the Diploma Examination Regulations of 1941

A problem for students of psychology in the twenties or thirties was the fact that there was really no qualification for professional activity outside the university. The only possibility was a doctorate, for which psychology was often not recognized as an independent subject. By the end of the thirties the growing demand for professional psychologists created pressure for academic training that prepared for these professional activities. The contradiction between the two was overcome by introducing an examination to confirm the ability or entitlement to exercise a profession. A prerequisite, of course, was some consensus about the scientific training necessary.

The DPO for psychology in 1941 were the first of their sort and the result of a lengthy process. In this chapter we will look at the possibilities of obtaining qualification in psychology before 1941, at the suggestions made to adapt purely scientific education to meet professional needs, at the interests that led to the fruition of these plans at the time, and at the concept of preparatory qualification that found expression in the DPO.[1]

In the early days psychology was always studied together with other subjects, forming at most a specialty area. The general regulations for doctorates required three subjects to be examined orally, with psychology often not even qualifying as a subject in its own right. In 1935 Rupp expressed the opinion that not many students chose psychology as the major subject for their doctorate because it did not offer any secure career prospects (Rep 76/37, f. 44). In order to have a backup some students sat for qualifications as a high school teacher in their minor subjects (Z, f. 91).

Before the expansion of Wehrmacht Psychology in the mid-thirties, a student of psychology did not have any clear career prospects outside the

---

1 The only previous detailed discussion of the DPO in the literature is in Däumling (1967), which considers their background and preparation, taking Kroh's essays as source.

university. Psychology was studied as a scientific discipline. There was hardly even general agreement about what a curriculum should include as the essentials of the subject. Students had to become acquainted with the fields of research and the scientific approach of their particular professors. The dissertation was often a small contribution to the development of the theory of a school or to the research work of an institute. It was not intended to show that the examinees knew how to apply the science of psychology, but that they were able to make their own scientific contribution to psychological research.

Even this qualification was in doubt when psychology was not itself a subject for doctoral theses. Marbe (1921, p. 208) had included the demand for the introduction of a doctorate in psychology as one of the measures with which he hoped to improve the position of psychology in the universities. The Executive Committee of the Society for Experimental Psychology had then addressed this demand to all facilities where psychology was not yet an independent subject (p. 210). Subsequent developments, however, showed only slight improvement.

In the following overview we will not consider the technical colleges, where as a rule it was possible to obtain only a doctorate in engineering (Dr. Ing.). In 1935 only the technical colleges in Braunschweig and Dresden had an independent psychology course for a qualification as doctor of cultural sciences (Dr. Cult.). At the Stuttgart Technical College it became possible after 1932 to submit psychotechnical topics for a doctorate in technical sciences. As at all other colleges, the precondition was an engineering diploma. Since 1923–4 psychotechnics had also been an option at Stuttgart for mechanical engineers and electrotechnicians, and was a compulsory subject for teachers at trade schools. At the Charlottenburg Technical College a Dr. Ing. could be obtained with a psychotechnical thesis.[2]

The university regulations offered various openings to obtain a doctorate on psychological topics and also posed obstacles. In 1926 it was possible to obtain a doctorate in psychology at only eleven of twenty-three universities; two of these required philosophy as an additional minor. Two other universities had options in "pedagogy and psychology."[3] At ten universities only philosophy or pedagogy could be taken. Eight of these altered their regulations during the period under consideration to make psychology a major subject: by 1934 Berlin and Bonn, by 1938 Halle, Heidelberg,

2 *Studium der Psychologie*, 1935, p. 17; I B Ebert (1978, p. 18); Giese (1933, p. 19); *Ind. Pst.*, 1 (1924), p. 59; Giese (1922, p. 40).
3 Psychology: Breslau, Frankfurt, Göttingen, Hamburg, Jena, Kiel, Cologne, Rostock, Würzburg; with philosophy: Freiburg, Giessen; pedagogy with psychology: Leipzig, Tübingen. The overview is based on Ammon (1926) and on an evaluation of doctorate regulations in the Federal Archives, the Geheimes Staatsarchiv and the Zentrales Staatsarchiv. Inquiries were also sent to university archives, but they often did not have the relevant records. Some archives did not respond. This means that it might be possible to obtain more information for Leipzig and Berlin.

Königsberg, Marburg, and Munich, and after 1938 Münster. Greifswald kept the old regulations – there was no professorship for psychology there; the sources for Erlangen allow no statement to be made. The tendency through the twenties to establish doctoral procedures for psychology[4] continued into the thirties. This progress in academic institutionalization augments the picture outlined in Chapter 2. If we were to investigate the minutes of faculty meetings, we might well find that the arguments used about doctorates in psychology were similar to those used for professorships. It is thus probable that the professional activities of psychologists in the Wehrmacht also influenced the reformulation of doctoral regulations. When the military head of Wehrmacht Psychology, von Voss, sent the new career regulation to all university philosophy faculties in November 1937, he also included a request that "it be arranged that students of your faculty also find the opportunity to study with a full professor of psychology, and to do their doctorate in psychology" (UAL Phil B I, 14[37], I, f. 59). When the Ministry of Science and Education also sent around the regulations (25 January 1939), the accompanying letter included the sentence "Where psychology is not yet included as a major subject for doctoral examinations in the doctoral regulations, I declare my agreement to appropriate alterations to the regulations" (UAT 148). This was probably seen as a signal to pursue the independence of the subject vigorously. Before this, the ministry had not intervened concerning psychology, although, having abandoned the idea of centralized doctoral regulations, all universities were required to draw up new ones in 1937–8, which required the approval of the ministry (REM/785, f. 219). These doctoral regulations seem to have remained in force for the duration of the Third Reich.

The problems involved in becoming a doctor in psychology (the "promotion") reappeared on the way to becoming a lecturer (the "habilitation"). The regulations for habilitation do not usually name subjects, but the postdoctoral thesis was linked to an application for the right to lecture (*venia legendi*) in a subject. As long as a professorship that a psychologist hoped to attain would be classed as philosophy, the psychologist was obliged either to apply with his psychological thesis for a *venia legendi* for philosophy, or to present a large proportion of philosophy in his thesis or additional work. This could lead to considerable conflicts in the faculties. In Leipzig in 1929 and 1936 there were disagreements over Karlfried Graf von Dürckheim, Arnulf Rüssel, and Johannes Rudert. In all these cases philosophy professors complained that the theses did not, for example, display sufficient philosophical thought. Felix Krueger, as spokesman for psychology, responded that psychological theses would have to be accepted for philosophical habilitation as long as psychology was not an independent habilitation subject. Otherwise,

4 Göttingen introduced psychology as a subject for doctorates in 1921, Würzburg also in 1921, Jena in 1925. Rep 76 Va, Sekt. 1, Tit. VI, I D, Bd. 5, ff. 173–4; Marbe (1921, p. 209); UAJ, communication, 6 April 1981.

he threatened, he would apply for an alteration of the regulations.[5] Dürckheim, Rüssel, and Rudert gained habilitation in philosophy. But in the case of Albert Wellek in 1938 the *venia legendi* was granted only for psychology, since Wellek was "emotionally determined" and did not think as a logician or philosopher (UAL PA 1046, f. 51). Thus between 1936 and 1938 it had become possible to gain habilitation in psychology in Leipzig.

The unclear situation led to a wide variety of terms being used to describe the permission to teach. In the case of Adhemar Gelb in 1919 it was for "philosophy with special consideration to psychology"; for Walther Poppelreuter, "pathological work psychology"; for Walther Blumenfeld in 1920, "psychology"; for Hans Keller in 1924, "philosophy, especially psychology"; for Adolf Busemann in 1926, "experimental psychology and pedagogy"; for Anneliese Argelander in 1926, "psychology"; and for Philipp Lersch in 1929, "philosophy and psychology."[6] Thus, for the habilitation, teaching areas were chosen that did not correspond to clearly defined subject demarcations. This may also have encouraged psychologists to push for the full institutionalization of their subject.

Proposals for professional qualification before 1940

The first calls for a professional qualification in psychology were raised after the First World War by some academic psychologists involved in applied work. They were accompanied by the suggestion that academic studies be oriented more toward practical tasks. Walther Poppelreuter seems to have been the first to use the term "diploma psychologist" (Giese 1919, p. 420). Giese also felt that the term was sensible – as a parallel to "diploma engineer" – even though it had no firm legal standing (1922, p. 5). Suggestions increased during the blossoming of psychotechnics between 1920 and 1923. At that time psychologists were working with the brain-damaged or at vocational guidance offices. In line with his own specialty, Poppelreuter (1921) wanted to see the course affiliated with medical studies. After practical training at one of the four institutes for the brain-damaged, a qualification could then be obtained with a doctoral thesis in psychology and a specialist examination. Elsewhere, Poppelreuter (1921) suggested that psychologists at the labor and vocational guidance offices be trained at a practical-psychological institute, take an exam, and have one year of practical training. For studies at universities he recommended teaching "human assessment" (Poppelreuter 1923, p. 78, note).

At the forefront of the reform discussion were not so much questions of syllabus, as questions of a professional qualification. The aim was to obtain

5 UAL PA 95, f. 45; PA 97, f. 21; for more on the Rudert conflict, see Geuter (1984).
6 *Hochschulkorrespondenz* No. 169, 25 July 1931; *Z. päd. Ps.* (1919), p. 429; (1920), p. 314; R. 21, appendix 10010, f. 4909; *Z. päd. Ps.* (1926), p. 206; UAJ information, 6 April 1981; R. 21, appendix 10012, f. 5931.

state approval for academic training in psychology as a flanking measure for the first hesitant steps into the areas governed by public law. Karl Marbe wrote that he heard the bureaucracy's call for specialized psychologists getting louder and louder. His reply was to suggest an examination similar to that for chemists, and considered a diploma as the formal title (Marbe 1921). Giese also felt that the state would have to regulate the examination of psychologists "since this profession has now matured to the level of public offices" (1922, p. 32). Hildegard Sachs (1923, p. 27) welcomed Poppelreuter's demand, since from a social policy standpoint, incompetent handling of aptitude testing should be prevented. All these authors were from practical psychology, even though Giese, Marbe, and Poppelreuter were teaching at universities. In the universities a doctorate or a habilitation was a mark of scientific competence. For practical work, however, graduates had no legally effective way of substantiating their claim to having a competence in psychological questions that others did not have. They were concerned to obtain offical recognition of their real or supposed competence.

The first comprehensive plan for studies intended to produce this result was outlined by Giese (1922) in his studies and career guide for psychology and psychotechnics. A state solution, he felt (p. 36), would resemble the plan of study he suggested. In the first two semesters he envisaged courses in experimental psychology, philososphy, physics, anatomy, and physiology. In the second semester quantitative methods in psychophysics would be introduced, and in the third semester would begin a four-semester cycle in applied psychology consisting of (1) pedagogy, (2) vocational counseling and aptitude testing, (3) object psychotechnics, and (4) advertising and organization. New subjects also envisaged for the fourth semester were differential psychology (types, correlations), psychophysics of work processes, psychiatry, and medical psychology. For the fifth semester there was social and forensic psychology, for the sixth semester the history of psychology, sociology, and other social sciences. The final semesters were to be accompanied by psychological exercises and research work. Folk psychology, psychology of language, parapsychology, and the psychology of religion were also suggested as topics but seem to have been intended for briefer treatment. The focus of both theoretical training and applied subjects was on qualifying students for the psychotechnical work of evaluation, selection, and measurement.

The VdpP was also concerned about establishing a formal claim to an area of competence. In 1927 a general meeting passed a resolution favoring the introduction of an internal or state examination for practical psychology. The Executive Committee added that it would undertake all efforts necessary to ensure the introduction of a state examination with state licensing.[7] Early

7 "Verband der deutschen....," 1927; the introduction of a state exam was also discussed at the sixth conference in 1928 in Hannover; see Rupp (1929).

in 1930 the first approaches were made to the Prussian Ministry, though without results.[8] Around this time there was cooperation between the VdpP and the DGfPs to develop an examination for practical psychologists. The Executive Committee of the DGfPs had realized, at the latest after the manifestation of 1929, that it would be useful to recommend psychological institutes as the places that should "ensure adequate education and training for the formerly unknown occupation of practical psychologist" (*Kundgebung*..., 1930, p. ix). William Stern (1931, p. 78), at that time DGfPs chairman, was of the opinion that universities should establish all the courses for training in this profession, which could not be left to the psychological laboratories at the technical colleges.

Whereas Stern said that people from other professions should only receive additional psychology training alongside the regular students, this aspect had pride of place for Moede, whose Charlottenburg Institute lived on courses for engineers from industrial firms. Just as Moede saw psychology as the science of the possible, so he favored directing training entirely toward industrial efficiency. In 1935 he presented the ministry with a "Memorandum on the State and Reorganization of Psychology and Psychotechnics in German Higher Education" (REM/821, f. 1–30). On the one hand he complained that many psychotechnical practitioners in industry and administration had not been academically trained and, on the other hand, that university teaching in psychology did not meet their needs. The ministry must act to clarify what knowledge practitioners needed to be taught. One could not expect the professors to come to any agreement on the matter. The ministry should create a new subject of "psychology and psychotechnics or psychology and its applications" (REM/821, f. 23) and introduce a state examination. It is interesting to see how Moede wished to structure this new university subject. He rejected majoring in psychology, which Giese had always demanded, because psychologists in industry, commerce, and administration had always had to operate outside their professional field. Instead, Moede suggested a new profession of work and performance engineer – someone who studied engineering, majoring in psychotechnics and labor technology. The new subject should also be open for other professionals, such as officers, lawyers,

8 As Walter Moede wrote in 1941, there were negotiations with von Rottenburg from the ministry about a draft for an internal examination of the Verband (REM/821, f. 150). This could not be traced in ministerial or institute records, but a note was found which indicates that Moede was going it alone in an attempt to secure psychotechnical training at his institute. This seems interesting in the context of the differences on training. For 5 April 1932 the ministry register shows that a file was consulted for the "introduction of academic qualification for psychotechnicians to conduct psychotechnical aptitude tests at the Institute for Industrial Psychotechnics at the Technical College Charlottenburg (Memorandum of Prof. Moede)" (REM/2289, f. 69). C. A. Roos reports that in 1931 there was an unsuccessful initiative by Wilhelm Hische, who invited four or five people, including Roos, to bring about standard appointment requirements for psychologists at labor offices (Z, f. 163).

economists, doctors, and teachers. This would have led to approving the sort of training Moede operated in Charlottenburg.

Moede's recommendation was doomed to failure. Industry was already appointing diploma engineers who were specialists in psychotechnics or labor technology or was having engineers trained in the new methods; no new examination system was needed. The recommendation did not meet the demand for a specific professional group with specific knowledge. At the same time it did not meet the interests of the university professors who could not be won as allies. It took the expansion of Wehrmacht Psychology to create a different situation and create the demand for fully trained psychologists. There a theoretically and methodologically broader conception of psychology found practical application, which was more interesting for the university psychologists at the time. This had favorable effects on the subsequent coalition between psychology in the universities and Wehrmacht Psychology.[9]

The assessor examination of Wehrmacht Psychology as a model for the DPO

The regulations governing civil servants called for a state examination for psychologists in the Wehrmacht (see Chapter 4). The career requirements of Wehrmacht Psychology (1 April 1937) specified that applicants for the three-year preparatory service have a philosophical doctorate with psychology as the major subject. After three years they would sit before a Wehrmacht psychologist examination to establish their "suitability as a tenured civil service official of Wehrmacht Psychology."[10]

The structure and content of this assessor examination already anticipated parts of the DPO, and Simoneit was later to refer to it as a forerunner (1972, p. 76). The examination was to consist of a written dissertation, a practical psychological investigation, and an oral examination. First a dissertation topic "from psychological studies of symptoms, methods or typology" (III, Para. 13) was set, which was to be written up in six weeks – formally the same period as for lawyers. This hurdle was followed by a psychological evaluation of three officer candidates as part of a normal selection session at a Wehrmacht Psychology test station. On the third selection day, the trainee had to evaluate written intelligence tests and formulate an assessment in a six-hour sitting. If successful, the trainee was subjected to an oral examination. This was to cover the assessment, general methods of writing

9 Noticing the change in climate, Moede altered his suggestions. At the sixteenth congress of the DGfPs in 1938 he called for a uniform introduction to psychology in the universities and of a profession of practical psychologist. But he named only industrial psychology as a major field of activity (Moede 1939b).
10 III, Para. 10, Vorschriften für die Ergänzung der Beamten des höheren Dienstes der Wehrmachtpsychologie, 1 April 1937, UAT 148.

reports, general psychology and its history, practical characterology, psychological-characterological typology, psychological theories of symptoms, and psychological diagnostics, especially theories of expression. Questions could also be asked on philosophy and weltanschauung, the Wehrmacht, and regulations affecting civil servants (III, Para. 16). According to Mierke (1942 p. 33) the navy psychologist examination covered slightly different areas: the assessment, general psychology, typology and methodology, psychology of expression, practical characterology, psychology of functions, and race and kinship studies. Race and kinship studies were still not examination topics in the army in 1942 (Simoneit 1942a, p. 11).

There was an examining board that consisted in 1940 of the head of army personnel, Lieutenant General Bodewin Keitel, the inspector for aptitude testing, Lieutenant General Hans von Voss, Max Simoneit and Rudolf H. Walther as representatives of Army Psychology, and Philipp Lersch as a university professor (Simoneit 1940b, p. 67). The examination board for Navy Psychology also consisted of two leading naval psychologists, two leading military officers, and a university professor (Mierke 1942, p. 32).

When the first examination had been held in 1940, Simoneit wrote that this deserved the attention of all of psychology. It had "received state recognition such as is necessary in Germany at the conclusion of the development of an academic professional group to a standing as senior servants of the Reich (*Reichsbeamte*)." Thus, due to the "specialist services of psychology for the state," formal confirmation had been given of the "conclusion of a stage in the history of psychology" (Simoneit 1940b, p. 67). One could now proceed with renewed energy to transform university teaching in the interests of Wehrmacht Psychology.

These interests included a closer link to practical activities and a standardization of teaching at the various institutes. It is therefore not surprising that it was Simoneit who was a driving force in working out the DPO. But a draft was not produced until the interests of Wehrmacht Psychology had come to coincide with those of academic psychologists, and the Executive Committee of the DGfPs took upon itself the task of formulating it.

Quacks and monopolies

The Executive Committee of the DGfPs became active in February 1940, according to Oswald Kroh (1941, p. 13), who had taken over the duties of DGfPs chairman following the death of Erich Jaensch (12 January 1940). The change of policy may be related to this changeover. At the sixteenth congress in 1938 Jaensch (1938c) had still favored an ideological "study of the soul" (*Seelenkunde*). Now two practically oriented professors, Lersch and Kroh, were in the committee of three heading the DGfPs, alongside Sander. The committee no longer pursued merely the policies of an academic discipline but those of a profession. It was active in the professionalization of the teaching of psychologists and the organization of professional activity, concluding, for example, an agreement with the NSV on the work of psychologists in educational counseling (see Chapter 8). They obviously felt that the time had come to use the success of Wehrmacht Psychology to improve the situation of psychology at the universities.

Academic psychologists wanted to strengthen the entire field of psychology, both discipline and profession. An important and effective way of doing this was to acquire a professional monopoly for academically trained psychologists by having formally regulated academic training declared a precondition for acceptance to certain occupations. This would have the effect of excluding others from such activities. The criticism of laymen and quacks damaging the image of psychology had been heard repeatedly through the twenties and thirties, but there had been no course of legal redress.[11] Such people could be charged with unfair competition only if they made obviously false claims or promises in public (Weber 1929). Kroh felt that a state certificate would effectively distinguish academic psychologists from "pseudopsychologists" as well as from those who "in the past have had some quick training to carry out certain methods of investigation for industry" (1938, p. 11). Kroh never tired of emphasizing that the DPO had excluded the quacks and the dilettantes.

Psychologists had adopted the strategy of the physicians in the nineteenth century, which had been based on their struggle against quackery (Clever 1980). Physicians had made important progress in the Third Reich with the establishment of the Association of General Practitioners (2 August 1933) and the Reich Physicians' Chamber (Reichsärtzekammer) (13 December 1935) as well as the Health Practitioners Law in 1939 (see Chapter 4). As in the case of doctors, psychologists were probably not motivated by a real fear of quacks. It was a convenient tactic to lend credibility to their demands. This is also the view of C. A. Roos, who belonged to the commission responsible for drawing up the DPO (Z, f. 164). This draft was thus the result of a combination of interests. Wehrmacht Psychology was interested in a state examination; the academics were interested in strengthening their discipline, and all were interested in state approval of a professional monopoly.

The commission to draw up the DPO

The commission that presented the ministry with the draft of certification regulations met for the first time in February 1940 at the German Institute for Psychological Research and Psychotherapy at the invitation of its leader, Matthias Heinrich Göring. Roos, the only survivor at the time of the interviews, told of his impression that Göring did little more than act as host (Z, f. 164). The commission represented the various groups interested to a greater or lesser degree in a regulation concerning the training of psychol-

---

11 Giese (1922a, p. 49); Stern (1931, p. 78); cf. Hinrichs (1981, p. 247).

ogists. The psychotherapists were represented by Göring and Gustav R. Heyer; they were interested in the full professionalization of the "attending psychologists" (see Chapter 4); present for the university psychologists were the acting Executive Committee members of the DGfPs, Kroh, Sander, and Lersch; for Wehrmacht Psychology, General von Voss and Simoneit; from the Ministry of Labor, Walter Stets and Mr. Krechel; and from Labor Front Albert Bremhorst, chief of Amt Bub, and Roos, as well as Arthur Hoffmann from a teacher training institution.[12]

The influence of the various groups in the commission varied with their interest in the DPO. As chairman of DGfPs Kroh also chaired the commission and led the negotiations with the ministry. Roos names Simoneit as the other driving force in the commission. The two had developed the outline of the examination regulations, whereas the contributions by labor psychologists Roos and Stets had next to no influence. Hoffmann cannot have played an important role since Roos could no longer remember his participation. Roos did recall two arguments for professionalization from the commission proceedings. Simoneit openly expressed his concern to regulate the careers of civil servants in Wehrmacht Psychology, and there was the opinion that "with the introduction of a course the academic standing of psychology would also be strengthened and the subject would be strengthened at the university" (Z, f. 165). There are, however, signs that there was considerable tension between Wehrmacht psychologists and academic psychologists, above all between Simoneit and Kroh. The academic psychologists seemed to be afraid of complete domination by those from the Wehrmacht. In a letter to the author (18 November 1981) the senior civil servant at the ministry, Harmjanz, wrote that Simoneit had initiated the examination regulations but that the academic psychologists had taken the matter up as soon as they got wind of it. First Kroh and then the two other committee members, Sander and Lersch, had come to him "because all three dared not undertake anything openly against Mr. S[imoneit]." However, since it was in the interests of all the participants, everything was done in the commission to bring the common cause to a conclusion.[13]

12 This list is from Heyer in "Jahresbericht 1940....," (1942), p. 11. It is doubtful whether Bremhorst took part. He is not named by Roos. When I spoke with Bremhorst I did not know about the list, but he did not raise this point himself. Roos also differs from Heyer in not naming Krechel or Hoffmann, but including the university psychologist Tumlirz. In a letter to the author dated 18 January 1981, the senior civil servant Harmjanz denied that the latter took part. Heyer's list is given preference here because it was written at the time, whereas Roos was recalling events after forty years. This does not reduce the great value of his comments, especially regarding the atmosphere and the constellation of forces in the commission (Z, ff. 163ff.).
13 According to Roos there was "a common wish to make use of the opportunity and bring the matter to a good conclusion. Thus one can assume the formation of a consensus in the commission: they did not want it to come to a controversy at all. Right from the start, therefore, it was approached very pragmatically" (Z, f. 164).

On 24 September 1940 Kroh submitted a draft of the DPO to the ministry on behalf of the DGfPs. In the last sentence of the accompanying letter, which he omitted in an otherwise exact repetition of the text on 1 November 1940 (REM/821, ff. 107–8), he named the partners that the psychologists claimed for the presentation of their draft. The DGfPs was acting in agreement with the Inspectorate for Aptitude Testing in the Army High Command, the Amt Bub of the Labor Front, the Reich Office for Vocational Counseling and Labor Supply, and the German Institute for Psychological Research and Psychotherapy (f. 89). A further interesting aspect of this text is that it contained in summary the arguments with which it hoped to convince the ministry of the professionalism of psychology. The distillation of the six points reads like a list of essentials for professionalization:

1. German psychology is in the middle of a positive development; previous criticism directed against it has been refuted; various fields such as Wehrmacht, politics, propaganda, business and so on, demand psychologically effective methods and thinking.
2. For selection, psychologically well-trained personalities are required everywhere; the Wehrmacht especially makes use of them.
3. The situation of the *Volk* demands the economical use of human resources and the means of production; psychology has the scientific means at its disposal to meet the needs for selection and guidance of the work force.
4. There is a shortage of psychologists on the labor market.
5. Too many dilettantes are active in the field of psychology as a result.
6. The demand has to be met by well-trained efficient recruits.

This meant it was necessary among other things to introduce "an examination clearly raising the trained psychologists above the dilettants which pays more consideration to the needs of practical psychology than hitherto usual in the doctorate. The examination must have the character of a state examination since most psychologists find application in state and Wehrmacht posts" (f. 88).

The draft regulations and Moede's alternative

The draft passed onto the ministry in 1940 laid down a basic structure of psychology teaching and a subdivision of the subject the effects of which are still felt in Germany. It envisaged a diploma degree with an intermediate and a final examination, a written intermediate dissertation, two written examinations for the finals, and three six-week periods of practical training during studies.[14] The very first paragraph states that the examination should not judge scientific qualification but suitability for a professional occupation, corresponding to the level of training in scientific fundamentals, as well as

14 "Entwurf. Diplomprüfungsordnung für Studierende der Psychologie"; REM/821, ff. 92–9.

the aptitude and personality of the applicant (I, Para. 1).[15] It was in line with the emphasis on practical diagnosis, that the written examination for the finals should test the examinee's ability "to develop a detailed notion of character and performance on the basis of the range of material provided, using scientifically proven means, and to provide an evaluation assessment," and furthermore test the ability "to provide sufficient information [about] the tasks and methods of applied psychology."[16] These written evaluation examinations were modeled on the aptitude evaluations in the Wehrmacht Assessor Examination. The list of places where it was possible to do the practical training, which was revised several times, provides an idea of the areas that were expected to employ psychologists:

Children's homes, young people's centers of the NSV, kindergartens, schools, and colleges
Special schools, as well as welfare and reform institutions
Vocational counseling centers of labor offices
Apprentice training, deployment and evaluation posts, and personnel offices in industry and commerce, psychological institutions, and evaluation posts of the Labor Front
Courts, prisons, and youth correction camps
Psychotherapeutic counseling centers
Psychiatric clinics
Race-hygiene counseling centers
Educational counseling centers of the NSV
Population policy centers (III, Para. 1, 2, c)

In 1944 some local offices of the counseling service of the Studentenwerk were added (*DWEV*, 1944, 10, pp. 263–4). Proof of attendance at inspection tours of these institutions was required; aptitude testing posts in the Wehrmacht could also be visited, although they had not been mentioned for practical studies.

The reduction in the significance attached to scientific qualification in the DPO compared with the doctorate was expressed by the fact that the only independent scientific work required was an eight-week dissertation at the intermediate stage; no dissertation was required for the finals. This may have been a reaction to the relatively short course length of six semesters enforced by the war. For the intermediate examination, after four semesters, the commission suggested four psychological and two additional subjects: General psychology, developmental psychology, characterology and hereditary psychology, and the psychology of expression, as well as philosophy and weltanschauung and biomedical auxiliary sciences. The last covered

15 Cited here and later from the examination regulations published in the gazette of the RMWEV, since these recommendations were adopted ("Prüfungsordnung...," 1941).
16 *Durchführungsbestimmungen*, III. 2.

biology, physiology, and general psychopathology. This was extended for the degree to include "medical psychology." The finals were in psychological diagnostics, applied psychology, pedagogical psychology and psychagogics, as well as cultural and ethnic psychology.

Moede presented his own recommendation to the ministry, which was oriented more toward industrial and business psychology. Writing in his capacity as secretary of the virtually defunct VdpP, Moede told the ministry (3 March 1941) that the draft regulations appeared to be intended for teachers of psychology, in other words, they were oriented too much toward science and not enough toward occupational activities (REM/821, f. 150). In contrast to Kroh, Moede felt that the examination plan was "too one-sided toward characterology and the psychology of expression." As much attention should be paid to the "experimental and exact tendency" (p. 163). On 22 March the VdpP conferred on the DPO with representatives of the ministries of Labor, Trade, Aviation, and Transport, the Reich Chamber of Commerce, the Reich Group Industry, the Reich Institute for Vocational Training, the Military Doctors' Academy, the NSD, and the labor offices of Brandenburg, Lower Saxony, and Berlin. According to Moede, Dr. Krechel represented the Ministry of Labor; he was on Heyer's list as belonging to the DPO commission. Moede was able to gather an impressive number of representatives whose authority he used to back up his proposal, which he presented in the form of a report to the Science Ministry.

The intermediate examination was to be as follows:

1. Psychology: (a) general and special psychology, including experimental psychology, personality and social psychology; (b) developmental psychology: hereditary psychology and psychology of life stages; (c) characterization with special emphasis on theories of performance and expression
2. Natural sciences: statistics, physics, anatomy, physiology, psychiatry, and biology
3. Human sciences: ethics, philosophy, and adult education
4. Industry and business studies: occupational studies, management and organization, economics, and industrial law (all at basic level)

The finals were to cover two main subjects, "*Kennzeichnungslehre*" (symptomatology and diagnostics) and "practical psychology" (aptitude studies, basic labor studies, accidents, as well as marketing, sales, and advertising) and three further options from the course (REM/821, f. 169). This was against the interests of Wehrmacht and academic psychology and favored the interests of psychology in industry and business. Characterology, ethnic psychology, and pedagogical psychology were not listed as independent subjects; the psychology of expression was put together with the study of performance under a new heading of "characterization." Applied psychology was limited under the concept of practical psychology to the concerns of industry and business. As in Giese's plan in 1922 natural sciences that had not found a place in the commission draft (statistics, physics, anatomy) were included. Neither the Executive Committee of DGfPs nor the ministry supported this proposal. Kroh carefully guarded the representation of the existing sections of the subject, probably not least in order to unite the various practical and theoretical interests in the cause of the DPO. The ministry adopted the suggestions of the commission in all decisive points. A number of alterations were made to the formal legal clauses, of which one is perhaps of particular interest. The commission had recommended requiring a declaration from students registering for the intermediate examination that they did not have "non-Aryan" parents or grandparents, particularly of "Jewish blood" or "Jewish confession" (REM/821, f. 93). The ministry dropped the clause.

The logic of the examination subjects

Kroh had written to Moede that the DPO must consider all approaches and tasks and form "an ordered summary of general minimum demands," doing justice "to the present-day level of psychological thought and work" (REM/821, f. 158). Was this really the case, or was everything existing in psychology at the time just sorted together under certain headings? Kluth wrote that the basis of a discipline is formed by "a canon of common, teachable theoretical and material propositions" (1968, p. 680). This is not to say that these propositions have to be valid, only that they must be generally accepted. But even this does not seem to me to be necessary for a discipline that will be formally confirmed as a major field by the establishment of examination regulations. For the DPO in psychology there was instead an agreement about the subdivisions of the subject, partly on the basis of systematic considerations, partly on the basis of fields of professional activity, and partly as a result of historical developments special to psychology. These categories did not reflect the dominance of research in general psychology, especially sensory psychology and theoretical research, nor the role that typology and race psychology had played in the thirties. The accent had been shifted to certain areas of knowledge and a certain understanding of areas of psychology that seemed to be useful for the practical work.

This claim may become more plausible if we look at the earlier development of subfields of psychology that can be seen in the structuring of bibliographies or in introductory texts. In the twenties there was little homogeneity. August Messer (1927) listed various approaches in psychology, including characterology, social psychology, mass psychology, cultural psychology, psychotechnics, developmental psychology, psychoanalysis, individual psychology, Couéism, parapsychology, and others. For Hans Henning (1925) areas of applied psychology were developmental, child, animal, and pedagogical psychologies, among others. A textbook first published in 1912 by Theodor Elsenhans

(1939) was structured in terms of the functions of perception, thought, emotions, and will and separated attention, memory, work, and consciousness as biological functions. Following Fritz Giese's plan to revise this work, Friedrich Dorsch and Hans Gruhle added chapters to the new edition in 1939 on "development of the mental," "personal totality" (theories of type, characterology, gender typology), and "mental community" (mass psychology, race, *Volk*, kinship, psychology of the Germans). There is no sign of a developing consensus on subfields that could have been transferred to the examination subjects in any of the textbooks. The situation was no different with the structure of the annual reports on psychological literature in the *Zeitschrift für Psychologie*. In the twenties the various areas of psychological functions, such as perception, emotion, attention, and motor functions, occupied the most space. In 1931 three applied areas appeared for the first time: psychotechnics, pedagogy, and legal psychology. In 1933 the psychology of education (*Erziehung*) was added; in 1934 characterology and the psychology of expressions. From 1935 until the DPO the literature report had three main headings: general psychology, developmental psychology, and applied psychology, each with a number of subdivisions. Characterology was among thirteen fields of general psychology, along with theories of types, expression studies, sensory perception, motor activity, emotion–drive–will, and voice–speech. Applied psychology included questions of military, business, educational, and medical psychology.

Neither these headings nor the subjects of the DPO reflected a consensus on the inner structure of psychology. It is possible to discern some logic in the structure of subjects for the intermediate examination since general, developmental, and characterological psychology all deal with mental reality in terms of its general laws, its development over a lifetime, and its individual peculiarities. But the fourth subject, the psychology of expressions, can no more be integrated into such a logic than hereditary psychology can be linked to characterology. The extent to which the subject names were used to bring in topical areas of interest becomes clearer on examining the procedural regulations. For the examination on characterology and hereditary psychology, the examinee had to show "that he has been thoroughly instructed on the basic laws of the structure of the character as well as the basic constitutional and racial forms of personality, with attention paid to connections with hereditary psychology and race psychology" (Prüfungsordnung ... 1941, II, para. 2, 1, 4). First of all it is noticeable that in the subject of examination characterology (*Charakterkunde*) the basic laws of the structure of character (*Charakteraufbau*) are given pride of place. Lersch (1938) had included this as one of six areas of characterology, the others for him being basic forms (types), conditions of the origin and development of character, as well as diagnostics and case studies. It was in line with the emphasis given to it in professional activities that diagnostics was an independent subject, in contrast to Lersch's theoretical structure, and that characterology concentrated on the area Lersch had termed the most significant area for diagnostics, namely, character structure. Thus, the subject was specified in accordance with Lersch's book (see Chapter 3). The reasons for specifically mentioning aspects of race psychology are clear; constitutional and hereditary psychology would possibly have been included even without political motives.

In the bibliographies animal psychology was a separate section. In the procedural regulations it was not to be found under biology but as a part of developmental psychology, which required not only knowledge of the laws of the developmental stages of life (particularly childhood and puberty), but also comparative psychology of "primitives" and animals. This corresponded to the developmental theory of Heinz Werner and the psychology of Krueger, who believed that in children, primitives, animals, and the mentally ill diffuse holistic emotions dominated – the original mental state of man. Developmental psychology thus distinguished between the developed, occidental, civilized adult psyche as the subject of general and characterological psychology and such atavistic forms.

The structuring of subjects for the diploma finals shows how systematic views coincided with those of the institutional fields of professional activity. When Stern first coined the term "*angewandte*" (applied) psvchology in 1903 he meant the "science of the psychological facts that come into question for practical applications" (Dorsch 1963, p. 9), whether the application was in pedagogical, business, or military areas. The DPO now separated "pedagogical psychology" and "applied psychology" as independent subjects and defined in the procedural regulations concerning the examination on applied psychology that an account was to be given "of the various areas of application of psychology, their methodological development and their results to date." Pedagogical psychology was to cover "folk education."

If, following Kroh, applied psychology was to include all those areas in which the tasks of psychology are taken from "not essentially psychological areas" (Kroh 1938, p. 4), then this ought to include pedagogical psychology, the application of psychology in education. Why, then, was this made a subject in its own right? I believe the reason is to be found in the institutional application of psychology as well as in the special position it occupied in the universities. Before the DPO, pedagogical psychology was a field concerned with the development of practically relevant knowledge, but this was not applied professionally by psychologists themselves. This changed only during, and above all after, the Second World War with the introduction of educational counseling centers and with the creation of school psychologists. Thus, when applied psychology was taken to mean virtually labor and military psychology, this corresponded to the areas of professional activity of psychologists – just as military, transport, occupation, industrial, and forensic psychologies did not arise from the structure of the science at the time but from the social-institutional areas of activity of psychology. A second reason for the separation of pedagogical psychology is probably that it was jointly developed by two disciplines (pedagogy and psychology) that were establishing themselves, so that experts on both sides viewed it as an independent research area between and for them both. This hypothesis seems reasonable in view of the long-standing discussion among pedagogues and psychologists about the independence of pedagogical psychology as a field of research between the two disciplines, which had begun in 1917 when Aloys Fischer referred to the task of pedagogical psychology as the "scientific investigation of the psychological side of education" (Hillebrand 1959, pp. 54–5). The leading light for the development of the DPO, Kroh, was both a pedagogue and a psychologist, and at least after the DPO he regarded pedagogical psychology as a discipline sui generis. In 1938 he had still assigned it to the applied areas of psychology (Kroh 1938, p. 5). In 1944 he was of the opinion that it distinguished itself from general psychology by its "pedagogical point of view," nor could it be regarded as an application of psychology (Kroh 1944a, p. 163). It therefore seems quite plausible that pedagogical psychology was taken out of applied psychology "at the instigation of Kroh" (Däumling 1967, p. 256). The reason Kroh himself gave, that this was because all psychological counseling and examination was really done to help (so that practical psychology was always pedagogical), sounds in contrast like an attempted justification, although it may have a modern ring to some ears (Kroh 1943, p. 36).

The inconsistent definition of applied psychology, arising from the field's professional and academic development, left its mark. The examination subject acquired the status of a scientific subdiscipline, even though it was only a heading. An older standard textbook on applied psychology (Dorsch 1963) covers the psychology of vocation, industry, transport, the military, and sport but not questions of pedagogical or clinical psychology. "Applied" and "pedagogical" also advanced to definitions of chairs. Finally, in the seventies clinical psychology dominated the concept of applied psychology in the examinations, where the regulations still contained the term. Occupational and business psychology, which had hitherto lived royally under the roof of applied psychology, were then relegated to a peripheral existence (on the impact of the headings on the discipline, see also Chapter 7).

The mention of clinical psychology of course raises the question of why this was not to be found as a separate subject in the examination regulations. Or, rather, why had it not yet advanced to a distinct field of psychology in 1941? In part the reason may have been that psychology had not yet developed methods specifically for clinical problems capable of competing with methods in other sciences, or outside science. It was only with the development of behavioral and client-centered therapies that psychologists were able to establish a claim for their own professional clinical work. The fact that psychologists in Germany were not active in counseling in the health system earlier (in contrast to the United States) is due to the near monopoly of physicians, who tried to keep psychologists out of their sector. In 1942, through pressure from the medical fraternity, general psychopathology was removed from the list of biomedical auxiliary sciences for the intermediate diploma examination (see Chapter 7 and Dorsch 1971, p. 36).

It is not easy to decide what concept of minimum requirements for professional qualification is actually reflected in the DPO of 1941. Kroh (1943, p. 34) referred to the following subjects as relevant to practical work: applied psychology, psychodiagnostics, the psychology of expression, and characterology. From its selection he appears to have been thinking of the practical work of characterological officer selection. For the new occupation of educational counselor, developmental and pedagogical psychology were certainly of greater importance. For the selection of military specialists, a knowledge of quantitative methods of sensory psychology was needed. Both occupational activities and research in psychology addressed a whole range of widely differing problems, so that different parts of the examination regulations seemed important for different areas of professional activity. However, the examination regulations were an expression of the fact that the qualification of future Wehrmacht psychologists was given a higher priority than the requirements of psychologists for labor administration and business. The emphasis on characterology and psychology of expression in the intermediate examimation corresponded to the interests of Wehrmacht Psychology and to its assessor examination. As Moede's suggestions demonstrated, an orientation toward industrial psychology would have favored quantitative methods or the psychology of work and performance, with physics as an auxiliary science. Even though Kroh was of the opinion that internal agreement on the basics of the subject had made the examination regulations possible, it seems to me that the choice of subjects and their interpretation in that situation also owed much to the external demands on psychology.

The DPO had not defined a canon of generally accepted knowledge but had defined a canon of subjects that were to become teaching areas. The internal agreement was an agreement on the structuring of fields of psychology. A house had been built that looked fine from the outside, no matter what went on within. In any event, whoever wished to enter the house in the future would have to come to terms with the architecture. It would take more than the wish for a better view or more elbow room to justify building alterations. Even given the cooperation of the architects, there was the need to get ministerial planning permission for changes. However, since the bureaucrats did not know the internal conditions, they tended either to leave well enough alone or to come up with new regulations that were along the right lines but that did not suit conditions in the house.

The question of professional ethics and
the selection of students

In the sociological literature the establishment of professional ethics and institutional means of controlling responsible professional activity are often named as basic aspects of professionalization (see Chapter 1). Napoli (1981) has described how important the discussion about professional ethics was for the professionalization of psychology in the United States. In Germany such ideas are to be found only in isolated cases such as the reaction of Rupp (1930) to a suggestion by Moede regarding the removal of undesirable personnel from firms (see Chapter 3, note 6). But industry was not the main area of professional activity.

In Germany, psychology did not develop as a free profession but as a civil service profession. Thus, the question of ethics was left completely to public institutions, and responsibility was felt only for the choice of appropriate recruits. It is therefore not at all surprising that in the DPO we do not come across the question of the ethics of psychology but rather that of the "ethical psychologist." The examination should test not only the ability of the students but also their character. Thus, the problem of setting up moral or ethical standards for the members of the profession was given an individualistic twist and became a problem of selecting morally acceptable psychologists. Selection according to character replaced the socialization of prospective professionals in the norms and values of their group. According to the regulations, the examination should ensure "that the applicant as a personality fulfills the demands that his future profession will place on him in terms of human values, sense of responsibility, and readiness to act for the state and the folk community" (I, 1, 2). After each examination, an impression of the examinee's personality was to be noted in the records (II, 2, 7f). Nevertheless, there was no clause stating that a negative impression of his or her personality could lead to an examinee's not passing. It only offered the option of refusing acceptance when "doubts existed as to the character and political attitudes of the applicant" (II, 1, 4b). All these clauses had been included in the commission's draft.

In reply to an engineer's criticism of the personal suitability of psychologists (Schoor 1941), Kroh commented that the concentration on practical tasks as a result of the new DPO was enough to force students to abandon an autistic approach self-oriented to studying psychology as a means of dealing with their own problems. Above all – and this was a new development over the DPO – applicants would in the future be sorted on admittance or during the first semester at the latest, according to "personality values and bearing"; this was the result of an agreement reached by the psychological institutes (Kroh 1942, p. 7). This practice was established in the examination regulations only after the war. The Bavarian regulations of 1948, for example, determined that students had to pass an entrance examination held by the director of the local institute. He was entitled to advise applicants against studying. If they still wanted to study, they could do so but would have to reckon with not being accepted for the diploma examination. Arnold (1948, pp. 20, 29) felt that characterological aptitude tests were urgently necessary to keep out cranks, outsiders, screwballs, know-it-alls, or illusionists clinging to "weird ideologies" or "social phantoms." The Professional Association of German Psychologists would not want to do without such aptitude tests.

Ministerial consultations and the approval of the DPO

In 1931 the Prussian Ministry did not react to the suggestion to introduce an examination. A decade later only three-quarters of a year passed between the submission of the draft by the DGfPs and the promulgation of the DPO. The decisive reason for rapid approval must have been the growing demand for psychologists and the need to find a legally watertight arrangement for the civil service career ladder in the Wehrmacht. An additional reason may have been that the new draft had wide support and did not reflect the interests of just one group, as had been the case with Moede's memorandum in 1935, and probably with the draft of 1931 as well, although the latter no longer exists. There are signs that the ministerial official responsible, Heinrich Harmjanz, did not take kindly to Wehrmacht Psychology; nevertheless, he supported the cause of the DPO. Indeed, in *Zeitschrift für Psychologie* Kroh expressed his "unreserved gratitude for the considerateness he had shown" (1941, p. 14).

After the draft had gone around the ministry and met with approval, Harmjanz put it to the minister on 9 January 1941, with a summary of the reasons such regulations were necessary. The (rather puzzling) antipathy toward Wehrmacht Psychology is apparent:

> The acquisition of the academic degree of "Diploma Psychologist" has been prepared here in order to put a stop to the numerous activities of quacks in the field of psychology. The underlying cause was the waxing influence of Wehrmacht Psychology. Here, in order to satisfy the need for "military psychologists" there had been a flood of appointments of "psychologists," extraordinarily well funded by the Wehrmacht and wearing a uniform who were gravely endangering the scientific development of psychology as a whole. The Wehrmacht's "Test Stations for Aptitude Testing," which were shooting up everywhere, and the Wehrmacht psychologists manning them, 80 percent of whom had no or minimal previous qualifications, had led to such a shift in the image of German Psychology that it looked as if in Germany Psychology = Wehrmacht Psychology. This dire situation had to be countered. The only remedy was an examination order issued by the Ministry to regulate the academic career of a "diploma

psychologist" for the first time. It would then be forbidden to pursue "Psychology" in order to earn money or to be employed as a psychologist without being in the possession of a certificate as "diploma psychologist".... The introduction of this academic degree of "diploma psychologist" met above all with intense interest from practical spheres, namely: NSV, Labor Services, Labor Front, youth care, trade school offices, special care institutes, and juvenile court judges. (REM/821, ff. 114–15)

The omission of Wehrmacht Psychology from this list corresponds neither to their interest nor to the role of Simoneit in preparing the DPO. Perhaps Harmjanz sided with some of the leading academics in their differences with the leader of the largest professional group. As former Wehrmacht psychologists reported, Simoneit let the university professors feel that when they were doing military service, they had to subordinate themselves to military aims. After the war – as a sort of retaliation – he was not awarded a professorship. However, it is doubtful whether this is the full explanation for the intraministerial memorandum. In his supplication to the ministry, Kroh had explicitly cited the agreement with the Inspectorate for Aptitude Testing. It is also questionable whether the opinion of Harmjanz was shared in the ministry or whether there were not others who wished to further precisely the interests of the Wehrmacht. Perhaps it is one of the quirks of history that the interests of Wehrmacht Psychology achieved domination because arguments were directed against it. Harmjanz's line of argument could indicate that the committee members of the DGfPs saw the new examination as a way of preventing Simoneit from employing so many nonpsychologists (particularly teachers) in Wehrmacht Psychology, and that they wished to produce a large number of formally qualified psychologists as soon as possible. Simoneit may, in turn, have pursued employment policies only because of the lack of suitable applicants, so that his interest in the DPO would have been just as great because it would enable him to take on trained psychologists without having to change his personnel policies right away. The common interest would then have been the point that Harmjanz emphasizes: replacing nonexperts with experts. The fact that Wehrmacht interests were also viewed sympathetically in the ministry is shown by the introduction to the DPO, which begins, "The growing demands placed on psychology by state, Wehrmacht, and business make it necessary to place the training of expert psychologists on a new footing" (*DWEV*, 1941, 7, p. 255). This was not just a phrase, since the introductions to the various examination regulations passed around at that time frequently named specific institutions.

Wehrmacht Psychology urged the ministry to adopt the draft. In a letter to Harmjanz dated 12 February 1941, General von Voss expressed his worry that the DPO had not yet been passed: "It is only to be hoped that the opponents of psychologists, who of course exist, have not attempted to intervene" (REM/821, f. 155). If von Voss suspected Nazi circles after the experience with the Amt Rosenberg, then he was mistaken. On 7 May 1941 Kroh rang Bechtold in the Brown House in Munich and was assured that at that time there were no objections against the DPO (f. 161). A ministry official, Heinrich Nipper, had already noted on 21 January 1941 that no problems had arisen with the deputy führer (f. 112), who nevertheless waited several months before writing to the ministry on 7 May 1941, making some remarks on the institutions at which practical studies could be absolved and on the formal details related to registering for examinations (f. 180). After the deputy führer approved some alterations on 29 May proposed by the DGfPs Executive Committee, Kroh felt that the "path to the introduction of the examination regulations" was cleared (f. 181). On 16 June 1941 the DPO was promulgated. It went into force retroactively from 1 April.

The diploma in psychology and the rationalization of university courses during the war

The reasons for the ministry's approval of the introduction of the examination regulations need not necessarily be sought in the development of psychology at the time. Perhaps it was not the reasons named by Harmjanz but a general rationalization of university courses during the war that tipped the balance for the edict on the DPO in 1941. There are various factors supporting this assumption. Beyerchen (1977, p. 174), for example, sees the efforts to introduce the diploma for physicists in connection with the alarm caused by the general lack of new scientists since 1938 (cf. Mehrtens 1979, p. 437). From September 1939 to the spring of 1941, the two semesters had been replaced by three terms to accelerate studies. In 1940 the senior government official in the RMWEV, Hans Huber, wrote in *Erziehung und Wissenschaft im Kriege* (Education and science in war) that curricula, course, and examination regulations for individual subjects had become more and more related to practical occupational requirements over the preceding years (1940, p. 17).

The development and promulgation of the DPO fell in a period when the ministry was vigorously pursuing a policy of orienting individual curricula to the relevant occupations. In November 1940 new plans had been passed for all engineering courses at the technical colleges, to come into effect on 1 April 1941, aimed at unifying courses and making them more effective (Nipper 1941). Ebert talks of examination regulations paring down the course requirements in general subjects, in tune with the wishes of large industrial firms (I B Ebert 1978, p. 12). At the same time examination regulations for diploma foresters (*Holzwirt*) and a diploma for geologists also came into force. On 1 November 1941 diplomas were introduced for geophysicists, meteorologists, and oceanographers, a year later for physicists and mathematicians. In the case of geologists the new course was to meet "not only the demands of the university, but also the needs of the Wehrmacht and the Reich Center for Soil Research" (*DWEV* 1941, 7, p. 41). The idea behind the diploma for foresters was to provide an opportunity of a regular course in higher education (p. 205). For geophysicists, meteorologists, and oceanographers, duties in the Wehrmacht, business, and transport were cited as reasons for the reorganization of studies (p. 413); for mathematicians and physicists it was also necessary to standardize teaching at technical colleges and universities. All these new examination regulations aimed to increase occupational orientation and tighten and standardize curricula. Curricula were also concentrated on the individual subject of the diploma. As the diploma replaced the doctorate, it was no longer necessary to study additional subjects.

The promulgation of the diploma regulations in psychology was possible because, in contrast, for example, to pedagogy, the discipline had managed to make itself useful to "state, Wehrmacht, and business." It is hard to decide the extent to which the precise timing of the decree was determined by the general rationalization of the curricula in addition to the requirements of Wehrmacht Psychology. In general it can be said that the interests of the university and Wehrmacht in a diploma coincided with the interests of the ministry in an occupational orientation and in rationalization of education. During the Second World War the technocratic logic of efficiency that the Nazis generally pursued replaced the old idea of a broadly based scientific education with that of occupational training – an example of Nazi "modernization."

Diploma and state examination

The career system for civil servants in the Wehrmacht really required a state examination, not a diploma. With the Assessor Examination, Wehrmacht Psychology had already created a sort of second state examination. But why was no regular first state examination introduced like the one for teachers or lawyers, and why was the diploma, formally only a university examination, nevertheless referred to as a state exam? A superficial answer would be that the matter was generally dealt with in this way. All the exam regulations named were for diplomas; all of them had landed on the desk of Franz Senger at the ministry.[17] But going further back we find a certain model for this.

The diploma examination as first state examination existed initially for engineers. In 1902 the state examination for construction leader (*Bauführer*) was replaced by the diploma for engineers, which had been introduced in 1899 (I B Ebert 1978). The general examination regulations for diploma engineers of 1941 explicitly state that the diploma examination is valid as a first state examination for those subject areas "in which a transfer to higher administrative service is possible" (Nipper 1941, p. 6). In this case, however, in contrast to psychology, the participation of government representatives in the examination was foreseen. The geology regulations also referred to the diploma examination as the first state examination; here six central examination offices were set up with one representative each from the RMWEV, the Reich Center for Soil Research, and the Military Geology Section of the Army High Command. They appointed the examination commissions in the universities (*DWEV* 1941, 7, pp. 41–6). Geophysics, meteorology, and oceanography had university examination commissions, although here the regulations also clearly stated that the diploma examination counted as the first state examination (p. 415). The DPO for psychology did not contain any such comment, but the new career regulations for Army Psychology referred to it as a state examination, and the synopsis of examination conditions issued by the RMWEV in 1943 classified the psychology diploma as "final (state) examination."[18] The diploma examination committees for psychology had only university representatives. Formally, they had to be nominated by the minister, but the local full professor was automatically chairman.

Nobody seems to have been interested in a different arrangement. Any other would have required a more comprehensive state system, either with a special license to practice like that granted to physicians, which would not have been appropriate in view of the fact that most psychologists were subsequently active as civil servants, or the organization of a second state training phase with a second state examination, all of which would have required new institutions and a new state examination apparatus. Thus, in contrast to the second state examination (which in Germany evaluates the practical work of the examinee in an institution like a school or a court), the university examination for the diploma remained in the final analysis an examination for a science and not for a profession, even though the DPO expressed a different intent.[19] The contradiction between profession and training remained, but the desired "state legitimation"[20] of the profession had been achieved.

18 *Zehnjahresstatistik* ..., 1943, p. 257. This also seems to have been the general impression created. In a newspaper article by Friedhilde Göppert on the Institute for Psychological Anthropology in Marburg, we read: "Our institute trains Diploma-Psychologists, a subject that has had a state examination since 1941. The establishment of a state exam is in itself a sign of the great importance that the professional psychologist will have in future, willing and destined to become the same as the teacher and doctor: supporter and helper" (*Oberhessische Zeitung* 18/19 December 1943, p. 5).
19 As Michaelis (1980, p. 41) rightly notes, according to the regulations of 1973 the academic "diploma" degree is "*in der Psychologie*" and not "*des Psychologen.*"
20 Kroh in a circular of the DGfPs, 15 January 1943 (UAT 148).

7

The Diploma Examination Regulations and their consequences

With the new diploma examination psychology became an independent teaching subject in the universities. For the first time psychologists appeared on the labor market as a professional group with state recognition. The DPO thus had consequences for the development of the discipline and the profession.

The effects of examination regulations on an academic discipline can be of two kinds. Above all they are apparent at the social level in the form of institutionalization at the universities. Psychology no longer had to justify its existence; it had legal backing for its complete autonomy. Much less direct and less tangible are the effects of curricula or new regulations at the cognitive level (Weingart 1976, p. 82). A system of knowledge may be a prerequisite for professionalization, but does this process then also regroup the existing knowledge systematically – in terms of the categories used in training? As far as the effects of the DPO on the development of the profession are concerned, our interest is in the revival of rivalry with other professional groups when psychologists were able to show formal qualifications (see Chapter 1). The resistance of physicians is instructive; they were able to force the exclusion of medical subjects from the intermediate examination.

A circular of the chairman of the DGfPs seldom displayed such high spirits as that of Christmas 1941. "With satisfaction" the society could look back on the war year of 1941, which brought with it the "obligatory basic course in psychology" and the "preconditions for the development of a united and recognized psychological profession." In addition the DGfPs could report "increased numbers of new members" (UAT 148). The military head of Wehrmacht Psychology, von Voss, and the head of the Army Personnel Office, General Bodewin Keitel, extended their thanks to the ministry for examination regulations of importance for the war.[1] The gratitude was

1 REM/821, letters from von Voss, n.d., f. 206; from Keitel, 12 December 1941, f. 282.

210

mingled with hope. In November von Voss and Simoneit toured the psychological institutes to gather information about their personnel, premises, and facilities. In a memorandum to the ministry dated 3 December 1941 the inspectorate complained of the inadequate facilities in the institutes. In Berlin, Freiburg, Göttingen, Münster, and Vienna psychology was not represented by a suitable professor (REM/821, ff. 248–9). The DGfPs was also interested in better facilities. An application was sent to the ministry to increase the number of psychological lectureships and to augment institute budgets; it was approved on both counts. The Wehrmacht had strengthened the "professional front" of psychology. Following the DPO Kroh also reckoned with a "considerable reinforcement of the scientific front of our subject."[2]

Examining boards and examination regulations

The new examination led to a number of measures for full academic professionalization, especially the establishment of examining boards for the intermediate examinations and the diploma finals. The occupant of the local chair for psychology was automatically chairman of both boards. When the DPO was decreed on 16 June 1941, the ministry asked for suggestions for nominations to the examining boards, and most professors answered before the beginning of the winter term.[3] The ministry obviously waited until all applications had been received before approving any. This led to complaints in 1942, especially from the University of Marburg and the Berlin Technical College that the boards had still not been approved, althogh the first students had registered for exams. On 4 May 1942 the applications were finally approved, with the exceptions of Greifswald and Hamburg, so that first intermediate examinations could take place. Heidelberg and Cologne had not sent in applications, so nineteen of twenty-three universities now had examinning boards (including Frankfurt and Rostock, although there were no chairs for psychology there). There were also boards in Graz, Innsbruck, Prague, and Vienna. At the same time examining boards were approved at the technical colleges in Berlin, Braunschweig, Danzig, Darmstadt, and Dresden. Heidelberg was included in October 1942, Posen in 1943, and Hamburg in February 1944.[4] Only Cologne, where the newly created professorship remained unoccupied (of which more later), Greifswald, and the Munich Technical College remained without examining boards.

The DPO had determined that for the finals examining board "persons should also be appointed under whose supervision the examinee had been introduced to practical psychology in the semesters between the intermediate

---

2 Circular Christmas, 1941, UAT 148.
3 R21/469, f. 1689 for the ministry's request. Account based on the detailed records in the file "*Prüfungsausschüsse Psychologie* Bd. 1, Okt. 1941," R21: Rep 76/912.
4 Information from the Archives of the University of Heidelberg, 27 August 1981, and the city of Hamburg, 21 April 1981.

exam and the finals" (Para. 3.3). The aim was to establish a closer link between academic studies and practice. But this was not realized immediately, since cooperation between institutes and practitioners scarcely existed. The directors of the psychological institutes discussed this problem in Munich on 16 October 1941. The opinion was that it would be possible to name those board members only "after they have proved themselves in the practical training of students."[5] The psychology professors reacted in different ways in their applications. Seventeen universities and technical colleges did not name practitioners. Danzig wrote that it would name someone later. Six universities asked for the postponement of such appointments; for example, Eckle from Breslau wrote that the aptitude and scientific competence of psychologists who trained psychology students needed to be established during a lengthy period of cooperation. In contrast, the universities of Bonn, Erlangen, Graz, Königsberg, Marburg, and Prague, and the technical colleges of Braunschweig and later Dresden named between one and four practitioners, in all two from school education, seven from vocational counseling, one engineer, one from the staff of the Labor Front, and six Wehrmacht psychologists.[6] At this time doctors were still appointed for the medical subjects in the intermediate examination. In the new regulations ten possible groups of institutions had been listed for the three six-week practical training sessions (see Chapter 6), but all the practitioners to be recommended came from the few areas that had long been familiar with the work of psychologists.

A further aspect of the transition from a scientific teaching subject to one preparing for a profession was that only diploma graduates were to be accepted for the doctorate in psychology. This went back to a general ministry regulation. On 23 December 1941 the ministry ordered that following the introduction of the diplomas in various fields the doctorate was no longer a final qualification, but was proof of the special scientific competence of the candidate. Acceptance for the doctorate examination would in the future require a diploma or a state teacher qualification. This also applied explicitly to the Ph.D. (REM/786, ff. 300ff.). The necessity of earning the profession-oriented diploma before taking the "scientific" examination led to so much dissension in the universities that the ministry was forced to retract parts of the decree. However, in contrast to other subjects, the diploma remained a prerequisite for the doctorate in psychology. Similar arrangements had already existed in other profession-oriented subjects, such as economics (since the early twenties) and chemistry (since 1 April 1939) (UAH Rep 21/6, 7). The peculiar situation of psychology as the only professionalized subject in the Philosophical Faculty – a faculty of nonprofessional scientific

5 Communication from Bollnow to the rector of Giessen University, 19 October 1941. For source see note 3.
6 Calculated from the applications in ministry files. For sources see note 3, for Heidelberg, note 4.

subjects – gave rise to the practice of adding a special clause to the faculty's doctorate regulations that in order to be accepted for a "promotion" in psychology it was necessary to have passed another exam, the diploma. When a student in Tübingen applied to take a doctorate in psychology, literary history, and art history in 1944, she was told by the institute that she would need a diploma in psychology and two further accredited semesters.[7]

The effects on academic appointments policy

In the preamble to the DPO the ministry had determined that psychology could be studied at those places of higher education at which psychology was represented by a "budgeted chair" and the other examination subjects were also appropriately represented. Since the universities that did not have such a chair also wanted to offer the examination, this led to a further strengthening of psychology's position in terms of chairs. Setting up a new chair no longer required efforts to prove the subject's scientific importance or practical use. A reference to the DPO was enough; it formed the central (though not the only) argument in such processes. It was now the duty of the universities to guarantee the training envisaged by the state. In Hamburg a new chair in psychology was set up, since William Stern's former chair had been lost. Freiburg, Cologne, and Münster received new professorships in the same year. The former chair of Wolfgang Köhler in Berlin was specified for psychology and reoccupied by Oswald Kroh. In Frankfurt efforts were made to revive Max Wertheimer's chair. In Erlangen the associate professorship was upgraded to a full professorship, but this was the conclusion of a long-term process, rather than a result of the DPO (see Chapter 2).

In Hamburg the educational scientist Gustav Deuchler had represented psychology since Stern's dismissal, in addition to the nonbudgeted professor, Georg E. Anschütz. Although in 1933 Deuchler was to be the replacement only "for the time being," nothing changed until 1941. On 12 November the dean wrote to the ministry that the faculty felt it necessary to reoccupy the chair for psychology due to the "importance of this subject" and because of the DPO. The rector supported the application with the single argument that meeting the requirements of the DPO made a budgeted representative of psychology necessary (SAH Ai 3/47). Both suggested the local head of the NSD, Anschütz. Despite approval by the Ministry of Science, the finance minister rejected the reestablishment of the chair, finally accepting the upgrading of Anschütz's budgeted teaching post to an associate professorship.[8]

In Frankfurt also the local party leadership took it upon itself to implement the ministerial decree. When the faculty and rector there applied to the

7 UAT 148, letter from Kohlmann, 15 December 1944.
8 Communication from the Hamburg City Archive, 8 November 1981. There was later a conflict at the university, obviously about Anschütz's regrouping as full professor. He was disliked as a Nazi official (see Giles 1978, p. 219).

ministry for the formation of an examining board for psychology under the chairmanship of Wolfgang Metzger, who held only a nonbudgeted associate professorship, the staff leader of the Nazi Party Gauleitung Hessen-Nassau intervened. In a letter to the curator of the university on 22 July 1941, he argued that it was necessary to start thinking about someone for the chair for psychology because this was required by the new diploma regulations. The curator could only reply that the chair was no longer included in the budget (Alt Ve 6 h). After Metzger left for Münster in 1942 the institute lay practically dormant. The teaching was kept going by the assistant, Edwin Rausch (see Pongratz et al. 1979, pp. 224ff.).

In Heidelberg an examining board could not be set up under the new regulations because psychology was represented solely by the *honorary* professor, Willy Hellpach. The rector and the minister of education in Baden tried to organize a budgeted professorship for him. In his application, the rector referred to the role of the University of Heidelberg in training new Wehrmacht psychologists, the use of psychology for vocational guidance, the needs of neighboring industrial areas, and the links between psychology and other disciplines at the university. Since Hellpach was a recognized authority in the field, it would be in the interests of the university "if Mr. Hellpach be placed, by the award of a budgeted professorship, in a position to organize psychological studies at the University of Heidelberg in accordance with the requirements of the new examination regulations" (Letter, 9 September 1941, R21: Rep 76/912). The application received the support of the State Ministry of Education (letter 25 November 1941 to RMWEV, R21: Rep 76/912). In the end they did not get the budgeted chair, but an examining board was set up on 30 October 1942 with Hellpach as chairman, and an institute was created in the winter semester of 1942–3.[9]

At the Dresden Technical College the Faculty for General Sciences, which had a chair for psychology and philosophy, wanted additional lecturers to spread the burden and thus to improve the opportunities for studying psychology. They sent in an application on 26 January 1943, and in a letter of 11 February the dean backed this up, pointing out the needs of industry for scientific psychologists. The local leader of the NSD supported the argument, adding that psychology was important for the training of technicians, who in the future would also increasingly be required to take on leadership tasks for other "tribes, races, and peoples" (R21: Rep 76/912). As in Frankfurt a party representative supported both the ministerial decree and the strengthening of psychology – further evidence against the hypothesis that psychology had to establish itself against the opposition of the Nazi Party in the Third Reich.

Psychology could become stronger only if new posts were set up or if it was given posts from other disciplines. This is what happened in Freiburg,

9 Information from the UA Heidelberg, 12 May 1981 and 27 August 1981.

Cologne, and Münster, although the chair in Cologne was not occupied. The Cologne Institute for Experimental Psychology was set up in 1938 under Professor Robert Heiss, who held a nonbudgeted chair. In order to examine for the diploma, the Curatorium set up a budgeted associate professorship in 1942, which it intended for Heiss. Matters dragged on, however, and as the faculty faced difficulties in providing for other subjects, it decided to use the still-unoccupied post for geography, since in the meantime Hitler had decreed that no new chairs could be set up for the duration of the war. Although the intended goal was not reached in this case, the new diploma regulations had set in motion activities to institutionalize psychology.[10]

In Münster and Freiburg, chairs were successfully occupied. There was no chair for psychology in Münster, although Richard Hellmuth Goldschmidt had set up a department for experimental psychology in the Philosophical Seminar in 1919. His right to teach was revoked in 1933 because he was a Jew. Psychology was taught only between 1935 and 1942 by Benno Kern as lecturer in applied psychology.[11] When Willy Kabitz, who held a budgeted chair for "philosophy of the human sciences with special emphasis on practical philosophy," reached retirement age in 1941, he applied to the RMWEV to accept his chair as a budgeted post for psychology in the sense of the DPO. The faculty recommended that the successor to Kabitz be appointed chairman of the examining board.[12] The university now did all it could to have a psychologist appointed to the post. On 4 November 1941 the rector wrote to the ministry that the university would like Kabitz's successor to chair the diploma examining board for psychology. He argued that pedagogy, which Kabitz had also represented, was no longer included in the examination regulations for teachers at higher schools (see Chapter 4), whereas the importance of psychology was growing. On 3 December 1941 he also added the needs for psychologists in the Wehrmacht, in business, and in employment offices. As a successor to Kabitz, only philosophers should be recommended whose work focused on psychology. With these arguments the university was also able to sway the ministry against some interests in the Nazi Party who were trying to get the pedagogue Erich Feldmann from Bonn appointed successor to Kabitz.[13]

10 For 1942 from the UA Cologne Zug 9–381, communication of the faculty, 12 March 1942. Information from the UA Cologne 14 April and 20 May 1981. For 1943: Rep 76/259, ff. 102ff. The university archives show that the professorship was approved but not finally included in the budget, whereas the ministerial files show that the chair existed, but in the end was used for other purposes.
11 On Goldschmidt, *Wer ist's*, 1935. On his cooperation with the Wehrmacht in Münster, see Chapter 4. Information on Kern from the UA Münster, 15 May 1981.
12 RMWEV to Rector, 14 October 1941 with a copy of Kabitz's letter, UAMs Neue Universität B I 11 spez. Bd. 3; Dean of Phil. Fac. to RMWEV, 26 September 1941, UAMs Kurator Dienst-Akt. Fach 45 No. 18.
13 Rektor to Curator, 18 December 1941, UAMs Kurator Dienst-Akt. F 8 No. 3 Bd. 5a; also, letter 4 November. Letter December 3 in UAMs Neue Universität B I 11 spez. Bd. 3.

In its list of candidates the faculty included only psychologists, with Robert Heiss in first place followed by Metzger and Hubert Rohracher. Heiss was favored because he had originally been a philosopher but also had experience in applied psychology. The rector also voted for Heiss: "That Heiss really is a qualified psychologist ought to be obvious from the fact that he is currently a senior official at the Air Ministry and trains Wehrmacht psychologists for the Luftwaffe." However, the ministry wrote on 29 November 1941 that Heiss was envisaged for a post in Cologne; therefore, at best Metzger could come to Münster.[14] The Amt Rosenberg did not see things in the same way as the university. The Science Office spoke to the party chancellery on 24 February 1942 in favor of Metzger, whose "tidy," "thorough," and "precise" scientific work made him preferable to Heiss (MA 116/10). This opinion may have been based on an unwillingness to strengthen Wehrmacht Psychology at the universities. It certainly conflicts with Metzger's later comment that in the shadow of the famous and courageous Cardinal von Galen one could "dare to appoint someone whose papers were not without blemishes" (Pongratz et al. 1972, p. 203), and that the chair in Münster was particularly independent of the party.[15] The leadership of the NSD had also emphasized to the party chancellery that in Münster a researcher was needed who could represent psychology as required in the DPO; Heiss and Metzger were especially well suited (NS 15/243, f. 233). The report on Metzger was positive in every way.[16]

The decisive impulse for establishing this "chair for psychology" had been the desire to constitute an examining board. The chance was offered because the DPO decree coincided with the retirement of a psychologically inclined philosopher, whose chair could then be redefined. The attempts inside the

14 UAMs Neue Universität B I 11 spez. Bd. 3.
15 Metzger to Donald Adams, 25 October 1946, AHAP M750, folder 13, Misc. Corresp.; Metzger also claimed that Rosenberg and Baeumler had worked against his appointment at Halle because he had studied under Wertheimer. But it was the Science Office of the Amt Rosenberg which in 1942 wrote that this should not be held against him "because the dominance of the Jews in psychology at that time was very great" (MA 116/10). In Halle other reasons had been decisive; there were no signs of intervention by the Amt Rosenberg (see Chapter 2).
16 "Metzger ... comes from the school of Gestalt psychology, whose basic tendencies he developed in an independent and productive manner. His book *Gesetze des Sehens* (1936) is an excellent scientific contribution, which has received attention and recognition.... [In his book *Psychologie*, 1941, he] presented a comprehensive, philosophically remarkable account of the basic concepts of the subject which aroused positive interest. Metzger's work shows a fortunate combination of energetic clarity of theoretical thought with down-to-earth use of experiments.... Metzger is regarded as a good teacher, who has proved his value in his Frankfurt research group.... In terms of character M. is regarded by the Frankfurt leadership [of the NSD] as respectable and clean. Politically, he belonged in the past to a liberalist, social-democratic circle in Frankfurt, but went through a genuine transformation in 1933, so that today he is felt to be politically irreproachable. He has belonged to the SA since 1933 and has been a party member since 1937" (Report of the leadership of the NSD, 21 January 1942, MA 116/10).

party to get a loyal pedagogue appointed to the chair smack more of nepotism than of systematic policy, all the more so since where the party had real influence on university affairs – Amt Rosenberg, the NSD leadership, and the party chancellery – the attitude had been positive and Metzger's appointment approved. Thus, in Münster psychology was able to strengthen itself, as elsewhere, at the expense of philosophy. Metzger's chair was the fifth after Jena (1923), Halle (1938), Hamburg (1942), and Berlin (1942) defined solely "for psychology." Under his leadership the Münster Institute for Psychology and Pedagogy was founded.

In Freiburg a similar situation arose in late 1941. Freiburg had no chair for psychology, and it was not financially possible to set up a new chair. On 20 October came the unexpected death of Martin Honecker, who had held the chair for "philosophy with emphasis on medieval philosophy" (UAF PA, Honecker). However, it was more difficult to get a psychologist appointed to the chair than had been the case in Münster. This was one of the two philosophical chairs – the other was occupied by Martin Heidegger. It had been transferred from theology in 1901 and was designated for the training of Catholic students according to a concordat with the Vatican. Seeing psychology represented, however, was more important to the university than honoring the concordat. On 18 September 1941 the rector had already recommended setting up an examining board, which Honecker was to chair (R 21: Rep 76/912). The examining board was approved; following Honecker's death his successor was to be the chairman. When the rector presented a list for the succession on 9 April 1942 he wrote to the state minister that he assumed that in the future it would be possible to make appointments for the chair without being bound by the concordat. Since both philosophy and psychology were to be covered, the faculty had listed Robert Heiss, Hans R. G. Günther, and Erich Rothacker, in that order. It was mentioned that Heiss and Günther had both published philosophical papers and had experience in Wehrmacht Psychology. The standing of Wehrmacht Psychology in the universities at this time is demonstrated by the fact that the rector wrote to Simoneit on 28 November 1941 to ask him for a comment (UAF Reg V 1/169). Heiss was appointed on 1 December 1942, after the chance to stay at Cologne had fallen through. Freiburg University also received its first psychological institute, renamed the Institute for Psychology and Characterology in July 1943. In 1943–4 the institute was provided with generous funds for a library and equipment, and an assistant's post was approved and occupied.[17]

Thus, in Freiburg the DPO also led to a new chair, this time at the expense not only of the philosophical chair but also of the philosophical training of Catholic theology students. This adds a certain irony to an argument advanced by Peter R. Hofstätter (1941) that psychology as a secular discipline

17 UAF Reg XVI 4/13 and PA Heiss.

should take over functions of pastoral care (*Seelsorge*) (see Chapter 8). The scales had once again tilted from philosophy to psychology. In the circular of the Executive Committee of the DGfPs of 15 January 1943, the members could read the victorious reports of the triumphant expansion of psychology: in Münster a full professorship had been established, and in Freiburg "a philosophy professorship previously linked to [Catholic] weltanschauung" had been transformed into a chair for philosophy and psychology; Heidelberg had been strengthened by the examining board and Hamburg by the new associate professorship. The successes in occupied areas were also welcomed: in Posen a full professorship had been occupied (Eckle), in Prague a new associate post for social and ethnic psychology (Hippius), and in Vienna one for psychology (Rohracher). "Thus the conditions for the development of our subject have been improved to a highly gratifying extent by the accommodating consideration of the Reich Education Ministry" (UAT 148).

Not only were new chairs occupied in 1942, but at the same time existing chairs changed hands, as in 1934 and 1938. Wilde came to Halle, Kroh went from Munich to Berlin, and Lersch from Leipzig to Munich; Rudert then stood in for the vacant professorship in Leipzig. In 1943 the chair in Breslau went to Wellek (who had previously been standing in for Wilde in Halle) after Eckle went to Posen. There were no more changes until the dismissals in 1945. It is worth while investigating the criteria influencing these appointments, for the situation in psychology had changed since the thirties, and changed again when Army Psychology and Luftwaffe Psychology were dissolved in 1942. Although this will be considered in detail in the next chapter, here I shall take an overall look at the development of appointments policies up to the end of the war.

The chairman of the DGfPs, Oswald Kroh, took the chair in Berlin on 1 April 1942. This was perhaps due to the interest that the Reich Ministry had in scientifically qualified teaching in the capital city. Whether or not this was so, the ministry negotiated with Kroh and Lersch to get one of them to come to Berlin. Kroh is supposed to have been reluctant (Thomae, Z, f. 87), but records on this were not available to me.[18] In contrast to Lersch, Kroh was a party member and chairman of the DGfPs, and it is possible that the ministry wanted to have its negotiating partner in psychological matters within easy reach.

Lersch took up Kroh's former chair for psychology and pedagogy in Munich on 1 October 1942, although he found an institute that was much less well equipped. At the top of its list the faculty nominated Friedrich Berger, who had studied under Kroh (see Chapter 2 on the events in Tübingen

18 The archive of the Humboldt University in (formerly) East Berlin rejected a request for information. The State Archive in Potsdam informed me that it had no relevant records. A communication of the NSD concerning the Munich chair confirms that Lersch and Kroh were the primary candidates for the Berlin chair (NS 15/243, f. 54246). In contrast to Lersch, Kroh was a party member.

in 1938). If Berger had scientific qualifications, they must have been in the field of education, but he was a party man and an SS man. Nevertheless, the NSD and the Amt Rosenberg supported neither him nor the third- and fourth-placed Sander or Pfahler, but favored Lersch, the only nonparty member on the list, calling him politically "impeccable."[19] The NSD leadership argued that Berger was less qualified than the others. His nomination must therefore have been the result of other nonscientific considerations, against the interests of the subject and even of the faculty. They recommended Lersch to the party chancellery. The Science Office of the Amt Rosenberg followed suit. Both emphasized his qualifications in psychology, although the chair was for psychology and pedagogy. The Science Office saw Lersch's strength in a "well-wrought characterology" (NS 15/243, ff. 54239–47). When Kroh went to Munich in 1938 there was no assistant post, though one was subsequently created. With Lersch's appointment a further post was gained, which had been taken from the Philosophical Seminar. Thus, below the level of professorship there were further examples of the institutional reinforcement of psychology during the war at the expense of philosophy (UAM O-1-15a).

The constellation of political forces and arguments in filling the chair of psychology in Halle was very different. In Munich Lersch, though second on the list, had been scientifically superior to Berger. The Halle faculty named Hans Keller, Werner Straub, and Kurt Wilde – none of whom was a leading psychologist in Germany. Probably Keller and Straub were included only because they really wanted Wilde, but as an instructor he could not head the list for a full professorship, especially since at the age of thirty-one he would have been exceptionally young. The faculty's evaluation clearly favored Wilde. In the application of 11 March 1942 the faculty even declared itself prepared to see the post reduced to an associate professorship in order to get him (UAH Rep 4/898). The main reasons were his research at Halle into hereditary psychology and his post as local deputy leader of the NSD. He also occasionally drew up reports for the Science Office of the Amt Rosenberg, which called on the party chancellery to use its influence to see Wilde appointed in the interests of the party.[20] Wilde was appointed as associate professor from December 1942, since he first had to go to the Wehrmacht; but Wellek stepped in immediately as stand-in for the chair.

The fact that at this time the commentaries were getting shorter and shorter makes it more difficult to judge the selection procedures. In some

19 NSD, 15 August 1942, to the party chancellery, NS 15/243, f. 54245. Lersch was, however, a member of the NSD (UAL Phil 684). In the fall of 1933 he had signed the *Bekenntnis der Professoren an den deutschen Universitäten und Hochschulen zu Adolf Hitler*...(Dresden 1933), along with Ach, Anschütz, Graf Dürckheim, Ehrenstein, Jaensch, Klemm, Krueger, Volkelt, and Wirth. But the list had included people like Theodor Litt who publicly attacked the Nazi race theories (Litt 1933).
20 NS 15/242, ff. 11, 14, 19, 24. See also Ash and Geuter (1985, p. 274).

cases only the names were given. In some cases, such as that of Wilde, the brevity reflected the concentration on a single argument. It may seem surprising that research into hereditary psychology reappeared as a criterion. The situation in Halle may have been a special one. But when Army Psychology and Luftwaffe Psychology were disbanded in mid-1942 (see Chapter 8), an important reference point for evaluating candidates for psychological chairs was lost. When a successor had to be found for Lersch in Leipzig, the Amt Rosenberg tried to push Hans Volkelt, who had been its man in psychology for years. But the faculty and the NSD did not find him sufficiently qualified scientifically. The faculty's list was headed by Pfahler, and the NSD emphasized once again that he was a leading representative of hereditary psychology (NS 15/243, ff. 54069–103). However, Pfahler refused the offer and the chair remained vacant for the duration of the war. It was represented by the associate professor Johannes Rudert, who had come back to the institute as successor to Otto Klemm in 1941 (UAL PA 95, ff. 64, 78). The last appointment during the war, that of Albert Wellek to Breslau, also shows how the criteria had shifted again. We saw how decisive the emphasis on preparation for an occupation and especially the activities of Wehrmacht Psychology was for setting up a chair in Breslau (see Chapter 2). Wellek had been rejected there in 1940; Christian Eckle, who had a background in practical psychology, was preferred. With the dissolution of Army Psychology and Luftwaffe Psychology everything changed. The faculty felt that Wellek met its requirements. It wanted to see psychology "continually in touch with the human sciences," and Wellek's approach promised to make a valuable addition. On the list of criteria, the competence of the future professor in practical psychology was now in second place. However, this time the faculty was prepared to attest to Wellek's abilities in that as well (Rep 76/131, f. 271).

The DPO, teaching, and subject matter

The effects of the new examination regulations on the cognitive structuring of psychology can at best be determined only in the long term, scarcely in the remaining four years of the Third Reich, which offered little scope for scientific development anyway. I shall therefore consider how, and at what levels, such effects might manifest themselves and illustrate them with the material available. First, the effects on teaching will be considered. At the time there was no fixed syllabus for psychology, and teaching at the individual universities depended largely on the professor concerned. The DPO now introduced a certain pressure to teach for the examination, without there necessarily being scientific consensus about the contents taught.[21] Did the DPO lead then to a realignment of the courses offered toward the new examination subjects?

21 This was not the case in sociology either; see Lepsius (1979, pp. 47–8).

For those who expected that it did, a count of the courses offered at various universities in certain semesters shows otherwise. Courses were evaluated at universities where psychology was taught relatively continually throughout the Nazi period. Six semesters were examined, three of them after the DPO and one directly before its imposition, in the summer semester of 1941. The choice of semesters and universities was influenced by the availability of the relevant lecture lists. The three universities with the most courses, Leipzig, Berlin, and Bonn, were included, as were seven other universities: Breslau, Giessen, Halle, Jena, Marburg, Munich, and Würzburg. All the courses of the incumbents listed in Table 5 (Chapter 2) and all other courses that were obviously declared to be psychological were included.[22] The following semesters were evaluated: summer 1934, summer 1938, summer 1941, summer 1942, winter 1943–4, summer 1944. In certain cases it was necessary to choose other semesters.[23]

The surprising result was that there were no obvious alterations. The overall number of courses and their distribution remained relatively constant, showing hardly any systematic alterations attributable to the DPO except for the diagnostics courses, which showed a marked increase. It would be going too far to look at the results for each university, but my impression is that the emphasis usually owed much to the interests of the incumbent professor. For example, courses on "statistics and quantitative methods" were given only by Wirth in Leipzig. There was strikingly little applied psychology. The evaluation of the lecture lists of the technical colleges would certainly show a different picture. Of the thirty-three courses in applied psychology at all ten universities fifteen were at Berlin University alone, most held by Rupp. This does not include courses held by Moede at the technical college but included in the university lists. In other areas there are also local peculiarities. Of thirty-three courses on social and "ethnic" psychology, nineteen were in Berlin. These were offered mainly by ethnologists such as Richard Thurnwald. The relatively small number may surprise those who would have expected a large number of courses on race psychology and such. If Rostock or Tübingen had been included, things would certainly have looked different (see Adam 1977, pp. 163ff.; Miehe 1968, p. 205).

22 The classification was not always easy; various titles such as "practical," "exercise," and "colloquium" did not clearly indicate whether they were theoretical or practical. When double titles were used ("characterology and expression") the first was taken for the classification. The headings were philosophy, general psychology, experimental exercises, statistical methods, personality (including typology and characterology), developmental and hereditary psychology, diagnostics (including expression and handwriting), pedagogical psychology, applied psychology, social and race psychology, and others.

23 In Giessen, the winter semester of 1937–8 was taken instead of the summer semester of 1938, when the chair was vacant and no timetable was issued. The availability of lecture lists also necessitated further exceptions: Giessen, third term 1941 instead of summer semester 1941; Jena, summer 1935 instead of 1934; Leipzig, summer 1943 instead of winter 1943–4; Munich, winter 1942–3 instead of summer 1942; Würzburg, winter 1942–3 instead of 1943–4.

Some of the details are perhaps more interesting than the overall figures. At some universities such as Giessen and Halle, the course topics show no orientation toward the professional practice of psychology. At other universities there were signs of innovation. In Marburg, for example, a tutorial was offered in 1942 on evaluation and observation, held at the educational counseling post of the institute. In Munich Lersch offered a tutorial in the summer of 1944 on educational counseling. These were directly oriented toward new practical needs (see Chapter 8). Other courses reflected the influence of the DPO, for example, by being named after examination topics. In the summer of 1944 Oskar Kutzner lectured in Bonn on "pedagogical psychology and psychagogics." The previous winter had seen courses offered in Berlin with names like examination subjects: characterology, cultural psychology, work psychology, pedagogical psychology, and developmental psychology, with applied psychology in the summer. This heralded the development of a new type of teaching course, the examination-related main lecture. Previously the main lecture had been on the theories of the full professor. A further innovation was the comment "only for psychologists" in the lecture list that Sander added to an "introduction to graphology" course in Jena, in the winter of 1943–4.

The obligation of the professors to make teaching relevant to the examinations might also have influenced the cognitive structure of the discipline at other levels. It provided a stimulus to arrange the knowledge as in textbooks, grouping it in terms of subject heading instead of in terms of problems. Textbooks and handbooks mark the claim of a science to maturity and integrity. I intentionally use the word "claim" here because in some cases the presentation of an "encyclopedia" of psychological knowledge seems more to cover up scientific insecurity. But this does not have to be the case; handbooks can take stock of the real core of available knowledge (Thomae 1977, p. 181). For us the relevant point is how much they also reflect the extent of professionalization, with the associated obligation to present some core of knowledge. It must be recalled that in 1958 and later the German *Handbuch der Psychologie* edited by Lersch, Sander, and Thomae still followed the structure of the DPO. Only the rather indeterminate subject of "applied psychology" was replaced by occupational psychology and in the seventies by clinical psychology as a response to new developments.

In 1922 Kafka had published his *Handbook of Comparative Psychology* in three volumes; the shorter general works at the end of the twenties and Giese's revision of the textbook of Elsenhans (1939) were largely compilations of theoretical psychology (see Chapter 6). The DPO encouraged plans to produce the first book that would cover all the areas of the DPO. In his foreword the editor Narziss Ach wrote that academic teachers were constantly being asked by their students to name a suitable textbook; practitioners also wanted to keep up on the state of things. "This textbook is thus to cover in particular those topics included in the examination regulations for diploma psychologists, but also to take into consideration the needs of other groups and at the same time to unify and modernize the face of German psychology" (Ach 1944, Foreword).

Four volumes were planned, of which only the third was published. They were (1) general psychology, (2) psychology of personality, (3) practical psychology, and (4) psychology of community and culture. We can deduce from a comment made by Kroh (1944a, p. 170) that developmental psychology was to be included in the volume on psychology of personality. A group of thirty had agreed to work on the project. Those involved with the third volume included the leading university psychotechnicians Herwig, Hische, Moede, and Rupp, the navy psychologist Mierke, the sport teacher Günter Scheele, and the deputy director of the Reich Institute for Psychological Research and Psychotherapy. J. H. Schultz.

The title and structure of this third volume show how examination subjects, fields of professional activity, scientific problems, and methods all had their influence on the shape of the subject, leading to bizarre and angular forms. "Practical psychology" might at first be supposed to cover applied and pedagogical psychology, yet there is no mention of the latter. One might also expect to encounter military psychology, which Simoneit regarded as part of applied psychology. After the demise of psychology in the army and Luftwaffe, however, this no longer appeared to be appropriate. The navy psychologist Mierke wrote about diagnostics in general, although he oriented himself on the procedures used in Wehrmacht Psychology. The volume also included contributions on psychology of commerce, psychology of work, diagnostics of visuality, diagnostics of technical abilities, vocational training and education, aptitude testing and judging individuals, traffic psychology, psychology of sport and physical traning, and medical psychology. Diagnostics thus fell under "practical psychology." But the contours of psychology become even more confused when we examine what was dealt with under the various headings. "Psychology of work," for example, was not the psychology of how a person works, but the investigation of relationships between work conditions and performance. One aspect of human labor became the subject "vocational training and education," whereas another aspect, "aptitude testing and judging individuals," which existed institutionally in the form of tests at employment offices, became a subject in its own right, with its own theories about inherited and acquired abilities. The logic behind these three chapters is less scientific or systematic than institutional: the problems of the deployment of labor in industry, their training by the Labor Front, and their selection by employment offices. The basic principles of diagnostics as a methodologically independent problem are set out in a chapter, alongside which other chapters deal with a special problem (diagnostics of visuality), with an area of industrial application, "technical ability diagnostics," and with "aptitude testing and judging individuals." Sander (1943, p. 21) had felt that the time was ripe for a "teaching edifice of German psychology." But the foundations were lacking, and it was hard to finish the interior when the builders did not know whether to base the construct on the unity of methods, a theoretical unity, a unity through institutionally defined fields of application, or a unity of examination subjects.

A final cognitive effect of the DPO that one would expect is further specialization. Establishing fixed examination subjects can lead to the outlook being restricted to one special field at the expense of others, diverting research along these particular channels. Characterology, which had previously been situated between the various disciplines (see Chapter 3) was now an area in its own right. In Stern's *"Personalistik"* the general theory of the person had included aspects of "general" psychology (see Revers 1960, p. 393). Later a personality psychology crystallized that was linked in individual models with general psychology, but that developed independently – especially where, for certain methodological reasons, it called itself "differential psychollogy." Clinical psychology is an example of an area that was ignored until the sixties. Where psychologists were active psychotherapeutically, they did not receive their training within the framework of academic psychology. The effect of specialization cannot be observed, however, in the short time before the end of the Third Reich. Nor did the social preconditions for it exist as yet in the universities. A single psychologist was still usually responsible for all fields. Only in the postwar period did a specialization of scientists' roles begin with the separation of theoretical and applied psychology.

The DPO created a new, clearer role for the scientist. When practical psychology had still been a matter for other professional groups or for university teachers who became practically active, people like Marbe and Stern were scientist and practical psychologist in one. We also meet this type of practitioner-researcher in the inspectorate of Wehrmacht Psychology. With the advent of the DPO the distinctions were more clearly drawn: here someone who provided triaining at the universities for practical work, there someone armed with a diploma who set about applying science. The institutionalization of the role of extramural expert and the training of new generations of professionals thus also contributed to the estrangement of the university teachers from practical matters. In the past university teachers had given advice, but now the practical psychologist was called on. At the same time, areas of practical activity included in the DPO could now become new areas of study for researchers far removed from all practice. The meanderings of diagnostics in irrelevant methodological disagreements in the sixties is an example. The DPO, which aimed to orient university psychology toward practice, thus created the preconditions for its separation. But again this effect of specialization and separation of roles could be investigated only over a longer period.

There are signs that certain scientific results were interpreted differently after the DPO and that attempts were made to construct a theoretical unity of psychology for didactic reasons. Wilhelm Witte (1957, p. 3–4) mentions that Hellpach's books were read by students after the decree of the DPO because they were the only ones that covered the new areas, in other words the new examination subject of social, ethnic, and cultural psychology. There was also a demand, not to be attributed solely to the DPO, for a theoretical construction encompassing the whole subject with its subdisciplines, rather than the autonomous theories such as holistic (*Ganzheit*) or Gestalt psychology. This may have increased the popularity of attempts at theoretical integration, such as the *Wesensfragen seelischen Seins* by Bruno Petermann (1938), the strata theory of Erich Rothacker (1938), and the anthropology of Arnold Gehlen (1940) (cf. Z. ff. 85–6).

Bornemann (1967, p. 226) has claimed that the DPO had a positive influence on the "inner integration" of psychology. If this is understood in more than a social sense, it seems doubtful whether this was really achieved. But the attempt was made. A good example was an essay by G. H. Fischer (1942a) in which he welcomed the DPO particularly because it sanctioned the synthesis of the psychological schools. The skepticism hitherto shown toward psychology had resulted from the fact that it had not been able to present a "self-contained system of completely established knowledge" to bear the load of psychological teaching (p. 1). With bold strokes he went on to make a broad sketch of psychology in which all the subdisciplines had their place – "phenomena and structure theories," "symptoms and functions theories," "hereditary and developmental psychology." There was even space for the great works from the history of psychology to reappear and add the appropriate veneer to the picture of unity. Fischer's conclusion was that here "the picture of a definitely self-contained body of knowledge could be gained, on which further research will be able to base itself with greater clarity of purpose, a picture that even in its present form already offers an adequate foundation for uniform training" (p. 7). The picture was thus not really drawn from life. Its function was to be a didactic one: the presentation of unity.

Kroh (1943, pp. 26ff.) constructed a logic of the development of psychology at the end of which was the "hard core of the subject" (p. 33): holistic outlook, stratification of the mental, anthropological perspective, structure-psychological interpretation of the mental framework (*Gefüge*), inclusion of a psychology of the acting person, phase theory in developmental psychology, the interpretation of expressions, and the view of man as a communal being. One cannot avoid gaining the impression that as a result of the external pressure to demonstrate theoretical unity and maturity a mixture of the theories of Kroh, Krueger, and Lersch was created to give outsiders the impression of a discourse that did not in fact exist.

The new professional group and its public reception

The DPO had affected not only the subject itself, but also the profession. Psychologists now appeared on the labor market as an officially recognized professional group. This must have been seen as a provocation by those who had previously carried out psychological activities but who had no official qualifications to show. Would psychologists in the future take over their activities? Practitioners, journalists, and psychologists commented on this question following the DPO. The psychologists regarded the DPO as "recognition of an officially protected professional body."[24] The aim must now be to "secure for the professional body of psychologists the same standing as tradition secures, for example, for the professional body of physicians" (Fischer 1942a, p. 2). Articles by psychologists appeared in the press about the new profession. In the journal *Das Reich* Hans Wenke wrote that the DPO not only determined the new teaching of the subject, but also "placed a new professional group on uniform foundations" (1941). E. Heinz (1942) wrote in the *Frankfurter Zeitung* that the diploma psychologist would now stand as a new professional alongside the diploma engineer, the diploma accountant, and the diploma chemist.

As Bornemann later wrote (1944, p. 37), questions about psychology were now discussed more frequently in the daily papers. Doubts were also raised about the competence of this new professional group. In a journal for industrial firms, *Der Maschinenmarkt*, an article appeared in 1941 that was very critical of the new diploma examination. The auther accepted the importance of selection and correct deployment in the war economy, but doubted whether such tasks were within the powers of diploma psychologists after a three-year course. First, he questioned the admission procedure for psychology students; he claimed that after the qualification they were lacking practical experience and training, just like accountants, and finally that the exam did not guarantee that they were suited for their profession. The "aptitude to recognize the vocational usefulness of other people" must be inborn (Schoor 1941, p. 22). Schoor, however, did not generally question the use of psychologists for allocating labor, and therefore, rather than replacing them with practitioners, he suggested that practitioners be involved in training the psychologists.

The editors of the journal subsequently received a number of letters on the matter. They felt that the discussion was important because in view of the discrepancy between growing demand and dwindling supply, the allocation of new workers was one of the most important problems for industry to solve. The daily papers picked up Schoor's arguments in various comments.[25] Nobody questioned the justification of the new profession or the need to employ psychologists in industry. All saw psychologists as aptitude testers and academically trained vocational counselors or as industrial "vocational planners." One paper even reported on the diploma psychologists

24 "Die neuen Reichsbestimmungen...," *Z. päd Ps*, 43 (1942), p. 29.
25 "Ein neues Diplomexamen," *Der Maschinenmarkt* No. 146, 27 August 1941. "Für und Wider eines neuen Diplomexamens," *Freiheitskampf*, the Nazi daily paper for the Gau Sachsen, No. 206, 27 July 1941; "Akademische Schulung für Berufsberater – Ein neues Diplomexamen: Dipl.-Psych.," *Magdeburgische Zeitung*, 7 August 1941. An article in the *Thüringer Gauzeitung* is also mentioned in a note in the *Maschinenmarkt*.

under the headline "Academic Training for Career Advisers." There was also agreement that completion of a university course was not the sole determinant of the suitability for practical activity. Therefore, it was necessary to establish aptitude for both the studies and the profession. Schoor's suggestion that practitioners be involved in the training was felt to be sensible, although the journal *Freiheitskampf* remarked that it would be difficult to concentrate personnel active in practical fields in the universities. A reader from Düsseldorf suggested introducing a sort of probationary period for psychologists. *Die sächsische Wirtschaft* published a very positive article defending psychology against psychoanalysis and individual psychology. For training and educational staff, a knowledge of youth psychology was useful.[26]

Kroh (1942) reacted to the mild criticism with the idiosyncrasy of a professional parvenu. In the journal *Arbeit und Betrieb* he published an article entitled "Why Diploma Psychologists?" aimed at quashing the objections. The arguments used here were different from those used to legitimate the subject to an academic audience (see Chapter 9). The necessity of using psychology arose from the "total responsibility" with respect to economizing with the labor force in war. The DPO emphasized practical training and thus took total responsibility into account. Proof of this for Kroh were the obligatory six-week practical training sessions in the diploma course and the introduction of such topics as expression psychology, characterology, diagnostics, applied psychology, and psychagogics. Contradicting the doubts aired in the press, Kroh also cited the selection of psychology students (see Chapter 6). The institutes paid the closest attention to "personality values and bearing" and rejected unsuitable applicants. In the pseudobiological terminology of the Nazis, G. H. Fischer (1942a, p. 3) countered the argument that the decisive factor for practical work was the personality of the psychologists and not their training. Before studies took place, there was a "preselection," and in the examination a "sieving out." Kroh dispelled one final reservation. Psychology did not intend to push others out of occupations in which they were doing good, practical work. The employment of psychologists should not be seen as a question of competition; they should be employed where the tasks made it necessary. The fears felt in other occupations about the presumptuousness of psychology were apparently not inconsiderable. Kroh felt obliged to repeat his argument in 1943 (p. 35).[27]

26 "Menschenkunde und Leistungssteigerung," *Die sächsische Wirtschaft*, 31 (1942), p. 440 (10 July 1942).
27 The discussion did not stop there. In the *Deutsche Allgemeine Zeitung*, 8 June 1943, an author named Muthesius questioned the need for any industrial psychologists at all. Industrial psychology was useful, but the two fields of selection and mental care of staff were original fields of the "*Betriebsführer*" (manager). Two weeks later a reply was published from the psychologist R. Walther, who argued that psychology was a discipline that demanded all the efforts of someone able to practice it. Its task was not mental care but selection. As disinterested observers with unprejudiced methods, psychologists were able to give an objective report, avoiding the "hopeless subjectivity of every personal judgment."

The rivalry of physicians and the deletion of medical examination topics

The institutionalization of professional roles for psychologists led to competition with other professional groups whose duties psychologists took over. In the period we are examining rival groups were primarily officers, engineers, vocational advisers, and doctors. The differences between psychologists and officers remained unaffected by the DPO. An institutional solution had been arrived at in 1937 with the regulations for psychological test stations (H. Dv. 26) and the regulations for the career ladder for psychologists. The DPO was important only for their position as civil servants. The conflict between the military and psychologists would flare up again when the usefulness of psychological selection was questioned (see Chapter 8). Signs of dissatisfaction among engineers were discussed in the preceding section. In view of the paucity of full-time professional psychologists in industry, it is probable that there were no major conflicts in this field. The same is probably also true of employment offices, and if there were conflicts they were not documented.

The greatest resistance to the DPO from a professional group came from physicians. In the ministerial files on the DPO the only protests are from the field of medicine, especially psychiatry. The examination regulations included a subject that was to become the bone of contention: "biomedical auxiliary sciences, which are fundamentals of biology (especially hereditary biology), physiology, medical psychology, and general psychopathology, in those parts relevant to studies in psychology" (*DWEV* 1941, 7, p. 256). This paragraph meant that physiologists and psychiatrists would in the future have to train and examine psychology students and belong to the new psychological examination boards. Wehrmacht psychiatrists further feared that the psychologists would now try to extend their influence to clinical questions. Thus, it came to a concerted action of Wehrmacht psychiatrists and representatives of psychiatry (the physiologists were obviously unafraid of competition), which was the case not only in Germany.[28]

Before the DPO, there had been a good relationship between psychologists and physicians only at the German Institute for Psychological Research and Psychotherapy. But this institute was also looked down on by prominent psychiatrists (Cocks 1985). The thoughts of developing psychology on the clinical side met with the disapproval of academic medicine, but they were expressed by the psychotherapists themselves. In 1938 M. H. Göring sug-

---

28 Other countries also experienced rivalry between physicians and psychologists. On the conflict in the U.S. Army see Samelson (1979); for the general situation in the United States see Napoli (1981), who reports on intrigues among physicians in the course of the first licensing of psychologists in Illinois at the end of the thirties; see also Reisman (1976, p. 301). Similar observations by Siguán (Spain) and Droz (Switzerland) are in Sexton and Misiak (1976, pp. 400–1, 420).

gested integrating psychological institutes in the medical faculties[29] in order to improve the combination of psychology, psychotherapy, and psychiatry. However, when in an isolated case at Strassburg University there was to be a link between psychology and medicine at a joint institute belonging to the philosophical and medical faculties, the medical dean opposed it and tried to get an institute under the control of physicians.[30]

The Third Reich was a successful period for physicians, bringing them among other things the long-desired Reich Physicians' Chamber (Güse and Schmacke 1976, p. 344), and their position was strong enough to repulse the attempted encroachment of psychologists into their domain. It is possible to follow this process in some detail since the physicians involved were employed in the army or the ministry so that files were left behind. It offers material that illuminates the long-standing conflict between psychologists and physicians that remains the decisive conflict between psychologists and other professional groups in Germany today.

Of the correspondence of physicians opposing the DPO, that of Max de Crinis and Otto Wuth is readily available. De Crinis, according to Cocks the most outspoken Nazi in the psychiatric establishment, was director of the Berlin University Nerve Clinic (Charité) and had been departmental head for medicine at the Ministry of Science and Education since 1941. Wuth was consulting psychiatrist to the Army Medical Inspectorate and in the second half of the thirties temporarily belonged to the Inspectorate for Aptitude Testing. The attempts of physicians in the Wehrmacht to repulse the advances of the psychologists were channeled through Wuth.

The first letter about the DPO to be traced is from Professor P. Nitsche from Berlin, a psychiatrist who was secretary of the Society of German Neurologists and Psychiatrists, from 1935 to 1939 and a member of the Reichsarbeitsgemeinschaft für Heil- und Pflegeanstalten (Reich Working Group for Mental Institutions) under Herbert Linden at the Ministry of the Interior. On 17 July 1941 he wrote to Professor Ernst Rüdin in Munich, who was head (*Reichsleiter*) of the Society of German Neurologists and Psychiatrists (Güse and Schmacke 1976, pp. 401–2), a supporter of forced sterilization and coauthor of the commentary entitled the "Law to Prevent Hereditarily Diseased Offspring" of 14 July 1933. Nitsche had heard from Wuth that the RMWEV wanted to create "the position of 'Diploma-psychologist' with 'medical'," by which he probably meant with the inclusion of selected medical subjects – "another group of quacks!" De Crinis had told him on the telephone that he had not been consulted in time, but that he wanted to fight it all the way. Nitsche requested Rüdin, as chairman, to send de Crinis a letter (for which de Crinis had asked) protesting against the fact "that these

---

29 Külpe had made the same suggestion in 1912; see Ash (1980).
30 For the source, see Chapter 2, note 45. The psychological institute that was founded never did get tied to the medical faculty (Z, ff. 84–5).

people should study psychiatry and neurology, since the academic teachers of these subjects could not be expected, and would refuse, to present psychologists – i.e., nonphysicians – to the sick, and to examine the psychologists in medicine."[31] Rüdin followed the advice and sent de Crinis a registered letter. He had heard of the plans, and as chairman of the Society of German Neurologists and Psychiatrists he wanted to express in advance his opposition to unqualified meddling in matters that were the responsibility of qualified doctors. He repeated the argument that university neurologists and psychiatrists could not be expected to present their patients to nonphysicians and asked that "the formation of a new group of nonphysicians, supposedly pretrained in medicine, be prevented" (REM/821, f. 257). In his reply de Crinis said that he favored the elimination of medical subjects from the DPO. The result of such an elimination would be to establish "that psychologists understand absolutely nothing about medicine, a fact with which we could then immediately confront them if ever they dared to reach out for our domain" (f. 256).

The physicians made use of their connections in the Ministry of the Interior, the stronghold of the state health bureaucracy that was strongly opposed to psychotherapy (Cocks 1985). On 15 October 1941 the Ministry of the Interior reported the psychiatrists' reservations to the RMWEV and requested the revision of the DPO. On 24 January 1942 they complained that they had still not received a reply. Soon afterward the RMWEV received a letter of protest against the DPO from the Breslau psychiatrist C. G. W. Villinger. Heinrich Harmjanz, responsible for psychology within the RMWEV, rejected these activities as the work of de Crinis, but was finally unable to prevent the removal of the passage from the examination regulations in August 1942 (REM/821, ff. 237, 250ff.).

Meanwhile, the psychiatrists Otto Wuth and Oswald Bumke, both university teachers and army psychiatrists, had become active. Their reservations about the DPO linked up with reservations against Army Psychology. On 23 February 1942 Wuth wrote to Bumke that the psychologists were worrying him, since they were encroaching "on psychiatric ground on a wide front." De Crinis and he had refused to teach these "lay people" clinical psychiatry with the presentation of patients, as the DPO envisaged. Rüdin had also sent Kroh a refusal. This meant that the psychiatrists at Berlin and Munich had taken a stand against the training of psychologists. In a report to the Breslau Medical Faculty on the training of psychologists, Villinger had also recommended that they not be admitted to clinical lectures. The argument that lay persons could not be admitted to presentations of patients was, however, only a tactical one. The DPO did not define the form of training, but only stated that "general psychopathology" was an examination sub-

---

31 German Documents among the War Crimes Records of the Judge Advocates Division, HQs, U.S. Army Europe, Microcopy T-1021, roll 11, frames 424–5. My thanks to Geoffrey Cocks for this information and the details about Nitsche.

ject, which thus also had to be instructed. The examination required "knowledge of the main forms of mental disturbance and the most important abnormalities" (*DWEV*, 1941, 7, p. 260).

Wuth clearly expressed his concern: securing the professional territory of psychiatry, the "threat" to which he described with unbelievable cynicism:

> I have made plain to him [de Crinis] the danger to German psychiatry that lies in the fact that on the one hand the psychologists and on the other the psychotherapists claim the whole field of psychopathy, "neuroses," etc. for themselves, while the mentally ill fall under euthanasia. It will be bad for our successors if the field is cut up like this.

Bumke had the same worries, but had not been so bold as to bar psychology students from his lectures. He replied that he would warmly welcome the appropriate general regulations from the ministry.[32]

Wuth and Bumke agreed that psychologists and psychotherapists in the Wehrmacht should not have anything to do with treating mental difficulties. Bumke declared that it would be a "catastrophe" if people suffering from the "epidemic" of shell shock were treated psychoanalytically. A psychiatrist in the Black Forest used forced drill as treatment. The psychotherapists of the Berlin Institute, with their psychoanalytical approach, were viewed with suspicion by Wuth because among other things they cooperated with psychologists. The psychologists in the Wehrmacht now regarded suicide as their domain and claimed that "childhood experiences (usually) lay behind it, and other such nonsense." The different etiological and therapeutic views linked together with professional territorial claims.[33]

On 25 July 1942 de Crinis and Harmjanz agreed that examining by neurologists and psychiatrists would be dropped immediately (REM/821, f. 253). The regulations were then altered on 20 August. The disputed paragraph now read, "biological auxiliary sciences in those aspects relevant to studies in psychology," replacing the term "biomedical auxiliary sciences" used variously in the regulations. This was commented on as follows: "With the alteration presented here the planned training of students of psychology in purely clinical subjects (general psychopathology), especially psychiatry and neurology, is canceled" (R 21/469, ff. 1711–12). The physicians had won. Although prominent Nazis had been at work, they had pursued the interests of their estate. Opinions in the party were divided. Erxleben from the Science Office of the Amt Rosenberg, writing in a letter of 2 January 1943 to the Nazi

---

32 Wuth to Bumke, 23 February 1942, H 20/480; Bumke to Wuth, 27 February 1942, H 20/480.
33 Wuth to Bumke, 3 March 1942, H 20/480. There had been a public argument about the responsibility of the so-called special cases in the Wehrmacht in 1938–9 between a senior psychiatric physician and Simoneit. The physician had treated psychology kindly but had firmly drawn the line between psychiatrists and psychologists on the question of special cases. See Simoneit (1939) and Tiling (1938, 1939).

Teachers' League, said that he "regarded the participation of physicians in the training of psychologists as a most essential matter" (MA 140).

Following the dissolution of psychology in the army and Luftwaffe, psychologists were being used for work in brain-damage stations, which led to territorial disputes between the professions flaring up again. Even toward the end of the war, the matter was important enough to psychiatrists to warrant letters and decrees. On 11 May 1942 the army medical inspector had forbidden the use of psychologists in special stations for the brain-damaged. Nevertheless, this had occurred. The order was thus repeated on 9 October 1944. Various doctors then inquired whether psychologists could not at least be used as auxiliaries without responsibility. Ernst Kretschmer, at that time consulting psychiatrist in Military Region 9, called the use of such auxiliaries a tried and trusted practice. The Army Medical Inspectorate had other ideas and declared the examination of the brain-damaged to be a purely medical matter (H 20/495). Kretschmer had at one time also taught psychology students in Marburg, on which Wuth commented to Bumke, "Obviously many still haven't grasped what it is all about" (23 February 1942; H 20/480). Wuth, de Crinis, and Bumke knew what it was about. As psychiatrists they were at least in agreement with the dissolution of Army Psychology and Luftwaffe Psychology in 1942. In their correspondence on the use of psychologists in the special stations they referred to this as evidence of the worthlessness of psychology.

Psychology had thus suffered its first setback during the war, 1941–2 being the apogee of its professional development in the Third Reich. At no other time had so many psychologists been active as full professionals, had the representation at the universities been so favorable, and had the members of the Executive Committee been so elated. Thus, the high point of German military successes coincided with that of the professional development of psychology. As we shall see in the following chapter, the first defeats of the Wehrmacht were to bring about the first major defeat for the profession.

8

The disbanding of psychological services in the Luftwaffe and the army in 1942 and the reorientation of psychology during the war

The expansion of Wehrmacht Psychology, a consequence of the expansion of the Wehrmacht itself, had essentially made the DPO possible in 1941. The consequences of the territorial expansion of the Wehrmacht during the war, especially the heavy losses in air battles and on the ground, and a new policy of officer recruitment led to the dissolution of the psychology services in the Luftwaffe and the army. The DPO was only three-quarters of a year old and chairs were just being established in the universities to ensure training in psychology when the orders went out terminating psychological testing. On 15 April 1942 the head of training in the Luftwaffe ordered that aptitude testing for flying personnel cease. In the future superiors alone would decide about appointments (LVBl, 1942 I, pp. 615–16). Thus, the central duty of Luftwaffe Psychology disappeared. Officer recruits were no longer to be tested psychologically either.[1] The army followed on 22 May 1942 with an order from the Supreme Command dissolving the Inspectorate for Personnel Testing and the personnel test stations as of 1 July 1942. This was also the end of aptitude testing for officers as well as "psychotechnical testing of other ranks" (HVBl, 1942 A, pp. 11–12). From July 1942 until the end of the financial year 31 March 1943, mopping-up operations were still going on at the former army testing stations and at the inspectorate. Navy Psychology remained intact until the end of the war, although psychological testing was no longer required for officer applicants there either, following a decree of 15 July 1942 (MVBl, 1942, pp. 732ff.).

---

1 The actual order could not be found in the archives. In the literature other dates for the dissolution are also given without further details. Former Luftwaffe psychologist P. R. Skawran wrote after the war that Göring had issued the order in January 1942 (I C Skawran 1, p. 61). Fitts (1946) cited an order by Göring on 11 February 1942, and Davis (1946) spoke of an order from the Supreme Command of the Wehrmacht on 15 December 1941. The certain fact is that the order for the Luftwaffe preceded that for the army.

There are probably more differing opinions about the dissolution of psychological services in the army and Luftwaffe than there are about any other issue concerning the development of psychology in the Third Reich. The reasons were not officially disclosed at the time, which gave rise to a number of rumors. Skawran (I C, 1, p. 61) had heard from a general on the staff of the head of training that the father of a candidate complained to Göring that his son had been asked questions about the bedroom secrets of his parents. Göring then ordered the dissolution. The case of Mölder received much attention (Z, f. 100). This famous fighter pilot had been classed as only conditionally suitable for flying (see I C Skawran 2). Pieper (1976, p. 166) felt that the dissolution of Luftwaffe Psychology resulted when Göring heard about this case. Hochheimer (Z, f. 56) remembers a contemporary rumor that Göring had been angered at the assessment of a nephew. In Army Psychology the dissolution was attributed to Hitler and his rejection of the use of academics. According to Kramp "the word had got round" that Hitler himself had ordered the dissolution of Army Psychology. He had suspected psychologists of being subversive in their field (Z, f. 110).

The reasons given by contemporaries and in the literature can be grouped into five categories. They cover the opposition of the Nazi Party, or certain leaders and organizations; the opposition of leading military leaders, or of the entire officer corps; the opposition of military physicians; the inadequate methodological-statistical results; and the altered situation for selection. The weight given to each reason seems to be governed more by the political or scientific stance of an author than by evidence. I will attempt to summarize these opinions and concentrate on the evidence for and against each hypothesis.

The most common argument advanced is the opposition of the party – for Luftwaffe Psychology that of Hermann Göring alone.[2] It is known that in the course of an intelligence test he was subjected to during the Nuremberg trials Göring called Wehrmacht Psychology "stuff that our psychologists played around with" (Gilbert 1977, p. 21), but there is no proof that he brought down Luftwaffe Psychology alone and on a whim. The Bundeswehr psychologist Martin Rauch (1977, p. 333) simply asserts that the Nazi leadership disbanded Wehrmacht Psychology, Gerathewohl (I B) speaks of the general mistrust of psychology in the Nazi Party. For Roth the "politicization" of the Wehrmacht was responsible (Pongratz et al. 1979, p. 268); for Arnold it was the conflict between psychological selection and nepotism (Z, f. 71) or between democratic psychology and dictatorial animosity (1970, p. 29). The conflict for Bornemann (Z, f. 147) was between characterology and Nazi

2 On Göring: Hanvik (1946), Pieper (1976, p. 166), Skawran in a letter to the author, 28 May 1979. Feder et al. (1948, p. 3) claimed on the basis of verbal information received after the war that General Milch disbanded Luftwaffe Psychology at the instigation of Göring, who in turn was moved by his cousin M. H. Göring. The latter is highly improbable.

ideology, and for Dirks (f. 25) it was about performance and ideological outlook as selection criteria. Some psychologists claim that it was Hitler himself who ordered the disbandment (Bornemann 1967, p. 226; Rothacker 1963, p. 108) or name – in addition to Hitler – Göring, Himmler, or Schirach (Rudert Z, f. 192; Simoneit 1954). Flik (Z, f. 150) and Rudert (UAL PA 92, ff. 92–3) point to the SS, von Voss (1949, p. 11) to the party chancellery. Elsewhere, Simoneit (1972, pp. 72, 106) says that Hitler's chief adjutant, Rudolf Schmundt, brought about the end of Army Psychology while he was head of the Army Personnel Office. However, the disbandment occurred under General Bodewin Keitel, who headed the office from March 1938 throughout the period of advancement of Army Psychology, being replaced by Schmundt only on 1 October 1942.[3] Thus, Mierke was also wrong when he told his U.S. audience that psychology had been upended after Keitel (Feder et al. 1948, p. 3). But this fit in well with the idea that Hitler and his close associates were the real opponents of psychology.

Some of these claims bear the hallmarks of a cliché among psychologists that psychology and National Socialism were incompatible; almost all presuppose an antagonism between the Nazis and Wehrmacht Psychology as an institution. It might already have become apparent that the first hypothesis, which ignores the encouragement given to psychology, is untenable; furthermore, in the Luftwaffe and the army it remained possible to use psychological methods after the testing stations had been disbanded, as we shall see. The second claim is not as easy to verify or falsify. Evidence of measures by a party organization against Wehrmacht Psychology are known to me only in the case of the attacks made by the Amt Rosenberg against the ideological precepts of Simoneit's psychology (see Chapter 5). It is true that teachers trying to avoid the obligation to propagate Nazi ideology in the schools could work in Wehrmacht Psychology. But this hardly made it a general "haven for anti-Nazis," as some would have it.[4] In September 1939, for example, 86 out of 170 military psychologists documented were members of the Nazi Party, although only two of the thirteen members of the inspectorate; a further twenty-three were members of affiliated party organizations (*WPsM*, 1939, 1, H. 9, p. 56). Later, as several psychologists report, urgent requests went out to army psychologists to join the party (Z, ff. 34, 55, 102, 150). Lieutenant Colonel Baumbach from the inspectorate remarked in *Wehrpsychologische Mitteilungen* (1940, 2, H. 4, p. 40) that membership in the NSDAP was desired. This may be related to the problems of legitimization encountered by Wehrmacht Psychology after 1938, but whether there was direct pressure from the party is hard to determine. Indeed, at least from 1934 to 1936 Army Psychology was called on to help select

---

3 In 1954 Simoneit wrote correctly that Keitel was still head of the Army Personnel Office; he had supported Army Psychology.
4 Flik, Z, f. 150; cf. Chapter 9.

SS-Führer candidates (Masuhr 1939; RH 12–2/101, f. 369). In 1935 there was a sympathetic article on Wehrmacht Psychology in the Nazi daily *Völkischer Beobachter*.[5] Finally, after Hitler's purge of the leadership of the army in 1938 and the subsequent attempts of the new commander in chief, General Walther von Brauchitsch, to bring the Personnel Office into line, Army Psychology received encouragement.

A special situation arose in the army when Hitler himself took over the command in December 1941. The order to disband the personnel test stations came from the Supreme Command, so it is not unlikely that Hitler himself was involved. Yet both Army Psychology and Luftwaffe Psychology reached their peak with Hitler as supreme commander of the Wehrmacht, which he had been since 1938 and with Göring as supreme commander of the Luftwaffe. If these two were now largely responsible for the disbandment, then what reasons did they have for doing so in 1942? The available sources do not show that there was a general antipathy toward Army Psychology and Luftwaffe Psychology within the Nazi Party, which then brought about their downfall. Nor do they offer any evidence to show that a party organization was involved in the disbandment.

Opposition to psychology in military circles is sometimes put forward as an explanation. The general problem discussed in Chapter 4 of the rivalry over officer selection had been exacerbated by the fact that sons of high-ranking officers had been assessed negatively. We have already heard of the reservations expressed about Göring's nephew. The army psychologist Gotthilf Flik reported that as a member of the Inspectorate for Aptitude Testing he had to give a second opinion (Z, f. 150). "The son of Reich Minister Rust, the son of General Fromm, and a son of Field Marshal Wilhelm Keitel were not assessed as their fathers had expected," Simoneit 1954, p. 141) later wrote. Fromm and Keitel had then caused problems for Wehrmacht Psychology. This is backed up by the fact that all the Wehrmacht psychologists interviewed knew about these cases (see Z passim). The U.S. prison psychologist in Nuremberg reported the following comment by Wilhelm Keitel about Wehrmacht Psychology: "Really, they even failed my son in the officer candidate tests because of some nonsense in a dark room; he also had to give a speech, during which his voice was not loud enough for the nonexistent audience. I cleared the whole stupidity out" (Gilbert 1977, pp. 32–3). Did he do it alone? At least he claimed it as a feather in his cap, which says much for his attitude.

Feder et al. (1948, p. 3), referring to information from navy psychologist Udo Undeutsch, named General Fromm as the driving force behind the disbandment of Army Psychology. Fritz Fromm, commander of the reserve from 1939 to 1944, had tried to collect pejorative comments about the value of Wehrmacht Psychology. The difficulty of evaluating such claims is perhaps

5 Fritz H. Chelius, "Zwanzig Jahre Wehrpsychologie," *Wochenbeilage zum Völkischen Beobachter*, No. 37, September 1935.

shown by Simoneit's (1954) reference to such inquiries by Hitler. We do not know who started the ball rolling. The rejection of relatives may have nurtured the skepticism of leading officers toward Wehrmacht Psychology. But it would be rash not to assume that the military top brass also undertook an assessment of Wehrmacht Psychology based on effectiveness. There certainly was a clear drop in efficiency, as we shall see later, and this led to the military making the decision on disbandment in 1942. The other problems of resentment among officers and the rejection of relatives occurred throughout the existence of Wehrmacht Psychology.

A third group who may have been responsible for the disbandment were the physicians in the Wehrmacht, the medical officers. Firgau sees the disagreement with the doctors as the main reason for the disbandment (Z, f. 13); Roth too emphasizes the "jealousy of the doctors" (Pongratz et al. 1979, p. 268). This allegation fits well into the general picture of the contradictions surrounding Wehrmacht Psychology. We saw in the preceding chapter that psychiatrists in the Wehrmacht tried to restrict the responsibilities of psychologists. Some doctors also wanted to see aptitude testing carried out by doctors instead of psychologists (e.g., Lottig 1938). The attitude of the Army Medical Inspectorate can perhaps be demonstrated by the fact that the ban on the employment of psychologists in the special wards for the brain-damaged (see Chapter 7) was issued after the disbanding of Luftwaffe Psychology and only eleven days before the order went out to disband Army Psychology. Nevertheless, there are no grounds for assuming that the resistance of the physicians was the primary cause for the disbandment.

In the Anglo-American literature after the war, the lack of validation of diagnostic methods used in the Wehrmacht was cited as the reason. According to Davis (1946, p. 6) psychologists were not adequately trained in scientific methods, were unable to provide sufficient proof of their success, undervalued the empirical verification of their conclusions, and pursued a subjective characterology instead of objective analysis of behavior. Dorsch (1963, pp. 157–8) and Hofstätter (1967, p. 326) followed this argument, but Arnold (1970, p. 31) contradicted it. What Hofstätter calls "diagnostic valency" was referred to as "*Bewährungskontrolle.*" This meant comparing assessments with troop evaluations and service certificates that had to be sent in to the inspectorate. Thus, the criterion of external validity was the active officers' idea of what made a suitable officer. From 1939 onward an additional criterion was the war itself. In *Wehrpsychologische Mitteilungen* psychological evaluations were now compared with individual war reports.[6] For the Luftwaffe the number of planes shot down by a fighter pilot became the measure of psychological character and performance diagnostics.[7] Since

6 *WPsM* 1939, H. 11, pp. 57ff.; 1940, H. 1, pp. 60ff., 64ff.; 1940, H. 5, pp. 93ff.
7 Skawran compares the evaluations with the "front reports," focusing on the kill rate. See I C 2, and I C 1, pp. 48ff.

it was not possible to control the so-called beta error, that is, the rejection of those who would in fact have proved suitable, controls always referred to the alpha error, that is, the acceptance of applicants who failed to prove themselves. The Army High Command recorded how many officer candidates failed. For example, of the cadets taken on in 1936 who became lieutenants in September 1938, 19.2 percent failed to be appointed officers. On 20 October 1938 Commander in Chief von Brauchitsch therefore demanded that more attention be paid to selection and training (MA RH 37/2378).

The figures published by Wehrmacht psychologists, who were naturally not unbiased, showed a good success rate Skawran quoted 80 percent correct predictions for the Luftwaffe (I C 2); for the navy Mierke (1942) found 92 to 95 percent agreement between training results and psychological evaluation. Figures published regularly in *Wehrpsychologische Mitteilungen* were similarly high. One is rarely told how these figures were arrived at or how they are to be interpreted (e.g., how was an agreement established, or what did the success rate of cadets in training say about the quality of psychological testing?), but it would clearly be wrong to say there was no statistical control.

This accusation cropped up only when the political situation changed. In 1941, when the U.S. Army was considering adopting German methods, Ansbacher and Nichols certified the quality of the controls in Wehrmacht Psychology (1941, pp. 46ff.); in 1949, in contrast, the neglect of validation and of objective methods numbered among the "transitory" aspects of German Wehrmacht Psychology for Ansbacher. The accusation resulted perhaps from the methodological prejudice of mathematical psychology. It is hard to imagine today that Davis (1947) really believed that Wehrmacht Psychology would have been able to defend itself against its critics if it had subjected its claims to experimental tests.

The fifth reason given is the declining effectiveness of psychological selection, especially for prospective officers and fliers. The former Wehrmacht psychologists Dirks, Munsch, Gerathewohl, and Renthe-Fink saw a link between the disbandment of Luftwaffe Psychology and the shortage of new recruits and reported that the selection rates were too low to satisfy demand.[8] In the army the demand was also greater than the supply; the possibilities for selection shrank.[9] Psychological selection and prognosis became virtually superfluous when prospective officers could be selected on

8 See Z, ff. 18, 26, 33, 100; I B Gerathewohl; I C Ge 5; Gerathewohl (1950, p. 1046).
9 See Z, ff. 14, 26; Mierke quoted by Feder et al. (1948, p. 2); Hofstätter and Wendt (1966, p. 181) write: "The German Wehrmacht saw itself obliged at this time [1942] to set the acceptance rates for most positions so high that even the best diagnostic methods could not effect a significant increase in the percentage of successes. But this meant that the selection process as such became superfluous."

the basis of their performance in the field.[10] Presumably these factors were decisive, so that the end of psychological services in the Luftwaffe and the army can be understood only against the backdrop of developments in the war and the personnel situation in the Wehrmacht. In addition to this the change in the military situation led to a change in the ideal type of officer required, which corresponded more and more to a practical and political selection than to a psychological one.

The Luftwaffe had not met the projected number of fliers in 1939 (Völker 1967, pp. 182–3). At the beginning of the war (August–September 1939) prospective pilots were accepted on the basis of their application and a medical checkup, without any psychological test at all (I C, Ge 5, f. 3; see Chapter 4, note 44). According to Gerathewohl the period until the spring of 1940 was a correspondingly hard time for Luftwaffe Psychology (I C, Ge 4, f. 9). The selection of fliers was, after all, its central task. During the preparations for the air battle over Britain in 1940, too many applicants failed the tests, according to Munsch (Z, f. 100). Gerathewohl also recalls that in 1939 the level among the applicants was still high, but that later the applicants were less qualified, "so that the psychologists rejected a greater proportion of applicants" (Z, f. 18). The commanders of the flying schools complained that they were not getting enough trainees. The situation became even more critical once the Battle of Britain began in August 1940. For the Germans there were 3,363 dead, 2,641 missing, and 2,117 wounded (Zipfel 1976, p. 188). There was thus a massive shortage of flying personnel in 1941, with the result that less suitable candidates were also accepted (Gerathewohl I B and 1950, p. 1046).

In contrast to the acceptance of officer cadets in the army, the commander of a flying school was bound by the psychological evaluation of candidates. At least from 1938 onward the precondition was "flying – psychologically at least conditionally qualified." Exceptions had to go through a complicated procedure (RL 5/870, f. 1; I C, Ge 2, f. 7). Every failed psychological test thus reduced the number of flying school trainees. Given the shortage of recruits, the tests stood in the way of an increase in numbers. Where they had previously chosen the most suitable candidates from abundant applicants and thus aided the expansion of the Luftwaffe, they were now an obstacle to further growth.

We can identify a structurally similar reason for the abandonment of tests for officer candidates in the army. Initially the recruiting situation was favorable: "The applications from gymnasium graduates for the officer corps – one in six in 1935 – rose to nearly one in three (!) immediately

---

10 Flik: Z, f. 152; E. Heinz in the *Frankfurter Zeitung*, 13 August 1942: "The officer recruits should no longer be selected solely by psychological testing, but by the experience and adequate performance in war itself."

afterward, but then leveled off at over one in five" (Grunberger 1974, p. 403). But by the time of the western offensive in 1940, there was already a shortage of suitable officers. Active soldiers recommended by the troops were sent to officer candidate courses in order to be promoted to officer rank after proving themselves (Müller-Hillebrand 1956, II, p. 40). Noncommissioned officers could be accepted as candidates for commissions after successfully completing special pretraining (*DWEV*, 1940, 6, pp. 129–30). Thus, the aptitude tests for officer applicants became less important.

By June 1941 the number of officers was satisfactory (Müller-Hillebrand 1956, II, p. 103). In 1942, however, the demand rose again. In the summer of 1941 the Wehrmacht had launched its attack against the Soviet Union. In December in the Battle of Moscow it had suffered its first major defeat, to which Hitler had responded by dismissing von Brauchitsch and taking over the High Command. The fighting and the cold winter caused the first heavy losses; by December 1941 the eastern campaign had already cost 24 percent of the original strength (Zipfel 1976, p. 196). The losses among officers had also been severe, and more and more sergeants were being promoted to officer rank (Absolon 1980, p. 252). The first defeats also "falsified" the prognoses of the psychologists. By 1941 a total of eleven countries had been occupied – the officers selected psychologically had "proved their worth" in the field; but now they were being defeated.

The war made it possible to do without an instrument to predict "performance," since it was possible to test "suitability" in battle. The guidelines for the selection of officers were altered correspondingly. In May 1941 the new regulations came into force (H. Dv. 82/3b). Officer candidates from the reserve could be commissioned if they had proved themselves in the field and were otherwise generally suitable. It was also possible for noncommissioned officers or candidates of other ranks to be recommended directly as officers following exceptional acts in the face of the enemy. On 3 September 1941 a new order was issued to accelerate special measures to promote soldiers to officer ranks (HVBl 1941 C, pp. 547–8). On 19 March 1942 the Army High Command declared that "the great demand for officer recruits requires rigorous measures to register, uniformly train, and supervise all soldiers of the army and reserve who come into question for war and active promotion" (HVBl 1942 B, p. 154). The decree named action in the face of the enemy, but not the *Abitur* school qualification or psychological tests, as a precondition. Thus, the war broke up the traditional system of officer recruiting. Two months later, on 22 May 1942, the order was given disbanding Army Psychology as of 1 July. The conditions for accepting officer candidates were revised on the same day. The army set up ten application offices (HVBl, 1942 C, pp. 340–1). The new leaflet entitled "The Active Officer Recruits for the Army in War" (June 1942), which replaced all former leaflets, established that "decisive for acceptance as an active officer during the war are actions in the field. . . . The previous psychological aptitude tests will no longer take place" (RHD 23/21, pp. 3, 7). Army Psychology had risen with the tests for officer candidates and fell with them.

The defeats of the Wehrmacht also changed the requirements for the ideal officer. When Hitler took over the command of the army in December 1941, he wanted to educate it in National Socialism (Messerschmidt 1969, p. 261). After the first defeats, it became more and more important for officers to be strong-willed and enduring, and able to improve the morale of the troops by ideological means. Ten days after the order to disband, on 1 June, Field Marshal Keitel told troop commanders that he wished to bring the supreme commander (i.e., Hitler) closer to the officer corps and thus to strengthen the "will to hold out" and the "belief in final victory" (Besson 1961, p. 84). In July 1942 the first specialists for military ideological leadership were introduced for all commands (p. 84). The attempt to impress the Nazi policy of holding out on the officer corps ended finally with the creation of National Socialist Leadership Officers in December 1942 (p. 79).

It was not general animosity from the party, officers, or physicians that put an end to Luftwaffe Psychology and Army Psychology. Rather, it was the loss of their military function in the war. This was due primarily to the change in the recruiting situation and the heavy losses, coupled in the army with the problems of troop morale, which led to a change in selection criteria. If Hitler himself was involved in the dissolution of Army Psychology, this was due less to a general attitude toward psychology and more to the fact that the military situation demanded radical alteration of the policy of officer recruitment.

In the smallest of the three organizations in Wehrmacht Psychology, Navy Psychology, the work went on. As of 15 July 1942 officer applicants were no longer examined at psychological test stations, and candidates were selected at the Navy Application Center in Stralsund (MVBl, 1942, pp. 732ff.) and later also in Vienna. The revised regulations, however, make no mention of psychological tests. Munsch reports that he had selected navy officer candidates until 1 April 1945 (Z, ff. 100ff.). It is not possible to determine why Navy Psychology was left untouched. There are no relevant records. One possibility is that specialist tests played a more important role than in the army and that there was not such a shortage of applicants there as in the Luftwaffe. According to Mierke, since the start of the war the emphasis had shifted to testing specialists and weapons specialists (Pongratz et al. 1972, p. 233). Navy Psychology had already been more concerned with the selection of technical staff than the army. Furthermore, there was no general selection of officers, but selection for either sea, engineering, signal section, administration, or medicine, all with different requirements. This may have made a need for psychological testing seem more plausible to the officers, and apparently they had more respect for psychology than did officers in the other services (Z, ff. 102–3). But this alone can scarcely have been the reason for sparing Navy Psychology. To find out more, it is necessary to investigate the recruitment situation. I know of no studies on this subject; at least as far as the officer corps is concerned none seem to exist (Salewski 1980, p. 229). Naming various admirals as saviors of Navy Psychology, such as Hasso von Bredow (Firgau, Z, f. 13) or Erich Raeder (Flik, Z, f. 152), is just as unsatisfactory as the theory of Davis (1946) that Navy Psychology escaped disbandment because it had little to do with character analysis and carried out more tests.

The dissolution of Army Psychology was finished by 31 March 1943. The apparatus went for the most part to the Army Medical Inspectorate. Psychotechnical test equipment was to be kept for the "technical aptitude testing of specialists."[11] Objective psychotechnical methods, which in the opinion of some postwar authors were a particularly sore point for the Nazis, continued to be used in the army after the end of Army Psychology, at the time of greatest Nazi penetration under the command of Hitler. The actual disbanding of Army Psychology was in the hands of the Medical Inspectorate. In a way Army Psychology was back where it had been during the First World War. Officers decided about commissions, and specialist testing was carried out by medical officers or soldiers using equipment developed by psychologists.

The development of Army and Luftwaffe Psychology demonstrates vividly the dependence of professional psychology on the demand situation, on the purposes to which psychology was put by the institutions employing psychologists. Wehrmacht Psychology was recognized as long as it contributed to the war effort. The significance of psychological prognoses of fighting performance vanished as soon as this had to be displayed in a real war. Actual fighting performance in the face of the enemy replaced psychological predictions. It is also clear that if professional psychological activity is geared to selection, it becomes pointless when the preconditions for selection no longer exist. Psychology thus experienced in the war how social preconditions determined its practical professional existence.

The profession after the dissolution

The dissolution of psychological services in the army and the Luftwaffe was a severe blow to the professionalization of psychology. The two largest employers, the army and the air force, where the role of expert had been securely institutionalized, had done away with its psychological experts. At one stroke a field of activity and research demonstrating the capabilities of the discipline had been lost. The survival of the smallest of the three service organizations could do nothing to soften the blow. The loss of a central "service function" can be understood according to Spiegel-Rösing (1974) as a cause of uncertainty about the status of a discipline. The loss of the

11 H 20/553.5, Allg. Heeresamt, 30 March 1943.

major field of professional activity caused more than uncertainty. A pillar of professionalization was demolished. But why didn't the entire building collapse?

Three reasons seem to be decisive. The DPO had provided a further pillar of professionalization. On 15 January 1943 in its annual report the Executive Committee of the DGfPs declared that "in a shattering year for the discipline [the DPO] had already fulfilled its consolidating function" (UAT 148). The demands of Wehrmacht Psychology had led to the reform of psychological teaching, but the dissolution of the one did not bring about the liquidation of the other. The state bureaucracy went on supporting the development of psychology according to the requirements of the DPO and not according to the actual demonstration of a need for psychology. We saw in the preceding chapter that the number of chairs increased. When the chairman of the DGfPs looked back over the year 1943, it led him to the opinion that "on the whole the grave shock that 1942 brought German psychology had been gradually overcome" (circular from late November 1943, UAT 148). As a second factor for this assessment the committee cited the gain of new areas of activity. This referred to the work in the NSV, where a new field of professional activity could be consolidated. The third factor was that the disbanding did not lead to an army of unemployed psychologists crowding the labor market because many psychologists were drafted into the real army. Since most of them were civil servants they would have had to be employed elsewhere in state service anyway. On 28 April 1943 the chief of army munitions and commander of the reserve reported that the "greater part" of the psychologists had been taken on "by offices of other Reich ministries" (BA-ZNS ES: Psychologen). Some idea of the areas in which they ended up can be gathered from autobiographical accounts: Pieper worked in a provincial administration as career adviser for war invalids, Hochheimer for the Employment Office in Berlin, Scharmann at an air force hospital for the brain-damaged in Vienna; Firgau went to the Reich Ministry for the Occupied Eastern Areas, Schänzle to the Reich Commissar for Pricing; Munsch was one of eight psychologists taken on by Navy Psychology; Flik went to the Military Academy; Zilian began to study medicine; Gerathewohl found a job as psychologist at Bavarian Motor Works; and Dirks, Renthe-Fink, and Simoneit went into battle. Some psychologists were employed at army colleges.[12] The Executive Committee spoke in January 1943 of the difficulties encountered in trying to place army and Luftwaffe psychologists in their profession (circular, 15 January 1943, UAT 148), but subsequent circulars in 1943 and 1944 mention the need for psychologists, especially in the field of education.

---

12 Sources: Pieper (1976, pp. 181–2); Pongratz et al. (1979); Z (passim). Fitts's comment (1946, p. 159) that *many* psychologists went to industry and career counseling is not accurate, at least for industry.

244 *The professionalization of psychology*

The DGfPs reacted to the dissolution with an active strategy to provide security for the profession (see Spiegel-Rösing 1974, pp. 31ff.). Internally they sent out a call for unity. In 1942 the Executive Committee passed a motion calling for the "cessation of disturbances to the unity of the subject."[13] Externally attempts were made to improve the image of psychology by publishing a series of articles demonstrating its relevance in the new situation. Furthermore, the committee now set out to find new partners who could further the application of psychology and to find new applied fields. Training in psychology was to be specialized to meet the needs of the areas of application then dominant. The work of the DGfPs became even more concerned with the direct value of psychology for day-to-day affairs. These various attempts to stabilize the subject, which will be examined in the following sections, were not at all the result of the disbanding of Army and Luftwaffe Psychology. They were also an attempt to secure a place for the newly established university subject in the academic and professional worlds.

The use of psychology in the National Socialist People's Welfare Organization during the war

During the war psychology was able to establish a professional foothold in the NSV. By 1942 the cooperation between the DGfPs and the NSV had reached a point where educational psychology could be presented as a field of activity and was an option for specialization. Since about 1940, before the disbanding of Army and Luftwaffe Psychology, this cooperation had been developing. As far as I know the first (female) psychologist, Hildegard Hetzer, was appointed by the NSV in April 1940 when it took over the Society for the Protection of Children Against Exploitation and Ill Treatment in Berlin. By the end of 1940 there was systematic cooperation between the DGfPs and the NSV. At Kroh's instigation the leader of the Office for Welfare and Youth Aid at the Head Office for People's Welfare at Nazi headquarters,[14] Althaus, recommended cooperation on the following basis (17 December 1940):

> You should name male and female psychologists able and willing to work in NSV Youth Aid, especially education counseling.... Their work will cover the following areas:
>
> 1. Counseling of those entitled to education and of children and youths

13 Letter from the Executive Committee of the DGfPs to Moede, Mathieu, and Engelmann, 30 December 1942, UAT 148 Schriftwechsel DGfPs.
14 Because it was affiliated with the Nazi Party there was a head office for the NSV at party headquarters, linked by personal connection to the NSV. The head of this office and "Reich administrator" of the NSV was Erich Hilgenfeldt, who was also in charge of the *Winterhilfswerk*, SS Brigadier and member of Heinrich Himmler's "Circle of Friends." See de Witt (1972).

in difficult cases and determining whether therapeutic pedagogical treatment is necessary

2. Counseling and examining
   (a) in adoption matters...
   (b) of children and youths with recognizable educational needs, but where the form of education has to be decided (foster family, NSV young people's homes or reform homes)
   (c) of difficult children and youths in the NSV young people's homes, in close cooperation with the wardens...
3. The training of men and women working in NSV Youth Aid in the field of psychology... (UAT 148)

When Kroh sent a copy of this letter to Pfahler, he commented "that here a new wonderful and important field of activity for psychologists, especially female psychologists," was opening up (UAT 148). Women psychologists had been left out by the professionalization in the Wehrmacht.

The use of psychologists in the NSV was made necessary by massive new problems in the field of education that the NSV tackled and the NSV's attempts to gain control over the system of youth care. The new problems were those of penury brought on by the war – youth criminality, gangs, destitution, and so on. Later there was an increasing number of orphans, whose upbringing had to be publicly organized. On 5 January 1943 Hitler decreed that all young children who could not be brought up by their parents were to be charged to NSV children's homes, unless as children of SS members they were sent to homes of the SS-Lebensborn.[15] As more women had to contribute to the war effort it became necessary to increase the already large number of kindergartens. When the bombing began, children were evacuated to foster families in the country, or to homes or camps partly organized by the Hitler Youth. According to Hetzer, psychologists handled the educational problems caused by the evacuation (Z, f. 157). All this formed the background for increased intervention by the NSV in so-called Youth Aid. In 1941 they were able to get the Ministry of the Interior and the Reich Chancellery to agree to transfer some responsibilities from youth welfare offices to the NSV.[16] Psychology profited from this switch of power from communal administration to the party. The NSV was the first institution in the field of education counseling to employ full-time qualified psychologists.

The first counseling centers in Germany had been set up by psychoanalytical therapists (mostly of Adler's school) – for example, in Munich in 1922 by Leonhard Seif, at the Vienna Psychoanalytical Laboratory in 1925, and in Wuppertal in 1929 by M. H. Göring.[17] The German Institute for Psychological

---

15 Letter of Reich Minister and Chief of the Reich Chancellery, Hans-Heinrich Lammers, 5 January 1943, MA 304.
16 Matzerath (1970, pp. 386–7); cf. de Witt (1972, pp. 258ff.).
17 Cocks (1985, pp. 53ff., 185ff.); Hitschmann (1932, p. 268).

Research and Psychotherapy continued this work. In November 1939 it set up its own counseling department (von Koenig-Fachsenfeld 1942). Where necessary, children could be given depth psychology treatment in institute homes, of which three existed in 1942. The branches of the institute in various towns acquired education aid centers. These were seen as providing opportunities for male and female "attending psychologists" (see Chapter 4). However, as with the treatment of adults, therapy could only be carried out in cooperation with the physician (von Koenig-Fachsenfeld 1940, 1941). Alongside this psychotherapeutic counseling was the counseling of the youth welfare offices. This was mostly in the hands of psychiatrists (i.e., physicians), with isolated psychologists such as the former Wehrmacht psychologist Robert Scholl in Stuttgart (Scholl 1939). Thus, in the development of educational counseling in Germany a given institutional form and certain patrons always corresponded to the employment of a specific professional group (Faust 1944, p. 41).

Psychologists had been active since 1941, when the NSV Youth Aid began counseling. The task of the psychologists was to evaluate difficult "cases" in addition to training the Youth Aid staff (Hetzer 1942). Among other things they also had to distinguish between those who were "worth the effort" and those "not worth the effort," as Hetzer writes (1940, p. 240; also 1942, p. 175). The counseling was not for anyone, but only for "genetically valuable" (*erbwerte*) young people. The factor deciding treatment was the value of the young person for the community, one of the things psychologists had to determine in their evaluation (Schott 1940). The practical tasks of counseling were to assign children to homes or foster parents, to name guardians, or to advise on custody (Hetzer 1940, 1940a, 1942; Faust 1944). Psychologists were also called on to provide expert evidence in court (see Thomae 1944).

The NSV Youth Aid also employed psychologists in Poland, where it was the first organization to establish itself behind the advancing German army. In a report it named among its duties establishing kindergartens, day-care centers, and juvenile vacation homes, as well as setting up a family care system (Fa 88/260). But the NSV in Poland was also given other duties, such as the relocation of so-called folk-German children in Germany, and the "Germanization" of Polish children. This Germanization constituted a crime against humanity, and since this seems to be the only point at which psychologists became linked to such a crime as professionals I would like to look in some detail at their role. Hrabar, Tokarz, and Wilczur (1981) have extensively documented the cooperation of the NSV with the SS and the Gestapo in stealing and selecting Polish children for German upbringing. Children of Polish women deported to work in Germany were sent to NSV homes; the NSV used force and subterfuge to get Polish orphans sent to foster families; finally they took control over children who, following Himmler's decree of 16 February 1942, had been separated from "politically particularly incriminated" Polish parents (1981, pp. 170–97; also see Łuczak

1966, pp. 223ff.). Were psychological methods used in the Germanization program? Were psychologists of the NSV involved? There are indications that this was the case; the name of a woman psychologist crops up in the literature. In order to decide whether a child became a German, the child was examined not only "racially" but also in terms of health, abilities, and character (Hrabar et al. 1981, pp. 187, 227). According to a decree of the head office of the Reich commissioner for the consolidation of the German *Volkstum*, Himmler, Polish orphans "whose racial appearance indicates Nordic parentage...should be subjected to a racial and psychological selection procedure. The children recognized as being valuable blood bearers for Germanness should be made Germans" (Sosnowski 1962, p. 376). This SS decree established a procedure in which a psychological investigation played a part in establishing "valuable blood bearers." First, the children were "racially examined" by the Race and Settlement Head Office of the SS, then by the Health Office; finally, they were psychologically examined at the children's home in Brockau "by Frau Professor Dr. Hildegard Hetzer (NSDAP, Reichsleitung, Head Office for People's Welfare)." The children were to stay there six weeks, during which time the home warden was to give a "characterological assessment" of each of them (pp. 376–7).[18] The Reich minister of the interior was consulted and a communication sent from the ministry to the local plenipotentiary in the "Warthegau" on 11 March 1942 ordering this procedure (Łuczak 1966, pp. 231 ff.). All the results went to the "representative of the Reich commissioner for the consolidation of the German *Volkstum*, who decides on sending children on to Lebensborn or to the inspector of home schools" (p. 233). This could be a matter of life or death. Of two thousand illegitimate Polish children evaluated with "racial and psychological methods" in June 1944, four hundred were *"ausgemerzt"* – murdered (p. 234).

At this time the psychologist Hildegard Hetzer no longer worked at the children's home in Brockau. Since the summer of 1940 she had worked in the NSV Youth Aid in Posen; in March 1942 (the month after the decree) she began in Brockau. But her contract was terminated in May, in her words because race considerations in the reports were insufficient for the Race and Settlement Head Office. According to information from the Poznan (Posen) Archive, documents on the Germanization action show that Hildegard Hetzer did not take part in the selection due to an excessive workload. Whether anyone else took over the work could not be determined.[19] On the work at the NSV in Posen, Hetzer reports: "The work of the psychologists in the NSV Youth Aid during the war consisted largely of choosing children for admission to homes and foster homes suitable for them. In the Warthegau there were a lot of resettled children, some of them orphans, and children

---

18 Checked against the original decree; Archiwum Głównej Komisji Badań Zbrodni Hitlerowskich w Polsce, Warsaw, Kolekcja fotokopii "X," Sygn. 611.
19 Wojewódzkie Archiwum Pánstowe, Poznan; communication of 12 April 1983.

248   *The professionalization of psychology*

brought in from the battle areas" (Z, f. 156). The NSV Youth Aid in the Warthegau was able to "withhold children from the SS." They were hidden and their tracks covered. "One could speak of 'underground work' in the interests of the children, which is too little known."[20] In view of the duties of the NSV in Poland, it is hard to imagine that the activities of psychologists consisted of such "underground work." However, the documents do not show that psychologists worked to enforce the decree either, so that we cannot conclude that psychologists participated in Nazi crimes in their professional capacity. If these accusations were to be pursued further, it would require a more detailed study of the work of the NSV in Poland.[21] On the basis of the sources examined, it is possible only to say that the SS intended that psychological methods be used in the Germanization action to determine "worthy blood bearers," which also implied condemning those who were "unworthy" to terror and destruction.

It is not possible to reconstruct exactly how many psychologists the NSV employed. According to Faust the plan "to place psychologists in charge of each *Gau* educational counseling center" was achieved in eleven cases (1944, p. 41). The psychologist Eyfert from the Head Office for People's Welfare spoke in a radio broadcast in November 1943 of *Gau* psychologists being employed in thirty-two of forty-five such districts (REM 821, f. 369). In Posen three psychologists were employed (Z, f. 157). In June 1944 the NSV

20 Z, f. 156. I found the text of the decree quoted in this section while finishing the manuscript. I sent the text to Professor Hildegard Hetzer, who wrote a longer statement on 6 November 1983 and provided additional documents. These included a report on her work as a psychologist for the NSV dated 4 December 1949 for the Hesse Cultural Ministry and various affidavits from 1948 and 1949 concerning her activities, which obviously figured in her de-Nazification, as well as her contract with the NSV and a letter terminating her work at Brockau. According to these documents she worked first at a resettlement camp in West Poland, the Warthegau, and came to Brockau in March 1942. It was said that the children there were German orphans who had previously been in Polish homes. She had not heard in Brockau of Polish children being taken away from their parents and had not helped to send Polish children to German families. The work was terminated in writing on 14 May by the *Gau* administration. Various witnesses speak of criticisms leveled against her by Dr. Grohmann, an SS leader of the Health Department in Lodz. She was regarded as politically unreliable by the party. According to her work contract she was not a party member when she began work for the NSV in 1940. These documents thus indicate that the investigations in Brockau for the Germanization program did not involve psychologists. But this says nothing about the work of all the psychologists for the NSV in Poland, about which only Hetzer was questioned. Thus, in a poorly documented book, in which the validity of witnesses' evidence cannot be verified, it is claimed that Hetzer participated in the medical investigation of Lidice children by the Race and Settlement Head Office of the SS in June 1942. One of the children who survived reported this to the authors. Hetzer is falsely described as a physician (Hillel and Henry 1975, pp. 217, 200-1). In their work in West Poland psychologists had duties that frequently brought them into contact with the SS, for example, in the resettlement and refugee camps. It is hard to imagine that Hetzer heard nothing of the stealing of Polish children, which according to Hillel and Henry was quite blatant.
21 This would involve further interviews and studies of NSV and SS records, particularly in Poland.

was looking for ten female psychologists to teach in schools for kindergarten staff and three female and one male psychologist for practical work.[22] In Berlin the Head Office planned to set up its own central institute in 1943 to provide scientific training for NSV personnel. Erich Hilgenfeldt offered Kroh the job of heading the institute, which was intended to become part of the High School of the Nazi Party.[23]

Cooperation with the NSV seemed to psychological institutes to offer new opportunities for practical activity. Hitherto, the link between practical psychology and scientific research had existed predominantly in Wehrmacht Psychology. Now it was possible to take on evaluation work or to attach practical facilities directly to the institutes. The Marburg Institute, for example, set up an education counseling center for the NSV in Kassel (Bröder 1943). The director, G. H. Fischer (1942, p. 8), reported in a speech that there were also plans to set up a kindergarten at the institute, as well as a counseling center and a youth hostel for observations. The Leipzig Institute held counseling courses for the NSV and helped to select expert staff. The institutes in Tübingen and Munich cooperated with the NSV and the Labor Front. In Münster, Professor Metzger was in charge of the *Gau* education counseling center. In Halle, Wilde worked for the Youth Aid, mainly as a speaker at training courses and as a consultant for difficult children. In 1942 he wanted psychology students to do their practical training in NSV children's homes.[24] The man in charge of the Head Office for People's Welfare at the NSDAP Reich Direction, Erich Hilgenfeldt, regarded the cooperation with psychological institutes as being so valuable that he wrote to the RMWEV opposing their closure, even in war, since then both cooperation between these institutes and the NSV Youth Aid and the training provided for the NSV Aid Program (Hilfswerk) Mother and Child would have to be ended (R 21/469). The NSV had thus become an important new coalition partner for psychology in the course of its professionalization. It provided a new occupational field, previously the domain of psychotherapists and psychiatrists, and thus created a new demand for psychologists. Furthermore, it made possible a form of practical orientation for psychological institutes that was to outlast the Nazi period.

The image of psychology after the demise of psychological services in the Luftwaffe and the army

In November 1942 the National Socialist Lecturers' League (NSD) held a seminar attended by the physician and head of the Science Office of the NSD

---

22 Letter from Kroh 12 June 1944, UAT 148.
23 Rep 76/37, f. 202; cf. Bollmus (1980); see also note 14.
24 Overview compiled from UAT 131/130, f. 216; UAT 148; UAM Sen 310, Lersch to Senate, 21 November 1944; UAL PA 95, ff. 70, 95; UAL Phil B I, 14[37] V, f. 92; Z, f. 188; UAMs Nachlass Herrmann, Nachrichtensammelstelle No. 3, f. 2; UAH PA Wilde, Wilde to Curator, 12 May 1942.

Gustav Borger, by Erich Hochstetter for the Amt Rosenberg, and by psychologists Kroh, Volkelt, Pfahler, Sander, Gottschaldt, Lersch, Eckle, Herwig, Bender, Thomae, Wilde, and Wenke.[25] Such seminars had been organized by the NSD since late 1941 with the approval of the Amt Rosenberg to bring individual disciplines into line ideologically (Kelly 1973, pp. 409ff.). The NSD planned to set up a psychology section.

According to Hochstetter's report made to the Amt Rosenberg, Lersch spoke at length about the practical activities of psychology in the wake of the dissolution of Army and Luftwaffe Psychology. The discussion was also dominated by "questions and possibilities of applied psychology." It was decided to publish a series of articles in the NSD journal *Deutschlands Erneuerung* to provide a portrait of the young subject (Kroh 1943, p. 21). Articles by Sander, Kroh, Wenke, and Lersch were published, and these were to be followed by an article on hereditary psychology by Gottschaldt.

This represented a direct response to the dissolution of Army and Luftwaffe Psychology. The articles were not a general attempt at ideological and political alignment, as Fritsche believes (1981, pp. 103ff.), nor a reaction to the restriction of university work due to the dismissals at the beginning of the Third Reich, as Seeger (1977, pp. 6–7) claims. It was an attempt by leading psychologists to document the theoretical and practical relevance of their subject in a situation where its usefulness had been called very much into question. Indeed, in individual press articles the opportunity had been taken for general rejection of diagnosis by psychologists (e.g., Döll 1943). The "crisis of usefulness" of psychology did not begin with more recent discussions in the United States (see, in contrast, Seeger 1977).

The various articles show how leading representatives of the field depicted their subject to outsiders. The readers in this case were not specialists, but the ideologically conformist "scientific core troop" for whom this journal was intended. There were only oblique references to the dissolution in the articles. Kroh, for example, wrote that individual cases should not be over-emphasized; rumors about mistaken evaluations had in part been proved false. He also refused to speak of a crisis in a science when "its external space for application is reduced, least of all then, when such a narrowing gives reason for self-critical examination of one's position and demands that the lines be drawn more tightly together" (Kroh 1943, p. 32).

Two years previously in the same journal Peter R. Hofstätter (1941) had proclaimed a "crisis of psychology," since nobody was prepared to register conflicting approaches in psychology or knew what a practical psychologist actually did. Hofstätter pleaded against scholasticism and for the practical usefulness of psychology. He envisaged psychologists carrying

---

25 Report by Hochstetter to Erxleben, 27 November 1942, NS 15/212, ff. 5ff. The report also mentions Klarer, unknown to me, and the psychologist "Wilke," probably mistakenly for Wilde.

out aptitude diagnostics and providing secular care (*Seelsorge*) for healthy people in need of advice, perhaps working at the local branch of the Nazi Party. This idea was picked up two years later by the Head Office for Science at the Amt Rosenberg. Erxleben, who was in charge there, pointed out that it would be wrong to conclude from the reduction of military psychology services that psychology no longer had any functions in the field of selection and counseling in general. He favored school and prison psychologists. Psychologists were more suitable than psychiatrists in schools, "because nobody wants to have anything to do with psychiatrists, because they run the danger of being caught out with some hereditary defect."[26] Thus, the advantage of using psychology could also be a tactical one.

However, this idea of providing secular guidance for "souls" did not catch on, perhaps because the field was already covered by the German Institute for Psychological Research and Psychotherapy. Kroh expressly opposed Hofstätter's article, above all because he had spoken of a crisis where none existed. Optimism was the key. Defending psychology against such "false views," or "misunderstanding about psychology" as Kroh called Hofstätter's article, the 1943 writers presented a theoretically coherent picture of psychology, which could claim its own domain among the other sciences and which had areas of practical activity.

Sander attempted to present the inner unity and theoretical coherence of psychological teaching. He offered Leipzig holistic (*Ganzheit*) psychology as a general teaching concept. The time was "ripe for a theoretical house (*Lehrgebäude*) of German psychology in which all the existing parts would be combined into a meaningful whole" (1943, p. 21). Wenke's essay examined the independence of psychology in the "web of sciences," and its relationship to the natural sciences, medicine, and the social sciences. Where Wenke assumed close links with the other sciences, Kroh placed the accent more on preventing those sciences from taking liberties by arrogating areas of psychology to their own subjects as special topics. He flatly rejected the thought of relinquishing an independent special course of studies in psychology or of attaching psychology as a special option onto other courses. In his extensive article Kroh defended the theoretical unity of psychology and maintained that it was an independent profession that could do without being hung onto some other, as E. Heinz (1942) and H. Döll (1943) had suggested in newspaper articles. Only graduates with full training in psychology were psychologists, and this was a precondition for being able to carry out diagnostics. Kroh obviously felt it was necessary to defend the image and the competence of psychologists as well as the subject itself. Kroh used the examination regulations to draw the dividing line between psychology and other subjects.

26 Letter from Erxleben to the Reich Administration of the Nazi Teachers' League, 1 December 1943. MA 40. Erxleben remarks, "The situation in psychology seems to me to be very well depicted in Hofstätter's essay 'The Crisis of Psychology.'"

Lersch, writing on practical fields of application for psychology, was more open on this point. He wanted both full psychologists and "hyphenated psychologists," that is, members of other professions who had completed an additional course in psychology. Lersch makes plain the main areas psychologists were considering for practical activity after the loss of Army and Luftwaffe Psychology. He named education (upbringing), vocational selection and counseling, work planning, and leadership. Both vocational counseling and work were traditional areas of psychological activity. The inclusion of education certainly reflected the intensification of links with the NSV. In addition to industry and advertising, leadership would find applications in national and international political affairs and in "psychological care" (*seelische Betreung*). In wartime, psychologists could teach officers leadership of populations in occupied areas. The four areas named crop up again in the supplementary regulations for the DPO in 1943.

The 1943 plan for the specialization of psychological training

A new field of activity had been established in the NSV, and a unified front had been displayed externally. The examination regulations were now to be supplemented to consolidate psychology in the area of training. As soon as the DPO had been decreed, Kroh had expressed the opinion that it would be necessary "to regulate the access of diploma psychologists to the various types of professional activity."[27] The new supplementary regulations represented a step in this direction. On 22 March 1943 the ministry decreed an addition to the DPO allowing a further examination in a special field after the diploma. It also became easier to acquire a diploma after the doctorate.

The initiative for the decree came from the DGfPs. At the beginning of 1943 the society completed its preparations and sent an application in to the ministry.[28] The draft was by Kroh, in consultation with Herwig, Lersch, and Sander (Kroh 1943a, p. 10). It was not based on special areas of the subject, as Lersch's essay had been, but on professional fields – an overlapping, but not identical approach. The supplementary examination, which could be administered with the rest or after them, included four fields, each divided into several subsections. Those named here were compulsory: for the examination as educational psychologist, general theories of education and social pedagogy; for the occupational psychologist (career counselors and aptitude testers), work and aptitude psychology; for the industrial psychologist, work and leadership psychology; and finally for the economic psychologist, the psychology of business (Kroh 1943a). The last was apparently intended for sales and marketing. When the DPO was still brand

27 Circular of DGfPs, Christmas 1941, UAT 148.
28 Circular of 15 January 1943, UAT 148.

new, Kroh (1941, p. 18) had spoken of trying to establish special fields with career regulations for industrial psychologist, economic psychologist, career counselor, and psychotherapist. G. H. Fischer (1942a, p. 3) had suggested setting up other career paths besides Wehrmacht Psychology – namely, in medicine, education, career guidance, and the furthering of the talented. The physicians had put an end to the dreams of creating an occupation of psychological psychotherapist, but the NSV had made it possible to establish the educational psychologist instead, something Kroh had not mentioned in 1941. Under the new regulation the local examination board could also examine in supplementary areas, where these were represented. The DGfPs wanted to prepare additional special areas. In April 1944 it suggested an examination as "psychological handwriting assessor" to the ministry, without success.[29]

The supplementary regulations were a reaction to developments, particularly the demise of military psychology services, which were no longer a special field, and to the increasing importance of educational psychology. The efforts of the DGfPs to secure these additions formed part of an active strategy to bolster the profession in wartime. They also fitted in with ministry plans to align the curricula and the examination regulations of all subjects more closely with the requirements of practical training (see Huber 1940).

Psychology in the "war effort of the human sciences"

The war led to an increasingly pragmatic orientation of the sciences in training, profession, and research. The natural sciences were particularly important for munitions technology, but the human sciences were also "called up" by the ministry. Their duty was to "develop the idea of a new European order . . . in a scientifically incontestable way and to prove this to be the truth and reality of the life of the European people" (Dietze 1940, p. 397). The Wehrmacht was bringing about this "reality." The human sciences, which according to Walther Frank were the creators of convictions, had to justify that reality by declaring the real to be right. Since 1940 meetings of the individual disciplines had been taking place for this purpose. Work groups had been set up, according to Dietze, with one for philosophy but not for psychology. Perhaps psychology had to become an independent subject at the universities before it could support the "battle of arms" with the "battle of minds" in this official form.

It was not until October 1943 that a "meeting on questions of the

29 REM/821, f. 379; cf. circular of 1943 November, UAT 148; Kroh to the RMWEV, 4 March 1944, R 21/469, f. 1718. A credit form from Marburg supplied by Professor Werner Traxel lists the following fields: (1) educational psychology, (2) psychology of career guidance, (3) industrial psychology, (4) economic psychology, (5) race-political folk studies, (6) psychology and pedagogy of folk care, and (7) constitution psychology.

deployment of psychology in wartime" was held in Weimar. Besides the speakers and guests, this was attended by representatives of the Ministries of the East and of Propaganda and other administrations as well as the heads of all university psychology institutes. The following topics were considered:

(a) Psychology of the peoples of the Eastern territories (*Ostraum*) (as an approach to a new folk characterology)
(b) The sociopolitical and sociopedagogical tasks of psychology
(c) Wartime leadership, preservation, and restoration of human performance capability
(d) The wartime use of applied animal psychology[30]

The Security Service (SD) of the SS had an "honorary" observer at the meeting, and fragments of the resulting report are preserved. They clearly express its function, as well as providing some information about the agenda:

> A central aim of the meeting was to make German psychologists familiar with the problems of the East, so that in the future a stronger participation of psychological research in clarifying certain eastern questions will be guaranteed. In view of the aims the meeting was begun with talks on the peoples of the East. (MA 641, Nos. 772283–4)

Professor Beyer and Professor Rudolf Hippius from Prague spoke first about their research on Eastern European peoples. Then Wilhelm Lejeune from the Institute for Work Psychology of the Labor Front spoke about experience in industry with Eastern workers. "From the Reich Ministry for the Occupied Eastern Areas two gentlemen then reported on their observations in Estonia, Latvia, and Lithuania" (MA 641, Nos. 772283–4), of which at least one was a psychologist.[31] The only other page of the report to survive shows that the psychologists attending discussed how propaganda could be disseminated more effectively to Eastern Europeans.

The *Ostforschung* (Eastern research) offered psychologists a new field in which to demonstrate their worth. The way they went about it is certainly typical of the short-term orientation of psychology during the war. Kroh (1941, p. 7) hoped that psychology could gain a further field of activity in selection, resettlement, and regional planning. The Marburg Institute was given a folk-psychological research assignment by the Society for Germanness in Foreign Countries (Fischer 1942a, p. 5 note). At the University of Posen Hippius led a study group called Aptitude Research in the Working Group for East Settlement from May 1942 onward, the aim of which was to increase

---

30 Circular of the DGfPs, November 1943, UAT 148; see also Kroh (1944, p. 189).
31 The former army psychologist Firgau was ordered to the East Ministry after the dissolution and spoke at the meeting; verbal communication, 10 April 1979; cf. Z, ff. 13, 227.

the economic efficiency of Nazi policies in Poland (Wróblewska 1980, pp. 243–4). They were to determine the willingness of various strata of the population, especially those of German origin, to settle in the annexed areas (Kalisch and Voigt 1961, p. 203). The circle around Hippius produced a number of empirical folk-psychological studies.[32] Since 22 January 1942 Hippius had worked for the Ministry for the Occupied East Regions.[33] The SS was also very interested in his suggestions and considered using aptitude testing for population planning in the Warthegau.[34] Hippius was granted a chair for social and folk psychology in Prague at the end of 1942, as well as being named codirector of the Institute for European Ethnology and Folk Psychology of the Reinhard Heydrich Foundation (BA:R 31/384).

This work included papers that from their content and language could have originated in periods other than the Third Reich (e.g., Hippius 1944). The function of the research, however, is clear. The race psychology of the first phase of the Nazi system, with its bombastic ideological concepts, was replaced by an empirical, methodologically sophisticated race psychology, which aimed to be an instrument for Nazi population planning. With the territorial conquests, psychologists saw a new opening for psychology (cf. Dirks 1942); they quickly adapted their research in the hope of demonstrating their practical value for war policies. The same also applies to the use of psychological tests on deported and prisoner-of-war workers from Eastern Europe (see Chapter 5). It was the last stage of such legitimation strategies during the Third Reich.[35]

The other noticeable feature of the Weimar meeting is the emphasis placed on "applied animal psychology." This also was new. Animal psychology as such had long been a specialty in psychology, but more for comparative investigations of theoretical interest than as an applied area. The mechanisms of professionalization also operated in this field. There had been an animal psychology group in the DGfPs with Robert Sommer as leader, although this made appearances only at congresses (Keller 1936, p. 12). In 1936 the German Society for Animal Psychology was founded with the growth in demand for animal psychological knowledge. The first meeting was attended by zoologists such as Friedrich Alverdes, psychologists such as Keller and Volkelt, and representatives of the Army Dog Section of the Ministry of War, of the Interior Ministry, the Breeding and Training Center for Police

32 See Hippius (1940), Hippius and Feldmann (1942), and Hippius et al. (1942).
33 "Curriculum vitae of Mag. Dr. Rudolf Hippius," written and signed in July 1942; Uniwersytet im A. Mickiewicza Poznan, Archiwum, Akten Reichsuniversität Posen, Sygn. 78/318.
34 Verbal communication from Mitchell G. Ash, who was able to inspect the relevant documents at the Berlin Document Center.
35 See e.g., I B Deuchler; the careful study by Michael Kater (1974) on the research group "Das Ahnenerbe" of the SS gives no indication of the involvement of psychologists in SS planning.

Dogs, the Propaganda Ministry, and animal breeders of the Reich Food Estate (*Gründungsbericht...*, 1936). According to Keller (1936), an applied animal psychology had been developed in Wehrmacht institutions. There was thus a link between research and practice that was to become closer with the demands of war, influencing in turn the development of research in the universities. On 1 September 1941 Werner Fischel was appointed the first lecturer in animal psychology in Leipzig. The faculty had used the argument of the army's interest in dog psychology. On 8 January 1941 the Army High Command had expressed interest in "both theoretical and applied animal psychology research at German universities." Early in 1945 Fischel was provided with dogs by the chief of dogs and messenger pigeons for the Reichsführer SS (UAL PA 1110, ff. 3, 87).

The DGfPs attempted to reorganize internally to meet the demands of war. In connection with the Weimar meeting, the society decided to set up study groups for folk characterology, performance leadership, occupational psychology, propaganda and advertising, school education, nonschool education, and applied animal psychology. Steps had been taken to set up a "section for applied psychology" in 1942–3. However, both plans seem to have got no further during the war.[36]

Although new areas such as Eastern research had attracted much attention, professional activity toward the end of the war actually centered on diagnostic duties: selection and allocation. At the Labor Front and in the factories, it was the selection of foreign workers or the examination of apprentices; at the employment offices, it was aptitude testing; in the navy, the selection of specialists; in the NSV, the diagnosis of children to determine how and where they should be brought up. There was also the determination of the employability of war invalids. In 1943 Professor Matthias Meier from Darmstadt wrote in a students' annual about the professional psychologist:

> Sought are teachers for psychology and pedagogy for teacher training institutes, youth leaders, psychologists for career management and factory leadership in the Labor Front, in the NSV, in the psychotechnical departments of employment offices, in the field of health pedagogy, etc. In all these areas psychologists have the duty to determine the aptitude of a pupil for a certain occupation, a soldier for a certain company, an applicant for a certain post, in fact to determine methodically and exactly the aptitude of a person for a certain position, and to provide expert guidance for the employment and promotion of people as members of the work force. (Meier 1943, p. 140)

As another example, Professor Christian Eckle from Posen wrote on 31 July 1944 about the application of psychology in war:

36 See the circular of 15 January 1943 and November 1943, UAT 148; none of those interviewed could confirm that the groups had actually worked.

The development of psychology in the last decade, reinforced by the war, is tending toward placing the research and teaching capacities of the subject increasingly in the service of the political-practical tasks of the nation....

This involves specifically the problems of occupational and economic life, of vocational and recruit guidance, work allocation and control, education and youth care with practical-diagnostic work and reports, which today are in part vital to the war effort. In addition there are psychological research problems, concerning the *Volk* and people in particular of the Eastern territories (*Ostraum*), the sociopolitical, socio-pedagogical, and folk-pedagogical tasks of psychology in wartime, the preservation and restoration of human performance capability, the use of animal psychology in war, etc.

Among the public institutions that continually – and increasingly – approach university psychological institutes with demands for recruits and with which working relationships exist and are initiated in part in Posen are Wehrmacht advisers, Nazi leadership officers, the staff of the Cadet schools, the agencies for war-disabled rehabilitation. Further, for the *Gau*, the Chamber of Commerce, and the Office for Technology, the Control Post for Occupational Competition, and the scholarship organizers of the Labor Front in Berlin, NSV offices, training and selection camps of the Hitler Youth, the children's relocation camps, the schools of the local authorities and the savings banks.

In satisfying all these and similar demands the nation can and should demand and expect the responsible cooperation of university psychological institutes.[37]

It was hardly worth all the trouble of drawing up such a long list of psychological achievements. Half a year later Poznan (Posen) was to be liberated from German occupation. Just as the war and its preparation had brought about a rise in fortunes for the professionalization of German psychology, so the war also ended this phase of the history of psychology in its own way.

Institutes burned down, such as in Leipzig on 4 December 1943. In the fall of 1944 nearly all students were drafted for the "Total War." Even professors were now called on to fight with weapons of steel. Pfahler in Tübingen, for example, headed the Volkssturm there (Adam 1977, p. 200). In this situation it seemed important to emphasize that psychology was really a natural science, and as such should also be kept going as part of the war effort. In the name of the DGfPs Kroh asked the ministry to treat psychology students as natural scientists. The ministry refused in January 1945 and denied that studying psychology was particularly important for the war.[38]

---

37 UA Posen, Files of the Reich University Posen, 78/81.
38 REM/821, f. 390; circular from Kroh, 4 November 1944, UAT 148.

It had not been of any use that the personal secretary in the SS Head Office, Dambach, had tried to convince the ministry that psychology was important for the war effort because psychological training was important for police and SS officers. Replying to an inquiry by Erich Hilgenfeldt from the NSV on the work of the psychological institutes, the ministry wrote in January 1945 that teaching was continuing at all universities with the exception of those that had been closed down or occupied by the Allies, at that time the universities of Bonn, Cologne, Münster, Kiel, Königsberg, and Strassburg, as well as the Aachen Technical College (R 21/469). It was only a matter of months before the universities collapsed everywhere.

9

Self-deception, loyalty, and solidarity: professionalization as a subjective process

The most important stages in the objective development of the professionalization of psychology in the Third Reich have now been examined, and the study could end here. However, several question remain to be considered: What did the professionalization demand subjectively from those involved, and what effects did it have? What can be said about motivation? The period under investigation lends these questions additional import.

A process of professionalization is initiated and actively supported by individuals; others then take part in it or are affected by it. In the course of the professionalization of psychology individuals began to see themselves as psychologists, members of a group, even though an unambiguous operational definition of membership was first provided in 1941 by the new academic qualification. Professionalization was a goal actively pursued by many, and yet at the same time also a subjectively formative process. It emerged as a movement to unite psychologists in the concerns of their subject. In the pursuit of their goals they were relatively blind to, when they did not actively affirm, the social and political context in which professionalization took place.

In this chapter I shall make some observations about this aspect of the professionalization process, based on the statements of participants. The aim is to try to find out something about the subjective motives for the increased attention paid to practical work, and how this was experienced and interpreted. This is a difficult undertaking. The only sources available are autobiographies and interviews with some psychologists. These are not only statements made from memory, but statements about experiences in the Third Reich which interviewees acknowledged or resisted in their own fashion. This is an interpretation of both the interviewees' past motives and their present view of things. It is therefore augmented by the interpretation of contemporary material.

The problem of interpreting these subjective reports is that National

259

Socialism is different for those of us who are studying it than it is for those who experienced it as part of their day-to-day existence for twelve years. They may be unable to provide answers to questions now that did not arise at that time. In some of the interviews, an inability to talk about identity and subjective motivation was encountered. We have no way of knowing the difficulties someone encounters in trying to remember harrowing details. None of the Wehrmacht psychologists said they could remember that the problem cases in the military, those "difficult to train" or of "sublevel character," who were in part examined by psychologists, could be transferred to punishment battalions or, if that failed to help, could be handed over to the police and taken to concentration camps. Yet this had been openly declared by the army, and there had been two articles on it in *Wehrpsychologische Mitteilungen*.[1] Nor do we have any way of assessing the extent to which the remembered past is a construction, except where things that are reported deviate from the written records, as was the case in some of the interviews. We know little of the hopes and bitter disappointments that determine one's view of the past. However, difficulties, pressures, and constraints are conspicuous in all the interviews and autobiographies – as are concessions, whether by the individual who included comments on race theory in an essay to get it published, or by a group such as the Wehrmacht psychologists who got involved with questions of race psychology, or by leading representatives of the discipline who included race studies in the examination regulations. But remarkably, in their recollection they really did everything right, cleverly maneuvering through adversity, making the best of a bad job. Only once did I experience a more intensive discussion of concessions made to censorship, and only once did I experience a self-critical admission of mistakes – from, of all people, a woman who had emigrated, and had set foot an German soil only once since 1945. The things that benefited the discipline or the profession were seen by many as being wholly good, no matter what the services, no matter in whose service. The attractiveness of working in the science or the profession seems to have been so great that it clouded the perception of contradictions and problems related to these activities. This must be a major subjective driving force for the professionalization of a discipline. Despite all limitations it seems to me that such subjective material offers the only way of finding out about such driving forces. It will be interpreted against the backdrop of the known objective conditions and forces behind professionalization.

Practical activity: attraction and loophole

Since Wehrmacht Psychology offered the most opportunities for practical activity in professional psychology, it elicited the greatest attraction for

1 AHM, 1939, p. 413; 1940, pp. 6–7; Grunwaldt (1941); Naegelsbach (1941).

psychologists. This was not so much a particular affection for military tasks and goals, but rather an interest in the practical use of their own science. At the end of the twenties the relevant circle of psychologists did not really identify themselves with military tasks according to Ludwig Kroeber-Keneth; they were more interested in their science and their own further training (Z, f. 198). Josef Schänzle also felt that Wehrmacht psychologists saw themselves primarily as psychologists and not as military personnel (f. 95). Most of them, in the judgment of Gerhard Munsch, were occupied with the idea of being able to apply their science (f. 103), though they were less concerned with the progress of the subject as a whole than with their individual opportunities to apply aptitude diagnostics, expression diagnostics, and characterology (see Schimansky, Z, f. 109). Siegfried Gerathewohl had an offer to stay on at the university but preferred the practical work in Wehrmacht Psychology (f. 18). Professors wanted to get acquainted with Wehrmacht diagnostics because it was the first large field of practical activity (see Rudert, Z, f. 193). Edwin Rausch saw his work in Luftwaffe Psychology mainly as an opportunity to have a look at diagnostic practice: "Getting to know the possibilities and the limitations of such examinations was a welcome augmentation for a theoretician who otherwise only experimented in basic research" (Pongratz et al. 1979, p. 224). This sounds sober enough in retrospect, but as Kroh said at the height of Wehrmacht Psychology, just after the new examination regulations had been decreed, even "the most extreme theoretician would no longer want to do without the beauty of practical psychological work such as the war has brought him" (1941, p. 5). As Wyatt and Teuber diagnosed in 1944, "Most psychologists, indeed, were only too glad to turn their minds to problems that might be of interest to the army" (p. 234).

Two further motives may have led some to enter Wehrmacht Psychology. It was secure work, and the diagnostic activity did not entail spreading Nazi views. Leonhard von Renthe-Fink reported that as a philosopher he had no chance of obtaining a habilitation in the thirties, so his teacher Rothacker advised him to take up something "more solid." When he heard that Wehrmacht Psychology was looking for recruits, he applied and was accepted (Z, f. 32). As a former assistant of Adhemar Gelb, as well as for other reasons, Wolfgang Hochheimer could not stay at the university in Halle, so he went to Wehrmacht Psychology (f. 53). Josef Pieper told of an acquaintance who "left the school service voluntarily because of political difficulties and went to 'Wehrmacht Psychology'" (1976, p. 160). Such events have led people to speak of Wehrmacht Psychology as offering a "hideout" to people at risk (Bornemann, Z, f. 147), or even as a "refuge for anti-Nazis" (Flik, Z, f. 150). The Wehrmacht psychologist Heinrich Lüderitz from Königsberg called it an "emigrants' station"; he himself had quit the school service. Gerd Schimansky, who reported this, added that this did not justify speaking of a "place for the persecuted," but only that people could get jobs there "who wanted to dodge the obligation to spread Nazi slogans (e.g., in the schools)" (ff. 108–9).

Memory may shift some of the facts a little, perhaps for peace of mind. Thus, one Wehrmacht psychologist described a "typical example of Simoneit's appointments policy" – a colleague who had problems at his school because of his anti-Nazi views, but who still got appointed (f. 28). Yet the colleague himself reports never having even applied for school service, to avoid political problems. He really wanted to become a writer anyway and summed up by saying Wehrmacht Psychology could not be referred to as a refuge for the politically persecuted (f. 110). The references to a refuge suggest that the persecuted might have been given jobs, but no such cases are known to me. Wehrmacht psychologists were mostly civil servants, and as such were covered by the revised Nazi civil service laws (see Chapter 2). It is probably more accurate to say that Wehrmacht Psychology offered an alternative in the case of political difficulties in teaching, where it was compulsory to spread the "correct" ideological outlook – in the schools more so than in the universities. In contrast to the universities, Wehrmacht Psychology had to account only for appointments to the corresponding personnel office, not to any party offices.

It is hard to say whether a similar statement can be made for the work of psychologists in industry, in view of the small number involved. Psychologists such as the former Communist Kroeber-Keneth found jobs there (see Chapter 4). We know that a number of physicists left the universities for political reasons and went to industry (Beyerchen 1977, pp. 168ff.). It was even possible for a psychologist like Martha Moers, who was not tolerated in teaching, to find employment with the Labor Front. Although such psychologists had better chances of finding a position in industry or the Wehrmacht than in higher education, thus also escaping being forced to spread Nazi ideology, this does not mean that their activities were not constrained by the political goals of the Nazis, even though one wished to think this was the case.

Practice as self-deception: oases, morality, and objectivity

The word "oasis" or "island" is often used by former military psychologists to describe Wehrmacht Psychology. Von Renthe-Fink says, "Just as the Wehrmacht as a whole was at that time a sort of oasis in the Nazi state so was Army Psychology" (f. 34), and Heinrich Roth writes, "In Army Psychology we lived on an undisputed island of undisturbed, scientifically based, psychological selection even in 1942" (Pongratz et al. 1979, p. 264).[2] Was this really still felt in 1942, even though at that time every army psychologist must have known what sort of war the officers and specialists were being selected for? It would be understandable if these psychologists had described

2 See also Rudert, who spoke of the "protective roof of the Wehrmacht," under which the psychologists "fled" (Z, f. 192).

their work as preferable to being sent to the front. But the talk of oases or islands indicates a certain self-deception; practical activity was seen only as it was directly experienced, without putting it in a wider context.

Maintaining this self-deception required a corresponding idea of the morality of one's actions. In Wehrmacht Psychology we find the common idea that the work helped people and that psychological methods as such were nonpartisan: "Morally we justified our work with the thesis that it would be to the advantage of every applicant to be allocated to the activity (officer, flier, radio operator, etc.) at which he was most successful" (Roth, in Pongratz et al. 1979, p. 265). Another psychologist said that having accepted that they could do nothing to oppose the war, some psychologists saw it as their duty "to allocate the examinee soldiers, in their own interests, optimally according to their abilities, since then they would be in less danger in the war" (Z, f. 150). How could one believe oneself to be acting in the best interests of the individual when, for example, validity criteria for the assessment of pilots were their survival and the number of enemy planes shot down, as in Skawran (I C)? Why is it not possible just to answer that in those days it was important for everyone in his or her profession or in the army simply to survive? Why, in contrast to the deeds of common soldiers, are professional activities always seen as having a humanitarian element? Probably this is a general trait of the ethical consciousness of professionals. As long as the ethics of the profession are limited to direct contact with individuals, it is possible to regard one's deeds as humanitarian, ignoring the overall conditions determining that contact.

The belief in the humane and unbiased character of psychological method lives on in the tales of the dissolution of psychological services in the Luftwaffe and the army. We saw in Chapter 8 where the reasons for the dissolution are to be found. Some former Wehrmacht psychologists, however, are certain that it was due to the democratic character of psychology: aptitude testing was inappropriate when ideological convictions counted more than ability, so the party was necessarily opposed to it (Z, f. 25); the democratic commission had frustrated the nepotism of the party, which led to the dissolution (f. 71; cf. Arnold 1970, p. 29); "objective personality selection" had a "democratic element" (UAL PA 95, f. 93). That drew the opposition of the party, so one would like to believe. Others believed that it was simply the good, practical work of psychology that led to the dissolution (Herrmann 1966, p. 274). Wehrmacht Psychology must have been politically subversive solely on the basis of its democratic methods.

Such a perspective obscures the question of the external demands for which psychology was functional or dysfunctional. If the methods were democratic there is no point in asking whether it was its functionality for the Nazis' aims that enabled psychology to flourish – the development of the Wehrmacht was, after all, such an aim. The fact that there was nearly no Nazification of the content of Wehrmacht Psychology (none was necessary for their purposes) can then be interpreted as a refusal to compromise with National Socialism, although its emergence had made the blossoming of the profession possible. If the ideological, the undemocratic, the inquisitory is always seen as being outside psychology then it cannot be contained in its methods themselves although in fact contemporary psychology was imbued with military ideology. By setting the context of use as extraneous to the procedure, one is not obliged to perceive external goals as determinants of psychological theory and activity. However, it was possible to set up the triad of psychology, client, and psychological method only in specific institutions with specific aims, and these alone made the development of professional psychological activity possible.

In occupational psychology we find the protective ideology of objectivity. The literature about engineers and natural scientists reports much the same. They were quite ready to take on jobs for the state as long as these could be carried out in a scientific manner (Ludwig 1979, p. 218). A female employee of the work psychology institute of the Labor Front remarked that psychologists in the institute did their practical work without paying any attention to ideology (Z, f. 49). An industrial psychologist commented that factory managers and engineers had refrained from all "ideologization" related to occupational psychology, as well as generally (f. 146). Politics was party ideology, but technical matters were a different matter. Even Albert Bremhorst, one of the heads of the Amt BuB of the Labor Front and a Nazi Party member, is of the opinion that the work of the office had nothing to do with politics and political ideology (Z, f. 185). But it was a highly political matter when the office's psychological institute inspected deported workers. The political ends here are dismissed as limiting conditions in order to justify past actions. Yet even in relation to this selection work, the big question of the ends being served is avoided in favor of the tiny hope that an optimum distribution "perhaps even contributed to an improvement in the situation of prisoners of war" (f. 48) – the same soothing tones again. Despite the ends that it served practical work somehow managed to be of benefit to the individual.

Passing the test

In addition to such protective ideologies, we find an identification with the goals set for psychology from outside. After the war psychologists defended Wehrmacht Psychology against the charge that it had not been effective. In their replies they accepted the criterion of military efficiency. Looking back, Wilhelm Arnold felt that the institutions of Wehrmacht Psychology had proved themselves to be practically useful (1970, p. 31). Karl Mierke saw confirmation of the "appropriateness of psychological selection" (in Pongratz et al. 1972, p. 233). Paul Robert Skawran went to great lengths to try to figure out the optimum contribution that psychology could have made to a greater German military victory. He tried to assess the selection of pilots using, among other criteria, the number of people killed. He criticized the dissolution of Luftwaffe Psychology as unwise for the successful continuation of the war, and before the new Federal Republic even had an army he had made a suggestion on the future role of psychologists in it (I C Skawran 1). Here we see an element of the pragmatic ideology of dominant science. The point is not the content and the aims – they are not seen as questions of science – but solely the rationality of the means. Science considers only whether it has tried to meet the goals that have been set. This means that individual experts are concerned to have a voice as authorities in their field and to prove their worth, whatever the aims may be.[3]

New loyalties

The opportunity to prove their worth created new loyalties for psychologists. Although those in the Wehrmacht saw themselves first as psychologists, they also felt that they had to serve military aims, as the "psychology corps" in the Wehrmacht. At the start of the war there was agreement in the inspectorate that those members who had not been at the front in the First World War should serve in the field (Z, f. 22). Only one did not do so, which still earns him criticism from an "old comrade" today. When an army psychologist fell in 1941, von Voss and Simoneit wrote in an obituary that he had "lived the life of an army psychologist to the highest meaningfulness." They were not content to express only their sorrow at the death of a colleague; he had fallen for the honor of the psychologists.[4]

The new loyalty also demanded obedience and gratitude. An army psychologist says that there was no question about continuing to do one's duty during the war (Z, f. 35). Even when psychological services in the army and Luftwaffe were dissolved, it seemed appropriate to react by following the higher purpose. In 1942 the *Wehrpsychologische Mitteilungen* stated that the replacement of the selection by psychologists with the selection by officers bound the "scientists who were affected to military obedience" and left no room for discussions "especially at a time when the gaze of the Germans is directed to higher aims than the relatively smaller worries of a certain professional group" (Oelrich 1942, p. 3).

Servility to one's own patrons thus continued even when they were dismantling psychology. Obedience was now shown rather than cheerful gratitude. This servility strikes one above all in the praise for von Voss on his departure from Wehrmacht Psychology (which was only temporary since he

---

3 See also the analysis by Ludwig (1979) on technicians and engineers in the Third Reich. For many of them, the desire to participate as experts, no matter where or for what purpose, was a mentality found throughout the NS period.

4 *WPsM*, 1941, 3 (H. 10), p. 127.

was recalled during the war). Not only Wehrmacht psychologists (e.g., Grunwaldt 1939), but also others (Rupp 1939) extended their thanks in the psychological journals. Von Voss had already been appointed an honorary member of the DGfPs out of gratitude for his efforts on behalf of psychology. The respect is still deeply ingrained: several former Wehrmacht psychologists when correcting my transcripts of our talks insisted on including that in the First World War he had been the last chief of staff of the Gardekorps and had been awarded the decoration *pour le mérite*.

Professionalization as belonging

The main subjective effect of professionalization was the creation or reinforcement of a feeling of belonging. Earlier hopes of increased support for psychology were realized with its practical use in the Wehrmacht and the recognition it received in the Nazi period. Thus, according to such spokesmen as Simoneit and Kroh there was in Wehrmacht Psychology the spirit of "cooperative work," in psychology as a whole a growing unity, and after the DPO excellent morale, or at least the attempt to create it. With the promulgation of the DPO Kroh felt that the "pressing discontent, which cast up innumerable fundamental questions," had been vanquished,

> first in psychological research itself, inasmuch as it increasingly demonstrated the basic principles on which all schools and tendencies agreed: aspects of holistic (*Ganzheit*), structural, community and developmental psychology in particular. The other point is that it was highly important that the various lines of research should be confronted with an area of cooperative effort, within which the usefulness of all methods directed at understanding and selecting people are reexamined repeatedly and conscientiously. This has been achieved by German Wehrmacht Psychology, and already one can say this is its historical contribution. (1941, p. 5)

Simoneit urged Wehrmacht psychologists to convince critics and doubters of the fertility of psychology in the military through their work together: "We need a real community of work and a real community of struggle for our idea" (1939b, p. 4). He was concerned with the interests of Wehrmacht Psychology as a whole. He speculated that this coincided with the interests of every individual psychologist, since if Wehrmacht Psychology proved its worth, it would legitimate the scientific usefulness of the subject as a whole and of each individual psychologist. He wrote: "Since belonging to the psychologists' corps of the Wehrmacht will in the future provide a decisive legitimation for every German psychologist we can hope that the entire body of German psychologists will grow together as a result of work and comradeship in military psychology in a way that must be described as unprecedented and unique in the history of psychology" (1940c, p. 2).

Coming closer together also implied shaving the edges off the differing scientific positions by means of the work and its regulations. Wehrmacht Psychology thus also functioned as a leveler. In Firgau's estimate this helped to bring down barriers of mistrust between the different psychological schools (Z, f. 13). Interestingly, Rudert reports that in his work for Army Psychology Jaensch made no use of his type theory (f. 193).

The work on the DPO brought together the top administrators in Wehrmacht Psychology and in academic psychology, as well as representatives from other areas. In the DPO Commission there was a firm desire for unity (see Chapter 6). Once the regulations had come into existence, Kroh hoped they would also have a subjective effect, since they provided a basis "for the first time for a professional status group (*Berufsstand*) of psychologists, and will thus also favor the development of a healthy feeling of standing (*Standesgefühl*)" (Kroh 1941, p. 19). The old feeling of belonging to an aristocracy that united professors (see Bleuel 1968, pp. 115ff.) was now obviously eclipsed by the feeling of belonging to a professional group. The professors underwent a subjective change. In the future they were to perceive themselves primarily as members of a professional group rather than in terms of a privileged social status. The Third Reich, which destroyed traditional social allegiances (Dahrendorf 1972, pp. 418ff.) and which did away with the outdated privileges of the old estates and favored the new professional groups of physicians or technicians, thus paved the way for "leveled-out middle-class society" (see Brückner 1978, p. 87) and may have helped the professors' subjective transformation into professional psychologists. Kroh symbolized this transformation and the consolidation of the psychologists in a professional group. After the war no other psychologist received such tributes as Kroh, not least because of his efforts for the DPO.[5]

Conformist restrictions on the rationality of scientific means

Thus, even after the war the history of psychology in the Third Reich was measured in terms of the benefits of this period for the professional standing of the subject itself. The question of whether psychology served National Socialism, however, is noticeable by its absence. The reflections of many remained bound up in the categories of experts, questioning the suitability of their actions but ignoring the purpose and goals for which they were acting.

In view of the relative irrelevance of psychology in the armed forces, and its irrelevance in society as a whole, one may understand why psychologists at that time harbored the illusion of being able to bring about a little bit of humanity. Today such arguments seem unconvincing. What was humane about a soldier becoming a radio operator or a driver or about a pupil

5 See Bornemann (1967), Bergius (1955), Kaminski and Märtin (1956), Lersch (1956), and Wellek (1955).

who had finished school becoming an officer or not? Why isn't it admitted that the "testing" of psychology was proving itself by selecting soldiers to conduct a bloody war?

Hardly any of the interviewees considered the possibility of subversive activity by psychologists in the Wehrmacht. One army psychologist at least mentioned that when investigating the "race components" of prisoners of war in occupied Eastern Europe he tried to diagnose a very high proportion of "Nordic" features. This provided the Army High Command with an argument against the SS campaign to kill these people (Z, ff. 57f.). Heinrich Düker told me that he wanted very much to work as a psychologist in a large factory so as to earn enough money to pay for supplies for people hiding from the Nazis in Berlin (ff. 40–1). Hardly anyone thought of even such elementary political evasion as Hellmuth Bogen, who changed his occupation at the beginning of the war in order to avoid having to work in Wehrmacht Psychology (Baumgarten-Tramer 1947). Apparently, after the war nobody asked why psychology as a subject had not been more immune to National Socialism. Instead, the psychologists considered whether they had fulfilled their professional tasks "dutifully," maintaining scientific standards. This alone is sometimes presented as an anti-Nazi attitude, as if applying rational scientific methods in itself was antitotalitarian.

The idea that applying psychology was humane as such, or at least neutral and uncommitted, preempts the question of its application and of the role of scientific experts. Of course, the psychologists of this period are not alone in seeing the question of the purpose to which a science is put as something affecting scientists solely as citizens but not as theoreticians. This view is central to pragmatic and positivist scientific approaches. When social science is reduced to social engineering, its theory remains impotent in the face of its abuse under dictatorships. If the application of psychology is seen as neutral or even humane in principle, an absolute loyalty to the state has already worked its way into the self-image of the science. Resistance is then a matter for the individual, critical *citoyen*.

But even as citizens, those psychologists who did not have to emigrate were largely conformist. The only ones to offer open political resistance were Heinrich Düker, a member of the International Socialist Kampfbund, who as a result of his illegal work ended in prison and concentration camps, and Professor Kurt Huber, a conservative Christian from Munich specializing in the psychology of music, who was condemned to death by the Nazi judge Roland Freisler on 19 April 1943 for his membership in the White Rose group.[6] For most other psychologists "normal" work went on as usual. Even in retrospect it is judged in frighteningly "normal" terms, in those of the prevailing conditions of professional activity. In an exceptional situation like the Third Reich the normal was the terrible, or as Hannah Arendt put it

---

6 See Huber (1947), Petry (1968, pp. 43ff.), and Weisenborn (1974, pp. 319ff.).

(1977, p. 26), "Under the conditions in the Third Reich only 'exceptions' could be expected to react normally." The German concept of loyalty – duty – which we encountered with the Wehrmacht psychologists, is well suited to the "normality" of experts who stick to their procedures without bothering about the ends these serve. If we wanted to find where psychology realized a little humanity in fascism, we would have to look not at its day-to-day routine, how it functioned, remained unaltered, or "proved its worth," but at the exceptions, the deviations, the narrow, subversive marginal paths.

10

Science, profession, and power

Two ideas about the role and development of psychology in the Third Reich are in need of correction. It was not true that psychology as a whole prostituted itself, or was called into service by the Nazis, nor was it the case that, as a result of its theories, it found itself in conflict with this regime, was more or less oppressed as a subject, or was able to secure its survival only in certain protected areas.

Psychologists did try to place their discipline in the service of organs of Nazi domination, but psychology contributed little to stabilizing that domination. It was socially much too insignificant for that. It was not systematically involved in the development of official propaganda; it had too little to offer for that. At that time in Germany it knew neither methods of propaganda nor research on its influence. As far as is known, psychologists were not used by the Nazis or the SS in persecution, torture, or murder. Whether the selection of soldiers by psychologists actually helped to improve the army's fighting strength is hard to say.[1] Even if the Wehrmacht may have benefited from the work of psychology, it was hardly dependent on it.

Some psychologists were persecuted by the Nazis, but psychology as a subject did not suffer particular oppression. The Nazis drove out Jewish scholars for racial reasons.[2] Among the emigrants, whom Wolfgang Köhler joined for political reasons, were the innovators of that period. Thus, as a result of Nazi policies, a number of theoretical developments in German psychology were interrupted, above all Gestalt psychology, Kurt Lewin's

1 According to Samelson (1979) the U.S. Army received almost no benefit from psychology during the First World War (see Chapter 1, note 46).
2 See the *Liste des schädlichen und unerwünschten Schrifttums. Stand vom 31.12.1938* (List of pestilential and undesirable writings), Leipzig (n.d.). This includes Adler, Freud, Fromm, and Reich; Horkheimer's *Studies on Authority and Family* is listed and banned, but the names Koffka, Lewin, and Stern are not to be found.

research, and the work of the Hamburg Institute under William Stern. However, if expulsion and emigration are taken as the only features of the development of psychology under the Nazis, it can easily lead to an exculpation of the entire discipline and those of its representatives who remained behind. Nor, as could be demonstrated, does the dissolution of Army Psychology and Luftwaffe Psychology in 1942 allow one to make the generalization that the party persecuted psychology, although many would like to think so.

The loss of leading theoreticians and theoretical stagnation in certain areas is one side of the coin; the other is the development of areas of practical activity and the improvement in the professional and university standing for the subject. Psychology had fulfilled the criteria of complete professionalization by 1941 (see Chapter 1). It had its own distinctive methods of diagnostics and models for the description of personalities, with the appropriate professional role. It was able to establish itself as a discipline in the universities, offering independent training with its own qualification, and it could improve its strength in relation to philosophy, its main competitor. Outside the universities, institutional areas of application were maintained, expanded, or gained, and in the Wehrmacht the first career was created that required professional qualification. Subjectively, the orientation toward application and the corresponding shift in teaching led the majority of psychologists to see themselves less as philosophers and more as professionals. The expert replaced the scholar. Thus, the period of the Third Reich was one in which the professionalization of psychology in Germany experienced a massive boost, certainly the largest it experienced up to the sixties.

What were the most important forces that provided this boost? Did Nazi rule have a particular influence on the type and tempo of professionalization in psychology? The foundations had been laid for this professionalization before the Nazis, inasmuch as German psychology had already begun to develop relevant practical methods and fields of application: it would have become a profession without Nazism. Nevertheless, Nazism favored the definite emphasis on the role of aptitude diagnostics and enormously accelerated professionalization. The most important factor there was the dramatic increase in the quantitative demands of the Wehrmacht that came with its expansion. Since psychology had been officially anchored in all the armed forces, when these expanded it automatically expanded with them. Thus, the interest psychology had in applying its knowledge met a demand that made possible the realization of this interest. The diagnostic methods that had been developed fit the practical requirements. Their use in the Wehrmacht allowed psychology to bolster its image in the universities and to convince the science administration to secure the status of the subject. This resulted in chairs being established. The science administration advanced professionalization primarily through the introduction of the diploma qualification in psychology. The interests of Wehrmacht psychologists in regulating careers and of academic psychologists in strengthening their subject united them in active efforts toward this end. Even under a dictatorial system the professional group was able to remain relatively independent and to achieve its own aims, though at the price of aligning itself with the aims of the system. When preparing the DPO there was thus a coalition in which the interests of the professional group, those employing the professionals, and the state apparatus all coincided.

This study has also shown factors that impeded professionalization, the most significant of which was the rivalry of other professional groups. Physicians, the epitome of a professional group, opposed the psychologists in the army when psychological tests were first introduced and later when they went to work in military hospitals. Physicians also blocked the inclusion of psychopathology in the diploma intermediate examination. In the factories, the engineers used psychotechnical methods for a time to the exclusion of psychologists. In the Wehrmacht, officers wanted to retain the task of selecting officer recruits. However, these were not the only hindrances to professionalization; internal rivalries also made joint efforts more difficult.

The results of this study contradict a common conception of the decline and fall of psychology under the Nazis. There was no fall in the sense of repression of psychology because of its content. In addition to the freezing of developments resulting from expulsions and repression, the fall of psychology lay more in its pliant conformism. Some psychologists attempted to trim their theories to provide ideological support for Nazism; a number of psychologists and their representatives were keen to meet the practical demands placed on their subject, without either questioning the political conditions governing the application of their knowledge or coming into conflict with their own ethical standards. They acted solely in terms of whether or not their actions served professional interests.[3]

Dahrendorf (1972) writes that due to the "modernizing" function of National Socialism, which destroyed traditional social ties and subjected people to the logic of the military–industrial megamachine, the resistance to the regime had to develop in a curiously conservative form, as an attempt to maintain traditional ties and to defend the "ancien regime." It is therefore not surprising that the reconstruction of postwar psychology represented to a certain extent a conservative restoration. Whereas the war had led to a practical, professional orientation, the professors now reverted to their traditional academic role. Psychologists remembered the importance of weltanschauung for their subject, but this time in a conservative sense (see Maikowski et al. 1976, pp. 23ff.). Lersch (1947) wanted to find a new meaning of life through psychology, wishing to confront the modernist rationalization of social life with the individual's contemplation of the inner world. But the professionalization of psychology in the Third Reich left its traces, and it was not possible to turn

3 Even after the war the period of the Third Reich was not seen as one of decline. The idea of a decline first cropped up when theoretical and methodological considerations led to the rejection of characterological and holistic approaches at the end of the fifties.

the clock back. Psychology was a profession: the DPO had formed a demarcation line against other disciplines; there were qualified psychologists who wished to exercise their profession; teaching was provided in the universities; and there were social expectations that psychology would be of use. It would be worth studying in its own right the way in which the formative influence of the Third Reich on psychology affected the postwar history of German psychology. Here I will content myself with a few observations on the traces of professionalization that we can observe in the history of West German psychology.[4]

Let us first look at the fields of occupational activity, where major employers disappeared with the Nazi system. Until the remilitarization of the Federal Republic there was no longer a Navy Psychology. The body for psychological educational guidance, the NSV, disappeared completely, as did the psychological institute of the Labor Front. With the end of Wehrmacht Psychology the main institution for large-scale practical research had gone. Psychologists had to adopt a new professional orientation. Right through to the sixties they persevered in the traditional channels, finding in some instances new institutional contacts. They worked in employment exchanges and firms, as well as in new educational counseling centers; and with the formation of a new army, they returned to military psychology. In 1953 a career for psychologists was set up in the psychological service of the Federal Labor Agency (Bundesanstalt für Arbeit). Clinical psychology was very slow to develop.

The Third Reich left academic psychology an ambivalent inheritance. Institutionally it could live off the progress it had made; theoretically it remained entangled in the work that had been pursued during the Nazi period. The theories and the research fields of the emigrants were largely ignored in favor of holistic (*Ganzheit*) psychology, characterology, theories of expression, and diagnostics (Maikowski et al. 1976, p. 33). The typologies still represented a large proportion of texts on personality theories. The concepts of selection applied in the Wehrmacht dominated ideas on diagnostics; after all, a whole generation of psychologists had experienced its socialization in Nazi Germany, many in Wehrmacht Psychology. This was still the situation in the fifties, although especially in schools, different methods of selection were expected. In the methodological debate of the fifties, in which some psychologists favored U.S.-style mathematical psychology, the older professors and former Wehrmacht psychologists stood their ground. Both sides, directly or indirectly, made references to the "test" in which psychology had proved itself in the Wehrmacht (see Geuter 1980, 1983).

Continuity in the universities was guaranteed by the continued tenure of professors. Lersch, Mierke, Rudert, and Heiss, all former (in some cases leading) Wehrmacht psychologists, occupied chairs for psychology. Repre-

---

4 It is, of course, just as necessary to examine the influence on East German psychology. There has been no study on this topic.

senting characterology and holistic psychology were Kroh, Rudert, Sander, and Wellek, and later the former Wehrmacht psychologists Undeutsch and Arnold. Kroh, Rudert, Sander, and Wilde, all former party members, left their chairs in the Soviet zone of occupation and subsequently found professorships in the western zones and West Berlin. Here the professors Anschütz, G. H. Fischer, and Pfahler had been dismissed for political reasons; Pfahler was later allowed to resume teaching in Tübingen.

In institutional terms psychology was able to preserve the successes it had gained in the Third Reich. It was an independent subject. Chairs for psychology were now a recognized part of a philosophical (arts and sciences) faculty and were included at universities newly founded after the war, such as Mainz. The decisive institutional inheritance remained the examination regulations, the DPO, which underwent only slight modifications in the occupation zones (Maikowski et al. 1976, p. 37). Some of the names of subjects were altered, and medical-clinical subjects were also reintroduced, but otherwise the basic structure influenced courses in psychology right through to the seventies. Thus, the continuity of academic psychology was preserved institutionally, personally, and in terms of content. The vanquishing of fascism, the end of the war, and the reorientation were "not a fundamental caesura" (p. 32). Where there were disagreements, as in the methodology debate or the split between the academic DGfPs and the Professional Association of German Psychologists (BDP), these were old differences, which did not indicate dissension in dealing with the history of psychology in the Third Reich. Nor does the experience of that period seem to have had an influence on the way psychologists approached the professionalization of their subject.

As it had done in 1929 (see Chapter 2), the Executive Committee of the DGfPs (following the general meeting of 2 August 1951) presented the ministerial bureaucracies and universities with a document entitled "Memorandum on the Dire Situation of Psychology at German Universities." The institutional state of psychology was bewailed; university chairs in areas that had been occupied by the Germans (Poznan, Prague, and Strassburg) were regarded as losses, and in marked contrast to 1933 (see Chapter 2) political dismissals were regretted, not only as a loss for the subject but also, in personal terms, for individuals. When reference was made to a "catastrophe" for psychology this did not mean the year 1933 but the year 1945. The enemy of psychology was not the Nazi system but the Soviet-occupied eastern zone. There was no hesitancy in citing the favorable situation of psychology before 1945 to justify demands for the improvement of the "catastrophic and completely untenable" state of things after 1945.[5]

When the first public rumors about establishing a West German army arose in 1951–2, the chairman of the BDP made an offer of cooperation to

5 Deutsche Gesselschaft für Psychologie, *Memorandum über den Notstand der Psychologie an den deutschen Hochschulen*, A 110.70, H. 4, ff. 192–6.

the Chancellor's Office in setting up a "Wehrmacht [sic] psychological service" (Mattes 1980, pp. 43–4). The BDP, university psychologists, and former Wehrmacht psychologists got busy, finally succeeding in setting up the Federal Army Psychology. The latest U.S. methods were employed, but as late as 1967 a journal of the Ministry of Defense was given the name *Wehrpsychologische Mitteilungen*, a link to the old days. An issue in 1972 with documents, essays, and a bibliography on the old Wehrmacht Psychology is unblemished by any criticism. Enthusiastic responses to the publication were received from practicing and academic psychologists, including the vice-president of the BDP.[6]

It took until the seventies before "productive unrest" in West German psychology led to a new orientation. By then a new generation had taken over in academic psychology; there had been theoretical and methodological developments (mostly adopted from the United States), new practical demands on the profession, and a vehement criticism of psychology during the post-1968 student movement. The framework of psychological theories, curricula, and examination subjects held as long as there were neither theoretical advances relevant to the profession nor a qualitative restructuring of the employment market for psychologists. The development of new psychological methods of psychotherapy (behavior and client-centered therapy) brought with it an expansion of opportunities for the profession in the clinical field. But in the universities the ties to practical activity were tenuous, and the emphasis lay on mathematics and methodology. Coupled with the expansion of the universities in general and the subsequent boom in psychology graduates a new fissure developed between scientific knowledge and professional practice. The old examination order reflected neither the development of clinical psychology nor the new requirements of the labor market. The psychologist was no longer the diagnostician for whom these examinations had been designed. Roles now included working on projects for educational innovation, or as adviser and therapist. But the scientists' demands for methodology in the curricula were not met by the old DPO either. A new framework introduced in 1973 attempted to do justice to everyone. Specialization was introduced in the second half of the course, replacing the idea of a common canon of knowledge for all students (an idea that some now wish were back).

In 1941 the situation had been clearer. The fields of professional activity were known, and psychology professors had an idea of what the student should learn for selection and classification. As authorities and norms were questioned in the wake of social change in the sixties, psychology faced new tasks requiring methods of counseling and therapy. Psychologists worked in a number of new areas, some far removed from academic psychology, and attempted to open up new opportunities and improve their legal standing. Today there is a new impulse of professionalization aimed at making permanent the gains made in therapy, counseling, and social work.

6 *Wehrpsychologische Mitteilungen*, 6 (H. 4), (1972), pp. 119–23.

Professionalization and the history of the social sciences

Professionalization seems to be an ongoing problem in the social sciences, although psychology now appears to be more successful, perhaps because it was more successful before. Sociology, political science, and pedagogy had no comparable success in the Third Reich – they were not among the subjects to receive examination regulations during the war. This is a good indicator of professionalization in Germany, because these regulations are passed by state authorities, which normally assess the practical relevance of a subject. It is instructive to compare briefly the histories of the different social sciences.

Before the Second World War sociology was a part of philosophy (social philosophy) or of economics (empirical social research). There was little sociotechnological knowledge that could be used for the solution of restricted social problems. There was no special university teaching for sociology and no special exam. According to Lepsius (1979) and Schelsky (1980) it was the problem of always being a subsidiary in the economics faculty that led representatives of the subject to set up a diploma course in the fifties. There was no occupational goal initially, but it was hoped that a supply of graduates would create its own demand. The first Diploma Examination Regulations were introduced in Frankfurt in 1955 and in Berlin in 1956 (Lepsius 1979, p. 46). The other universities then gradually followed suit.[7] As the number of students increased and the demand grew for a usable social science, sociologists adopted the strategy of the creation of professional demand in new areas such as development sociology and regional or town planning (see Matthes 1973, p. 52). Schmitz and Weingart (n.d.) refer to this as a case of "active professionalization."

The development of political science was somewhat similar. The first diploma had been introduced in 1930 at the Deutsche Hochschule für Politik in Berlin, which could not continue its work under the Nazis. In 1956 the restored college was granted permission to award the academic degree of diploma political scientist, and the first examination regulations were passed in 1962. The aim then was to establish political science as a career in the state administration (Hartung et al. 1970, p. 51), where law graduates had always been favored. The efforts from within the science were not all successful, perhaps because it had no theoretical tradition or research methodology applicable in governmental or administrative sciences. The professional diploma did not catch on.[8]

As a basic subject for teacher training, pedagogy enjoyed a traditionally strong position in the universities. It was to provide future high school

---

7 General regulations for the diploma examination in sociology were passed in 1968, but the procedures continued to differ. In addition to the diploma sociologist, there was the diploma social economist, the diploma social scientist, and the diploma economist (social sciences). At several universities no qualification is possible. See *Blätter für Berufskunde*, 3 IV A 01.

8 See van den Daele and Weingart (n.d.), pp. 76ff. and *Blätter für Berufskunde*, 3 IV A 02.

teachers with general pedagogical expertise. But pedagogy was not an independent subject of study, nor was there a corresponding profession – only the possibility of obtaining a doctorate. In the sixties it was hoped that educationalists would supply reform plans for the educational system, so general regulations for a diploma in pedagogy were introduced. It was hoped also that this would bring pedagogical training at universities and teacher training colleges into line as a step toward becoming comprehensive institutions of higher education. Although university teachers were naturally interested in this development (Koch 1977), on the job market nobody knew anything about the new qualification. In adult education, an area in which the new diploma pedagogues had high hopes of being innovative, a certification had been created in the view of Weingart (1976, p. 229) without a recognizable body of knowledge that could be used for solving specific professional problems and without an expert role being institutionalized. Thus, semigovernmental coordinating bodies had created a course for which there was no defined field of professional activity. There was no correspondence between scientific supply and the real demands of institutions. The rise in the number of unemployed graduates in the late seventies was correspondingly dramatic.

By contrast, psychology had already met with success in National Socialism. In contrast to that of the other disciplines, its professionalization was supported by the demands of the labor market and by the development of scientific knowledge that could be transformed into professional deeds. We can thus follow Schmitz and Weingart in speaking of a "problem-induced" professionalization, the success of which, however, was linked to institutions in the Nazi state and thus partially negated by the end of the war.

This short comparison of the disciplines demonstrates that different factors influenced professionalization in each case. There is no law governing the professionalization of the social sciences or their sequence, but there are certain factors that obviously govern timing, speed, and progression. These were presented at the beginning of this book in a model. The professionalization of a discipline in Germany is neither independent of its scientific knowledge nor solely the result of the efforts of an academic group. It draws its impulses subjectively from the hopes of representatives to improve the position of their discipline, but it will meet with success only if it has a basis of applicable knowledge. It may be pursued by the professional group itself, but will not come about without state support, nor without adaptation to the needs of the labor market. The interests of social centers of power and the attitude toward them adopted by the professional group are important for the professionalization of a discipline. Even the theoretical development of a discipline in such a phase depends on these two factors. Histories of theory that neglect these links do not encompass the full history of a science. The history of professionalization in turn cannot omit the history of theory, because it is no more independent of social centers of power than it is of the theories and methodology of the discipline. Just as the history of a professionalized science is incomplete without the aspect of its development as a profession, the history of professionalization of a science is incomplete if it neglects the function of its knowledge.

For a long time the history of psychology restricted itself to being a history of theories.[9] But science is no purely intellectual affair involving a group of scientists producing theories. Particularly when a discipline sets about applying its knowledge professionally, it becomes more and more entangled in a web of social ties. In such a phase the significance, for example, of the universities for the development of the subject changes, as recent historiography has noted. The classical German university was seen as a place where students were to learn to follow the scientific course of reason, so that state-supervised practical training for the older professions always followed graduation. Now the university was also to be an *école professionnelle*. The state appeared as guarantor of training standards, a fact that poses new questions. For a period when psychology was not pursuing professional goals, the history of theories may be able to tell us something about the relationship between the development of psychological thought and the development of theories in other sciences or general developments in society. But the history of professionalization shows us another dimension: the links between science, profession, and power. Such historical studies are concerned with the forces at work between two poles, the discipline and the academic group at one and the social powers affecting scientific development at the other. This does not make it "whiggish history," the history of the winners, a charge leveled at the classical history of theories (Stocking 1965) and more recently also at the history of professionalization (Kuklick 1980, 1980a); in contrast to the historiography criticized by Kuklick, it does not assume that sciences that professionalize themselves are mature in Kuhn's sense, progressing independently of external influences. Rather, it opposes a false idea of autonomy.

The history of a science and a profession is made by individuals against a background of existing objective social conditions. Perhaps it is therefore necessary to develop a new concept of the subject in the history of psychology, a new concept of the role of the scientist and of the professional. Older works explained the development of psychology with the thoughts of "great men" (e.g., Boring 1950), while the critical historiography that followed the student movement neglected the scientist as subject in its analysis of the links between social structure and forms of scientific thought. The reintroduction of the subject ought to lead to a new understanding of scientists; individually and as a group their actions are based on their intellectual and social interests while at the same time they are caught in a net of external interests.

9 There has since been much criticism of the way the theoretical position of authors has determined their historical views of the various schools or theories; see O'Donnel (1979), Weimer (1974), and Young (1966).

The network of power and Nazi science policy

What was this net like in the Third Reich? What were the forces acting among science, profession, and power? Franz Neumann's analysis of the powers dominating National Socialism (see Chapter 1) provides a way of describing them. The first seat of power in the Nazi system, the Nazi Party, had only one constant policy concerning science that directly affected psychology: the expulsion of Jews and political undesirables from the universities. The party obviously had no clear concept of psychology, but neither did it view the discipline with constant animosity. In general the Amt Rosenberg supported those psychologists who pledged to buttress Nazi ideology. Attacks from this office against Wehrmacht Psychology were directed against the ideological fundamentals of Simoneit's psychology, not its diagnostic methods. Simoneit's later claim (1972) that the SS opposed Wehrmacht Psychology is not supported by the fact that Wehrmacht Psychology methods were used experimentally to select SS candidates. The Labor Front supported sociotechnical work psychology, and the NSV employed psychologists for educational counseling. The Nazi Party was interested not only in ideological servitude but also in the practical efficiency of the science. Thus, various offices in the party supported the establishment of psychological professorships needed for teaching, without paying particular attention to the ideological background of the intended recipients.

The science administration supported psychology, especially as it turned increasingly to professional matters such as aptitude selection for the Wehrmacht, and carried on with its institutionalization at the universities, without meeting high-level resistance in the party. No case could be found in which an appointment foundered on objections from the party. This result supports the hypothesis of Kelly (1973) that the science administration held its own against the party in university matters, an area in which elements of the Nazi movement could be subdued very early, as can be seen from the repulsion of the ambitions of the Nazi Student Organization and its incorporation into the lecturers' organization. However, given the interpenetration of bureaucracy and party, such a hypothesis is also problematic. In the case of psychology, for example, it was an SS man at the top of the ministry who pushed ahead the examination regulations, for which the chairman of the DGfPs expressed his gratitude in the *Zeitschrift für Psychologie.*

In contrast to its importance for the natural sciences (Mehrtens 1980, p. 45), industry played a comparatively minor role in the professionalization of psychology. It was probably the second largest area of application of psychological methods and began at the end of the thirties to establish posts for psychologists, but it did not produce any major initiatives that affected psychology as a whole. When there were such initiatives that affected, for example, university psychology, they came from the Labor Front.

It was the Wehrmacht that did most to further the professional development of psychology. Its strong position and its pragmatic use of methods to build a strong army were the primary factors that made the growth of psychology possible. Psychology grew together with the Wehrmacht. The controversy about Wehrmacht Psychology shows that there were also different expectations, especially regarding attitude toward Nazi ideology. The expansion of the Wehrmacht, however (in which psychology participated), was in everyone's interests – those of leading industrialists, the state apparatus, and the Nazis alike. Inasmuch as military expansion was an integral part of Nazi domination, it was this dominance that accelerated the professionalization of psychology.

Behind the diverse interests of the various centers of power, there is an interrelationship and a logic, above all the logic of the effective application of science to the war economy and the purposes of war. The results of this study support the view, such as that held by Neumann, that behind the multiplicity of power centers in the Nazi system there was a common factor; the results do not support those authors who tend toward a polycratic approach, which allows no recognizable logic in Nazi rule. Psychology did not have to establish a specific form of existence in the midst of chaotic rivalry by means of a tactical coalition barring all logic except that of keeping to its own terrain. Rather, the use of and the support given to psychology had a systematic relationship to the interests of the power centers. In Nazism there was also a rationality of demands placed on science.

Nazi science policies were concerned with both the ideological and the practical utilization of science. It was more than anti-Jewish policies in the civil service, more than antirationalism or antiintellectualism, more than an attempt to impregnate the universities with their politics. Even among the highest leaders, including Hitler, there was a belief in the power of science and the usefulness of qualified intellectuals, a belief that grew stronger during the course of the war. In 1944 the Berlin professor of philosophy Alfred Baeumler, who worked for the Amt Rosenberg and had drawn up its critique of Wehrmacht Psychology, wrote in a much-quoted note that "in the field of science ... [there is] only one way ... to encourage the production of work with a National Socialist approach: the establishment of exemplary achievements" (Poliakov and Wulf 1959, p. 99). In the Third Reich the increased ideological conformity of large portions of scientific theory was accompanied by an increased implementation of rational-scientific methods, including those for the control of social problems.

The various sciences were certainly more suited for either ideological or practical tasks. Roughly speaking, it was German studies, ethnology, history, and the like that were best suited for the former task and natural sciences such as physics and chemistry that were best suited for the latter, although they were also seen as supports for a world outlook, and biological-anthropological subjects made important efforts to back up Nazi ideology (see Mehrtens 1980). We have also heard of the "war effort" of the human sciences (see Chapter 8). Psychology could take on both tasks. In the first phase of Nazi rule there were efforts to demonstrate the ideological usefulness of the discipline, and a primacy of politics in appointments to chairs. In the preparations for war psychology was then used increasingly by industry and the armed forces as a practical discipline. A similar conclusion is to be found elsewhere in the research on science in the Third Reich. Summarizing the results, we can say that initially, until about 1936, when the victory of the "movement" had to be established on all "fronts," the attitude toward science was dominated by weltanschauung, whereas in a second phase, beginning somewhere between the Four-Year Plan in 1936 and the political shake-up in 1938, science was valued above all for its practical usefulness in war preparations.

During the first part of their rule, the Nazis were concerned to mobilize the people ideologically for the "movement." The intellectuals were to adopt the ideology through "folk science." Oswald Kroh, later chairman of the DGfPs, wrote in a letter to the Amt Rosenberg in 1936 that the topics and aims of scientific research should express the "unreserved orientation of science to the realities and tasks of folk state life (*völkisch-staatlichen Lebens*)" (NS 15/216, f. 171). As soon as the requirements of industry and weapons came to dominate the sciences the concept of "folk science," with "German mathematics," "German physics," and "German chemistry," was no longer adequate.[10] These requirements began to dominate with the mobilization of all reserves during the Four-Year Plan. This transfer was clearly formulated by the Führer of university lectures in Berlin, Willig, at the meeting on science and the Four-Year Plan on 18 January 1937. The First Four-Year Plan of the universities had been fulfilled: "Cleansing of German higher education from liberalist and materialistic bonds.... Now the thing is to employ the universities and German science for the completion of the Second Four-Year Plan that has been declared by the Führer.... The total deployment of German science and technology for the fulfillment of the Second Four-Year Plan is the slogan for the next years."[11]

Mehrtens (1979, 1980) has identified a change in the ideological base of the natural sciences corresponding to this transfer: from the "folk science" that accompanied the expulsion of the Jews to "science as a service for the *Volk*" and "research as national duty." From 1936 on, "German science" no longer played any role in the support provided for natural sciences; research in industry was expanded. In 1937 the Reich Research Council was founded to advance science, particularly the natural sciences, in the framework of the Four-Year Plan, even if in the end the support was not on as large a scale or as generous as some scientists may have wished (Ludwig 1979,

---

10 See Mehrtens (1979, p. 431) and Mehrtens and Richter (1980) for essays on the individual sciences.
11 *Wissenschaft und Vierjahresplan*, 1937, p. 3.

pp. 216ff.). Symptomatic of the triumph of technologically effective science over inappropriate folk science is the Munich Religionsgespräch of physicists in 1940, organized by the NSD. In a five-point declaration the "Aryan physicists" had to acknowledge formerly deprecated achievements of theoretical physics, including quantum mechanics (Beyerchen 1977, p. 178). At another conference in Seefeld in 1942 the theory of relativity, which Heisenberg had already described as the obvious basis of modern physics in 1936 in the *Völkischer Beobachter*, was recognized as an established contribution to physics. Support was then available for effective nuclear research.

This shift in emphasis was not confined to the natural sciences. The study of psychotherapy by Cocks (1985) gives the same impression. When it came to the practical application of psychotherapy, the initial efforts to devise a new German "healing art of the soul" (*Seelenheilkunde*) were no longer relevant. Although psychoanalysis was ideologically unacceptable, it received official support in the end as a technique and therapy. Klingemann (1981) sees sociology in the Third Reich transformed to a provider of sociotechnical knowledge. Adam (1977) concludes that Tübingen University had completed the adaptation desired by the Nazis by 1936 and that from 1938 the ideological-political element in teaching decreased.

The explanation for this shift from the ideological to the practical in the sciences must be sought in the changing Nazi policies. In 1935 general conscription was reintroduced for the armed forces, which led to their rapid growth. The Four-Year Plan decreed at Nuremberg in 1936 set the goal of economic autarky, an element of the preparations for war. By 1938 the imminence of war permeated all policies (Mason 1978, p. 305). It was the year when Germany marched into Czechoslovakia and Austria. Hitler took over supreme control of the armed forces and replaced the War Ministry with the Supreme Command of the Wehrmacht. General von Brauchitsch became supreme army commander and Ribbentrop the new foreign minister. Broszat saw 1938 as a "caesura in the constitutional development of the Hitler state," the year in which the "totalitarian forces in the regime" won the upper hand over the forces of the authoritarian state and in which considerations of effectiveness became dominant (1975, pp. 363, 424, 438). Technocrats such as Fritz Todt, Albert Speer, and Himmler gained more and more influence in the party compared with the old ideologists such as Walter Darré, Robert Ley, and Rosenberg. Thus "in all areas of leadership and administration the ideologically sound, but useless party men were displaced in the course of time by educationally qualified (non) party members" (Kater 1983, p. 39). The "Crystal Night" in 1938 marked the transition from the persecution of Jews to their destruction. Around this time it also became plain that ideological methods were not winning over the working class and that – as a result of a labor shortage since 1936, according to Mason – the government was forced to take account of workers' stubborn desire for improved conditions; this resulted in political terror becoming a "means of education" at work (Mason 1978, pp. 300, 322). "The intensification of terror in Winter 1938–9 represented a reaction of all leadership groups to developments that, viewed as a whole, posed the threat of social disintegration" (p. 312).

Prewar developments were now backed up less by ideological mobilization than by terror and the deployment of all forces for war production and the army. Against this background science became orientated more and more toward actual usefulness, and scientists were eager "to be included in war activities," as Ludwig wrote about natural scientists and engineers (1979, p. 230). This was also the background against which German psychology professionalized itself. It was called on to regulate manpower problems arising from rationalization in social life, particularly in the armed forces and industry.

During the First World War scientists had raised their voices to support their country; during the Second World War scientists were employed to conduct war. The encouragement of psychology by the military and military developments was thus not specific to fascism, as was apparent in the United States during the First World War. Highly developed industrial societies concentrate their scientific potential more for war than for any other purpose. In other fascist countries, such as Italy and Japan, there was no corresponding rapid development of psychology during the Second World War. In the United States, however, as in Britain and Canada, the war led to a significant advance in the professionalization of psychology.[12]

Rationality and ethics

If psychology wishes to see its history as one of continual, positive improvement, it can maintain its historical identity[13] only by defining the Third Reich as a period of negation of development, as a decline, loss, or caesura. However, in terms of current criteria of professional success, there was progress. This progress became possible only through work for institutions of the most inhumane of political systems. This forms part of the history of German psychology. The development of psychology in the Third Reich is a challenge to the idea that the progressive use of normal science is welcome per se. It raises questions about the ethics of a profession that puts its own interests first and about the character of a science that can be functional in its normality even in periods of war and dictatorship. Psychology and National Socialism were not incompatible, as some believe. Nor did psychology remain the same, as some have claimed (see Chapter 1). There was some incompatibility, as far as it was not possible to subsume psychological theories under a new ideology without alterations and breaks. But

12 See Camfield (1975), Hearnshaw (1964), Napoli (1981), Samelson (1979), and Wright (1974).
13 For this term, see Graumann (1983).

for practical purposes, Nazified typology and race psychology were dysfunctional. The professionalization of psychology progressed in the Third Reich because with its "normal" theories and methods the discipline was able to satisfy the practical requirements of controlling the allocation of people. Its "technological" side met the demands arising from the Four-Year Plan of 1936, war preparations, and war. The professionalization of psychology in the Third Reich was not pushed ahead by theoretically atypical elements but by the "normal" psychology practiced before, during, and after the Third Reich.

This aspect is perhaps more disturbing than that of the ideological Nazification of psychology. It shows that psychology's past in the Third Reich is not yet purified when the taint of ideology is removed. The rationality of scientific methods provided no protection against Nazism's dictatorial clutches; the effective use of science was not anti-Nazi. Timeless criteria of the nature of science, which do not take into account the function of a science in a system of domination, are therefore inappropriate for measuring the fall of psychology in the Third Reich. This throws light on the ambiguity of some U.S. reactions to German psychology in the Nazi period. Because normal psychology was able to meet practical demands, the discussion in the United States often addressed the question of how to adopt these methods usefully.[14] When Ansbacher (1941a) concluded that the American Henry Murray and the German Max Simoneit had arrived at similar methods of investigating personality, he did not consider the significance of the fact that these methods were applicable to a fascist army.

There was another normality, perhaps an unexpected one. Centers of power were decisive for the professionalization of psychology, but the professional group itself had its own room for maneuvering. Its own interests formed a sort of constant force toward professionalization. Success was possible because the interests of the power centers were taken into consideration, because the profession conformed, compromised, and took the chances that offered themselves and made alliances so as to provide practical opportunities for psychologists. There was no incompatibility between their professional interests and the interests of the power centers. Professionalization was possible in a dictatorship.

Nor were the professional ethics of psychologists any protection against involvement in a murderous war. In the hope of furthering their own interests psychologists cooperated with army, state, party, and industry, and yet still seemed to believe that they were acting as reformers in a humanitarian sense, because psychology was humane and necessarily served the individual. Their professional ethics, which considered only the direct implications of the relationship to the client, with social conditions accepted as an external factor, left them helpless. It may be a trait of the professional expert in

14 See Chapter 1, notes 9 and 10.

general to go along with things as long as professional opportunities are provided. This was Beyerchen's (1977) diagnosis of physicists in the Third Reich.

The way experts worked to rationalize procedures whose goals they did not consider represents a part of the "banality of evil" in the Third Reich, a term Hannah Arendt used in her study of Adolf Eichmann (1963). Eichmann described himself at his trial in 1961 as an administrator who only did his duty and never wanted to do anybody any harm. He was interested only in making his own way, yet he nevertheless kept the extermination machine running. However, this unwillingness to "examine morality" or to decide personally not to participate is not only a problem of experts. Brückner (1978, p. 96) sees it as central to the German mentality.

The "examination of morality" was a different matter for psychologists in the Third Reich than for some physicians or members of the legal profession who were involved in murderous crimes. But they were not protected against being used by the Nazis (where they did not actively support their goals) as long as they saw themselves as being responsible only for the means, but not for the ends, not for reflection on the social implications. This positivist view of science has been accused by critical theorists of being caught up in "instrumental reason" (Horkheimer 1967; Adorno et al. 1970). Perhaps it helps to avoid being instrumentalized if a science opens itself, its functions, aims, and standards to critical discussion (Habermas 1970). Whether this is generally possible under existing conditions is questionable. In all societies to date efforts to arrive at an understanding of the professional expert and of science that include the question of emancipatory goals as an object of reflection are only to be found on the critical periphery.

Comments on sources

This research work drew mainly on the files of the Reich Ministry of Science, the Prussian Ministry of Science, the Amt Rosenberg, the army, and various universities.

Files on psychological institutes and professorships from the RMWEV are to be found in the Zentrales Staatsarchiv (REM) and in the Bundesarchiv (R 21), in the various university archives, and, for the Prussian Ministry, in the Merseburg branch of the Zentrales Staatsarchiv and in the Geheimes Staatsarchiv (both Rep 76). The files for some universities were destroyed or have been lost, in particular those for Danzig, Königsberg, Rostock, and Stuttgart. The university or state archives concerned report that no files relevant to the history of psychology exist for the technical colleges in Berlin, Darmstadt, Dresden, and Stuttgart, or for the universities of Königsberg (Olsztyn) and Würzburg. The files for the University of Breslau (Wroclaw) have also been lost, but here the ministerial records are well preserved. No information was received for Humboldt University in Berlin, the University of Greifswald, or the Hesse State Archives (for Marburg).

The files of the Amt Rosenberg are among the best preserved from this period. The lengthy exchanges are of interest for appointments, especially NS 15, MA 116, MA 141, and MA 205. Similar files kept in the Chancellery are much less complete.

The creation of the Reich Ministry of Science and Education (RMWEV) in 1934 means that a central file covers all aspects of the DPO from January 1935 to February 1945 (REM/821). It begins with Moede's memorandum. The ZStA/DM and the GStA seem to have no earlier files on activities concerning examination regulations. The ministerial files on Moede's Institut für industrielle Psychotechnik also contain no relevant material (1917–25 ZStA-DM; '26–'43 ZStA). Two files by Senger at the ministry are concerned with developments following the DPO: "Studium der Psychologie; Stillegung psychologischer Institute 1941–45" (R 21/469) and "Prüfungsausschüsse Psychologie Band 1, Okt. 1941" (R 21: Rep 76/912).

The files of the DGfPs and the VdpP have not been preserved, nor is the correspondence of leading psychologists available. The secretary of the DGfPs wrote on 8 August 1979 that they had no records predating 1948.

The last secretary of the DGfPs in the Third Reich was Albert Wellek, who probably lost the records during the war. On 14 December 1943 he wrote to the curator of the University of Halle that the files in Leipzig had been destroyed by fire (UAH/PA 16853). Records of the VdpP could not be traced at the Berlin Technical College, where Moede had taught; his adopted daughter did not reply to inquiries.

The activities of the board of the DGfPs can be reconstructed to some extent from (1) the ministry files about the DPO (REM 281); (2) the files of Amt Rosenberg on scientific societies (NS 15 alt/14 and 17); (3) a correspondence file of the Psychological Institute of the University of Tübingen (UAT 148). The last includes the best collection of circulars. It is possible that similar material exists elsewhere, although I received no replies to my inquiries.

Nothing in Germany resembled the Archives of the History of American Psychology at the time of writing, though an archive is being set up in Passau. A few letters were found in U.S. archives and in Tübingen. As far as I know, there is no unpublished correspondence of leading psychologists in the state or communal archives. The Spranger Material of the Bundesarchiv covers only the postwar period; the Generallandesarchiv in Karlsruhe has some of Hellpach's papers. But both were involved only at the margins of the profession. The widows of Kroh, Simoneit, and Wellek no longer had any papers or letters.

Despite intensive efforts, it was not possible, as a resident of West Berlin, to see the documents and the Nazi files of the Berlin Document Center, which is under U.S. administration. The Berlin Senate refused permission. As a U.S. citizen, Mitchell G. Ash was later able to inspect a number of documents. As I was then finishing the manuscript these could not all be considered, but they introduced no fundamentally new aspects (see Ash and Geuter 1985). Some personal files on public servants in the Third Reich, for example, in West German university archives, are still inaccessible.

For investigating psychology's application and professional activities, the most important sources come from the fields of military and labor psychology. An archive of all test material of the Wehrmacht Inspectorate did exist, but according to Simoneit (1972) it was destroyed by bombs in the Tegel Barracks. The Military Archive of the Bundesarchiv had nothing from the inspectorate, nor any files from the test stations of individual districts. For the period before 1933 the records of the Inspektion der Infanterie (RH 12-2/37 and 38) also touch on Army Psychotechnics and the Psychotechnical Laboratory. Annual reports of the head office of Wehrmacht Psychology exist for the years 1932–3, 1934–5, 1936–7, and 1938–9. Together with organizational plans, they are in the records of the District Command V (RH 53-5/41), the Inspektion der Infanterie (RH 12-2/101) and the Heeresgruppe C (RH 19 III/494 and 686). Material relating to the disbanding of Army Psychology can be found in the records of the Army Medical Inspectorate (H 20/553.5), which was responsible for the actual mop-up operation. Few files could be found in Luftwaffe or navy records concerning psychological services.

Printed material on military psychology was often difficult to track down. The *Wehrpsychologische Mitteilungen* had to be destroyed on order as service documents when Wehrmacht Psychology was disbanded (Z, f. 153). However, a complete set could be compiled from public and private collections, with the exception of "Ergänzungsheft 1." Some material on Wehrmacht Psychology collected by U.S. troops for scientific evaluation was burned in France (Z, f. 171), some disappeared in the United States. Many of the documents used by Feder, Gulliksen, and Ansbacher (1948) and unpublished reports by Wehrmacht psychologists written after the war could not be traced in the United States at any of the following institutions: Educational Testing Service, Princeton, New Jersey; Navy Personnel Research and Development Center, San Diego, California; Naval Historical Center, Department of the Navy, Washington, D.C.; Naval Military Personnel Command Library, Washington, D.C.; Navy Department Library, Washington D.C.; Navy Office of Naval Research Library, Arlington, Virginia; National Archives, Captured German Archives, Washington, D.C.; private library, Professor Heinz L. Ansbacher, Burlington, Vt.

Labor psychology is little documented in the files. The Bundesarchiv reported that the records of the Reich Ministry of Labor contain nothing on the employment of psychologists at labor offices. The index of the (incomplete) Labor Front files (BA: NS 5) shows no records for the Institute for Work Psychology and Pedagogy at the Office for Vocational Training and Works Management, but there are files on general activities in labor psychology. An organizational plan of the office is contained in the files of the Reich Organization Leader of the NSDAP (NS 22/236 and 238). The Institut für Zeitgeschichte has some printed material concerning the institute and the office. The 1944 annual report of the institute has been made available from a private source (Z).

The archives of firms were not evaluated. On inquiry the archive of the Badische Anilin und Sodafabrik replied that it did not possess the Kohlhof-Briefe on management training in labor psychology.

It was important, when trying to reconstruct everyday professional activities in industry and the armed forces, to be able to refer to materials used, such as instructions on test procedures. These could as a rule be found only in private collections (cf. I C and II B in the sources register).

It is possible that other relevant material may exist. It was not possible to inspect all university and state archives systematically. However, inquiries were sent out in an attempt to establish if there were further important sources.

Unpublished and printed sources

For citations from sources listed in I A, the signature abbreviation used below is given; for citations from I B, I C, and II B, the section of the sources and the name or the document is given. Periodicals from II A are named (or an abbreviation is used). Where the abbreviation from I A alone would be ambiguous, the specific archive is indicated as well. Where files are paginated, the file page is quoted (f.).

I. Unpublished sources

A. Archives

	Zentrales Staatsarchiv, Potsdam (ZStA)
REM	Reich Ministry of Science and Education
	Zentrales Staatsarchiv, Dienststelle Merseburg (ZStA/DM)
Rep 76 V	Prussian Ministry of Science, Art, and Public Education (files are marked Va for universities or Vb for technical colleges; the next figure denotes the institution (1 is general); the third is for the title; within one title the files are numerically ordered, and divided again into volumes; this way of

marking distinguishes these files from the files of the same
ministry in the GStA

Geheimes Staatsarchiv, Berlin-Dahlem (GStA)

Rep 76	Prussian Ministry of Science, Art, and Public Education
42a M I	Personnel information from the Amtsblatt of the Reich
	Ministry for Science and Education 1924–1944
	(alphabetically ordered cards)

Bundesarchiv, Coblenz (BA)

R 21	Reich Ministry of Science and Education
NS 15 alt	Amt Rosenberg, "the designate of the Führer for the
NS 15	supervision of the entire intellectual and ideological
	instruction and education of the NSDAP"
NS 37	NSV (National Socialist People's Welfare Organization)
Kl. Erw. 762	German Institute for Psychological Research and Psychotherapy

Bundesarchiv/Militärarchiv, Freiburg (BA/MA)

RH 2	OKH/General Staff of the Army
RH 7	Army Personnel Office
RH 12–1	Inspectorate of Education
RH 12–2	Inspectorate of the Infantry
RH 19 III	Army Group C
RH 53–5	Military District Command 5
H 4	
H 6	
H 10–4	OKH/Army Personnel Office
H 20	Army Medical Inspectorate
RL 2 III	General Staff of the Luftwaffe/General Quartermaster
RL 5	Luftwaffenwehramt
RL 17	Plenipotentiary of the Reich Air Ministry (RLM) for aviation
	industry personnel and commander of flight-technical
	preschools
RL 19	Luftgaukommandos

Institut für Zeitgeschichte, Munich (IfZ) (Microfilm Archiv,
MA)

MA 40	
MA 116	
MA 141	Files of the Head Science Office (Hauptamt Wissenschaft) of
MA 205	the Amt Rosenberg
MA 256	Reich Ministry for the Occupied Eastern Regions
MA 641	Files of the Reich Führer SS and Head of the German Police
Fa 88	NSDAP Main Archive
Fa 91	NSDAP Party Chancellery
Z (= ZS/A 37)	(Collected transcripts of the author's interviews)
	Zeugenschrifttum: Gespräche zur Entwicklung der
	Psychologie in Deutschland von den 20er Jahren bis 1945.
	Eine Protokollsammlung; with an appendix:
	Der Jahresbericht des Instituts für Arbeitspsychologie
	und Arbeitspädagogik der Deutschen Arbeitsfront, 1944

Staatsarchiv Hamburg (SAH)

A 110.70	Hamburg University, Chair for Psychology (Heft 4)

C 20/4	Minutes of the University Senate meetings
Ai 3/47	Hochschulwesen (University matters) II
	Archives of the History of American Psychology,
	Akron, Ohio (AHAP)
M 750	Adams Papers
	Wertheimer memoir
	New York Public Library
Werth. Corr.	Wertheimer papers. Wertheimer correspondence
	Library of Congress, Manuscript Division, Washington, D.C.
APA	Records of the American Psychological Association
	Universities and University Archives
	University of Frankfurt
	Files of the University Curatorium
alt Ve 6 h	Psychological Institute, 1916–51
PA Wertheimer	Personnel file Wertheimer
	University Archive Freiburg (UAF)
Reg	Files in the records office
UAF/PA	Personnel files
	University Archive Halle (UAH)
Rep 4	Files of the Rector's Office
Rep 6	Files of the Curator
Rep 21	Files of the Philosophical Faculty
UAH/PA	Personnel files
	University Archive Jena (UAJ)
C-753	Psychological Institute
D 941	Personnel file Sander
	University Archive Leipzig (UAL)
UAL/PA	Personnel files
Phil	Files of the Philosophical Faculty
	University Archive Munich (UAM)
Sen 310	Senate files, Psychological Seminar
O-I-15a	Faculty files, Pedagogical Seminar and Psychological Institute
	University Archive Tübingen (UAT)
131	Dean's files, Philosophical Faculty
117	
205	Files of the Rector's Office
148	General correspondence, A–Z, Psychological Institute

Documents from the following archives are quoted individually: the university archives of Erlangen (UAE), Göttingen (UAG), Cologne (= Köln) (UAK), Münster (UAMs), Posen, and Rostock (UAR); also the Zentralnachweisstelle of the Bundesarchiv, Kornelimünster (BA-ZNS), the Bavarian Hauptstaatsarchiv (BHSA), and the Główna Komisja Badania Zbrodni Hitlerowskich w Polsce, Warsaw.

B. Individual documents

Die Anwendung psychologischer Prüfungen in der deutschen Wehrmacht (1927–1942). Personnel and Administration Project, 2 b, v, 1948. Militärgeschichtliches Forschungsamt Freiburg, MS P-007.

Deuchler, G. Die Psychologie im Osteinsatz. Talk given in Berlin, 28 January 1943, at a course of the Institute for Industrial Psychotechnics. BA: R 21, Rep. 76, Abt. Wiss., Personalia.

Ebert, H. Historische Entwicklung der Diplom-Prüfungsordnung, 1978. University Archive, Technical University Berlin, Hochschulgeschichtliche Sammlung No. 482.

Gerathewohl, S. The Personnel [orig.: Personal] Aptitude Testing Committee of the German Air Force (trans. W. S. Sheeley), June 1945. Wright Patterson Air Force Base, Ohio, Aeromedical Research Library.

Pieper, J. Die Vorgeschichte der münsterischen Personal-Prüfstelle VI Ost (1917–25). Deutsches Literaturarchiv/Schiller Nationalmuseum, Marbach am Neckar, Reg, No. 26.2.

C. Unpublished private material

Professor Dr. Adolf M. Däumling, Bonn
A- and E-Stelle (L), Luftgaukommando VII, Aptitude testing post. Picture report on aptitude tests for flying personnel. Fliegerhorst Neubiberberg, April 1941. Papers and notes on test procedures for officer applicants.

Professor Dr. Friedrich Dorsch, Tübingen
Tests and test results from psychological aptitude tests at the labor administration

Dr. Walter Ehrenstein (Jun.), Konstanz
Curriculum vitae of Walter Ehrenstein (written after 1945. n.d.).

Professor Dr. Siegfried Gerathewohl, Munich
Ge 1: Paper on the organization and goals of the aptitude testing post. Enlisting and Demobilization Office of the Luftgaukommando XI, Hamburg-Rissen, 24 October 1940.
Ge 2: Paper on questions of practical selection work, 4 March 1941.
Ge 3: "Experience with Aptitude testing of Volunteers with Glider Training"; paper given at a meeting of the Enlisting and Demobilization Office of the Luftgaukommando XI in Hamburg-Rissen.
Ge 4: Report on a visit to the bombers' school at Bug a. Rügen (after June 1941, before June 1942).
Ge 5: "From the Practice of Selection Work at the Flier Training Regiments," n.d.

Professor Dr. Elisabeth Lucker, Essen
Documents on aptitude tests of the "Institute for Work Psychology and Pedagogy" of the DAF.

Lottmann, Werner. Auslese auf Grundlage der Begabung. Paper delivered at the 2nd Seminar of the Fachamt Eisen und Metall, 5 June 1942 in Wernigerode a.H.

Dr. Gerhard Munsch, Munich
Documents on diagnostic tests in Wehrmacht Psychology.

Professor Dr. Paul Robert Skawran, Pretoria
Skawran, Paul Robert. (1) "Das Eignungsprüfwesen der deutschen Luftwaffe während des letzten Weltkrieges." Manuscript, n.d.
Skawran, Paul Robert. (2) "Wissenschaftlich-statistische Verarbeitung der Erhebungen beim Geschwader im Zusammenhang mit den psychologichen Gutachten der Flugzeugführer." Part II of a report on visits to J. G. 5, 23 March to 22 April and 4–13 September 1941. Manuscript, n.d. 23 p.

II. Printed sources

A. Periodicals

AHM   Allgemeine Heeresmitteilungen (BA/MA, RHD 1/2)
AN   Allgemeine Nachrichten für die Oberkommandos der Wehrmacht,
       des Heeres und der Kriegsmarine (BA/MA, RWD 2)
BLB   Besondere Luftwaffenbestimmungen (BA/MA, RLD 1/2)
H.Dv. Heeres-Druckvorschriften (BA/MA, RHD 4/···)
HVBl   Heeresverordnungsblatt (BA/MA, RHD 1)
LVBl   Luftwaffenverordnungsblatt (BA/MA, RHD 1)
MVBl   Marineverordnungsblatt (BA/MA, RMD 2)
Anregungen-Anleitungen für Berufserziehung und Betriebsführung. Mitteilungsblatt
    der Deutschen Arbeitsfront, Amt für Berufserziehung und Betriebsführung (IfZ,
    Db 61.44)
Heeresverwaltungsverfügungen, Bd. 15, 1941 (BA/MA, RHD 25/49/16)
Merkblätter des Oberkommando des Heeres (BA/MA, RHD 23/...)
Wehrpsychologische Mitteilungen, 1, 1939 – 4, 1942:
    BA/MA: 1939 (2–4, 6–8)
    UB Giessen: 1940 (7–12), 1941, 1942; Ergänzungshefte 5–12 and 14–16
    Dr. Walter Kröber, Bonn-Beuel: 1940(4), Ergänzungshefte 4 and 13
    Dr. Gustav Spengler, Wuppertal: 1939 (1, 5, 9–12), 1940 (1–3, 5, 6), Ergänzungshefte
        2 and 3; Beilage 1 and 2, Luftwaffe
Wissenschaftliche Berichte und Erfahrungen aus dem Personal-Eignungsprüfwesen
    der Luftwaffe, Heft 1, 1941. Private collection of Professor Dr. Siegfried Grubitzsch,
    Oldenburg
Staff and lecture lists of the universities: Berlin, Bonn, Breslau, Giessen, Halle,
    Hamburg, Jena, Königsberg, Leipzig, Marburg, Munich, Rostock, Würzburg

B. Individual documents

Holzheimer, F. Schwierige Persönlichkeiten unter den Rekruten und ihre
    Behandlung. Wehrpsychologische Arbeiten No. 7; BA/MA, RHD 23/30, No. 7
Leistungsbericht des Amtes für Berufserziehung und Betriebsführung der Deutchen
    Arbeitsfront für das Kriegsjahr 1940. IfZ, Db 61.118
Prüfungsordnung für Unteroffizier-Bewerber des Heeres, 1.9.1943 (= Merkblatt
    5a/17); BA/MA, RHD 6/5a–17
Oberkommando des Heeres. Inspektion für Eignungsuntersuchungen. Anweisungen
    für Eignungsuntersuchungen. Berlin 1940. Library of the Psychological Institute
    of the Free University, Berlin
Inspektion für Eignungsuntersuchungen. Oberkommando des Heeres. Atlas D. Berlin
    1941. Library of the Psychological Institute of the Free University, Berlin
Inspektion für Eignungsuntersuchungen, Oberkommando des Heeres. Atlas E.
    Befehlsreihen-Protokolle. Berlin 1941. Library of the Psychological Institute of the
    Free University, Berlin

III. Written and oral communications

A. Written communications

Bonn, University Archive
Borsig Company, Berlin

Bundesarchiv, Zentralnachweisstelle Kornelimünster
Cologne, University Archive
Darmstadt Technical College, Dr. Marianne Viehaus
Erlangen, University Archive
Gdansk (Danzig), Wojewódzkie Archiwum Państwowe
Göttingen, Prof. Dr. N. Kamp
Hamburg, State Archive
Heidelberg, University Archive
Jena, University Archive
Kiel, University Library, Dr. J. Blunck
Munich, University Archive
Münster, University Archive
Olsztyn, Wojewódzkie Archiwum Państwowe
Poznan (Posen), University Archive
Poznan, Wojewódzkie Archiwum Państwowe
Rheinmetall Company, Düsseldorf
Rostock, University Archive
Stuttgart, Main State Archive
Würzburg, Commission for the History of the Julius Maximilian University

Prof. Dr. Geoffrey Cocks, Albion, Michigan
Prof. Dr. Heinrich Harmjanz, Burgdorf
Dr. Walter Jacobsen, Hamburg
Dr. Walter Kröber, Bonn-Beuel
Dr. Hans Märtin, Goslar
Prof. Dr. Wolfgang Michaelis, Augsburg
Prof. Dr. Paul Robert Skawran, Pretoria
Hertha Wellek, Mainz

Written information filed in ZS/A 37

Dr. Willy Kramp, Schwerte/Villigst
Dr. Waldemar Lichtenberger, Sobernheim
Dr. Egon v. Niederhöffer, Munich
Dr. Gerd Schimansky, Schwerte/Villigst

B. Oral information

Transcripts of talks with the following persons have been deposited at the "Institut für Zeitgeschichte" under ZS/A 37 (source cited as Z). Some additional meetings also took place. Oral information was also received from Ms. Kroh, Berlin, and Gertrud Simoneit, Berlin. (Asterisks indicate that the person is deceased.)

Prof. Dr. Heinz L. Ansbacher, Burlington, Vermont
Prof. Dr. Rudolf Arnheim, Ann Arbor, Michigan
Prof. Dr. Wilhelm Arnold, Würzburg*
Prof. Dr. Hans Bender, Freiburg i. Br.
Prof. Dr. Ernst Bornemann, Altenberge
Prof. Dr. Dr. Helmut v. Bracken, Marburg*

Dr. Albert Bremhorst, Bad Reichenhall*
Prof. Dr. Adolf M. Däumling, Bonn
Dr. phil. habil. Heinz Dirks, Meckenheim
Prof. Dr. Friedrich Dorsch, Tübingen
Prof. Dr. Dr. h. c. Heinrich Düker, Marburg*
Prof. Dr. Karlfried Reichsgraf Eckbrecht von Dürckheim-Montmartin, Todtmoos-
  Rütte*
Prof. Dr. Hans-Joachim Firgau, Munich
Dr. phil. habil. Gotthilf Flik, Munich
Prof. Dr. Erika Fromm, Chicago
Prof. Dr. Rainer Fuchs, Munich
Prof. Dr. Siegfried Gerathewohl, Munich
Prof. Dr. Dr. Erich Goldmeier, White Plains, New York
Prof. Dr. Hildegard Hetzer, Giessen*
Dr. Maria Hippius, Todtmoos-Rütte
Prof. Dr. Wolfgang Hochheimer, Berlin
Prof. Dr. Dr. h. c. George Katona, Ann Arbor, Michigan*
Theodor Katz, Stockholm, Schweden
Ludwig Kroeber-Keneth, Kronberg/Ts*
Prof. Dr. Erwin Levy, New York
Prof. Dr. Elisabeth Lucker, Essen
Prof. Dr. Dr. h. c. Wolfgang Metzger, Bebenhausen*
Dr. Gerhard Munsch, Munich*
Prof. Dr. Edwin B. Newman, Cambridge, Massachusetts
Dr. Engelbert Pechhold, Bonn
Dr. Leonhard von Renthe-Fink, Leverkusen
Dr. Carl-Alexander Roos, Bonn-Oberkassel*
Prof. Dr. Johannes Rudert, Heidelberg*
Dr. Josef Schänzle, Stuttgart
Dr. phil. habil. Elisabeth Schliebe-Lippert, Wiesbaden
Prof. Dr. Kurt Strunz, Brühl*
Prof. Dr. Dr. h. c. Hans Thomae, Bonn
Prof. Dr. Hans Wallach, Swarthmore, Pennsylvania
Prof. Dr. Frederick Wyatt, Freiburg

Bibliography

A. Literature up to 1945

Ach, N. (Ed.), *Lehrbuch für Psychologie*. 3. Band: Praktische Psychologie, Bamberg 1944.

Ammon, H., *Die philosophische Doktorwürde. Die Promotionsbestimmungen der deutschen Universitäten, nebst Erläuterungen und Ratschlägen*, Dessau 1926 (= Dünnhaupts Studien- und Berufsführer, Bd. 12).

Anon., "Eignungsprüfung bei Freiwilligeneinstellung," *Militär-Wochenblatt*, 1932, 116, pp. 1078f.

Anon., "Psychologische Prüfung der Offiziersanwärter," *Militär-Wochenblatt*,1932a, 116, pp. 1213–16.

Anon., "Psychologie und Offizier," *Militär-Wochenblatt*, 1935, 119, pp. 1850–3.

Anon., "Das Leistungsabzeichen der Deutschen Arbeitsfront," *Ind. Pst.*, 1936, 13, pp. 158ff.

Ansbacher, H. L., "German Military Psychology," *Psychological Bulletin*, 1941, 38, pp. 370–92.

Ansbacher, H. L., "Murray's and Simoneit's (German Military) Methods of Personality Study," *Journal of Abnormal and Social Psychology*, 1941a, 36, pp. 589–92.

Ansbacher, H. L., "Curtailment of Military Psychology in Germany," *Science*, 1943, pp. 218f.

Ansbacher, H. L., "German Industrial Psychology in the Fifth Year of War," *Psychological Bulletin*, 1944, 41, pp. 605–14.

Ansbacher, H. L., and K. R. Nichols, "Selecting the Nazi Officer," *Infantry Journal*, 1941, 49, pp. 44–8.

Anschütz, G., "Sinn und Aufgaben einer kommenden Psychologie," *Deutschlands Erneuerung*, 1941, 25, pp. 252–7.

Arnhold, K., *Grundsätze nationalsozialistischer Berufserziehung*, Berlin 1937.

Arnhold, K., "Arbeitspsychologische Aufgaben und Probleme im Amt für Berufs-erziehung und Betriebsführung der DAF," *Zeitschrift für Arbeitspsychologie*, 1938, 11, pp. 33–8.

Arnhold, K., "Psychologische Kräfte im Dienste der Berufserziehung und Leistungssteigerung," in R. Bilz (1941, pp. 105–27).

Arnhold, K., "Das Unterführerkorps im Betrieb," *Arbeit und Betrieb*, 1941a, 12, pp. 1f.

Arnhold, K., *Leistungsertüchtigung*, Berlin 1942.

Arnhold, S., "Die psychophysische Struktur bei Hühnern verschiedener Rassen," *Z. Ps.*, 1938, 144, pp. 10–91.

Arnold, W., "Leistung und Charakter," *Z. ang. Ps*, 1937–8, 53, pp. 48–79.

Baumbach, –, "Beamte auf Lebenszeit und auf Widerruf," *WPsM*, 1939, 1 (H. 1), pp. 30–32.

Baumgarten, F., *Die Charakterprüfung der Berufsanwärter*, Zurich 1941.

Becker, –, "Psychologie und Reichsanstalt für Arbeitsvermittlung und Arbeitslosenversicherung," *Soziale Praxis, Zentralblatt für Sozialpolitik und Wohlfahrtspflege*, 1930, 39, Sp. 68–70, 81–86.

Becker, F., "Industrieform und Menschentypus," *Zeitschrift für Arbeitspsychologie*, 1938, 11, pp. 65–76.

Becker, F., "Die Intelligenzprüfung unter völkischem und typologischem Gesichtspunk," *Z. ang. Ps.*, 1938a, 55, pp. 15–111.

"(Die) behördlichen psychotechnischen Einrichtungen in Deutschland," *Ind. Pst.*, 1930, 7, pp. 339–52 and 1931, 8, p. 190.

Bekenntnis der Professoren an den deutschen Universitäten und Hochschulen zu Adolf Hitler und dem nationalsozialistischen Staat, Dresden 1933.

Benkert, H., *Mensch und Fortschritt im Betrieb*, Wiesbaden 1943.

Bilz, R. (Ed.), *Psyche und Leistung*. Bericht über die 3. Tagung der Deutschen Allgemeinen Aerztlichen Gesellschaft für Psychotherapie in Wien 1940, Stuttgart 1941.

Bloch, E., *Erbschaft dieser Zeit*, Frankfurt/M. 1979 [1935].

Bobertag, O., "Der Kampf für und gegen die Psychologie," *Zeitschrift für Kinderforschung*, 1934, 42, pp. 190–9.

Boder, D. P., "Nazi Science," *Chicago Jewish Forum*, 1942, 1, pp. 23–9.

Bornemann, E., "Aufgaben der Arbeitspsychologie der Gegenwart," *Stahl und Eisen*, 1944, 64, pp. 37–47, 249–56.

Bornemann, E., "Bedeutung des Vorschlagswesens für die Eisenhüttenbetriebe," *Stahl und Eisen*, 1944a, 64, pp. 706–16.

Bramesfeld, E., *Psychotechnik als Lehrfach der Technischen Hochschule*, Schriften der Hessischen Hochschulen, Technische Hochschule Darmstadt, 1926, H. 4.

Bramesfeld, E., "Psychologe – Unterkommen in der praktischen Psychologie (Psychotechnik)," *Studium und Beruf*, 1932, 2, pp. 306f.

Bramesfeld, E., and E. Eberle, "Zur Psychologie des Praktikerurteils," *Ind. Pst.*, 1927, 4, pp. 302–6.

Brecht, –, "Was verstehen wir Soldaten unter Charakter und warum wiegen im Kriege die Eigenschaften des Charakters schwerer als die des Verstandes?" *Soldatentum*, 1939, 6, pp. 198–204, 260–7.

Bröder, P., "Persönlichkeitsaufbau und Gefährdungstypen, ein Beitrag zu Aufbau und Praxis einer Erziehungsberatungsstelle der NSV," *Zeitschrift für Kinderforschung*, 1943 , 49, pp. 325–35.

Bühler, K., *Die Krise der Psychologie*, Frankfurt/Berlin/Vienna 1978 [1926].

Bühler, K., "Ansprache des Vorsitzenden. Eröffnung des XII. Kongresses der

Deutschen Gesellschaft für Psychologie in Hamburg am 13. April 1931," in G. Kafka 1932, pp. 3–6.

Bühler, K., *Ausdruckstheorie. Das System an der Geschichte aufgezeigt*, Jena 1933.

Busemann, A. (Ed.), *Handbuch der pädagogischen Milieukunde*, Halle 1932.

Busemann, A., "Die Psychologie inmitten der neuen Bewegung," *Z. Päd. Ps.*, 1933, 34, pp. 193–9.

Busemann, A., "Die Frage des Aufstiegs der Begabten in neuer Sicht," *Z. Päd. Ps.*, 1933a, 34, pp. 259–65.

Chleusebairgue, A., "Industrial Psychology in Spain," *Occupational Psychology*, 1939, 13, pp. 32–41.

Dach, J. S., *Der erste Eindruck. Seine Bedeutung und Bedingtheit*, Berlin 1937.

Deutsche Wissenschaft, Erziehung und Volksbildung. Amtsblatt des Reichs- und Preussischen Ministers für Wissenschaft, Erziehung und Volksbildung und der Unterrichtsverwaltungen der anderen Länder. Vol. 1, 1935ff.

Dietze, H. H., "Bericht über die Arbeitstagung zum Kriegseinsatz der deutschen Geisteswissenschaften am 27. und 28.4.1940 in Kiel," *Kieler Blätter*, 1940, pp. 397f.

Dilthey, W., *Ideen über eine beschreibende und zergliedernde Psychologie. Gesammelte Schriften*, Bd. 5, Leipzig/Berlin 1924, pp. 139–240.

Dirks, H., *Lebenskraft und Charakter*, Berlin 1940 (= Die Lehre von der praktischen Menschenkenntnis, Teil 2, Bd. 1).

Dirks, H., "Einige Gedanken zur Methodik stammespsychologischer Forschung," *WpsM*, 1941, 3 (H. 4), pp. 51–4.

Dirks, H., "Einige Gedanken über die Bedeutung stammespsychologischer Arbeit und über die Möglichkeit der Beteiligung der Personal-Prüfstellen an dieser Arbeit," *WPsM*, 1942, 4 (H. 2), pp. 47–9.

Döll, H., "Psychologie und soldatische Anforderung," *Kölnische Zeitung*, 16. Feb. 1943.

Döring, W. O., *Die Hauptströmungen in der neueren Psychologie*, Leipzig 1932.

Eckstein, L., *Psychologie des ersten Eindrucks*, Leipzig 1937.

Eckstein, L., *Die Sprache der menschlichen Leibeserscheinungen*, Leipzig 1943 (= Z. ang. Ps., Beiheft 92).

Eckstein, L., *Rassenleib und Rassenseele*, no date (Ed.: Der Reichsführer SS, SS-Hauptamt).

Ehrhardt, A., and O. Klemm, "Rasse und Leistung auf Grund von Erfahrungen im Felde der Eignungsuntersuchung," *Z. ang. Ps.*, 1937, 53, pp. 1–18.

Elsenhans, T., *Lehrbuch der Psychologie.* 30 ed., fully revised by Fritz Giese; edited by H. W. Gruhle, and F. Dorsch, Tübingen 1939.

Engelmann, W., "Berufsberatung und psychologische Eignungsuntersuchungen in der Wirtschaft," *Technische Erziehung*, 1938, 11, pp. 262–8.

Ermisch, H., "Psychophysische und psychologische Untersuchungen an verschiedenen Hühnerrassen," *Z. Ps.*, 1936, 137, pp. 209–44.

"(Die) erste Tagung für angewandte Psychologie der Gesellschaft für experimentelle Psychologie in Berlin," *Praktische Psychologie*, 1922–3, 4, pp. 28–32.

Farago, L. (Ed.), *German Psychological Warfare*, New York 1942 (Reprint: 1972).

Faust, F., "Ueber Organisationsformen und Arbeitsweisen in der Erziehungsberatung," *Zeitschrift für Kinderforschung*, 1944, 50, pp. 34–44.

Ferrari, C. A., "Industrial Psychology in Italy," *Occupational Psychology*, 1939, 13, pp. 141–51.

Fischer, A., "'Der praktische Psychologe' – ein neuer Beruf," *Der Kunstwart und Kulturwart*, 1913, 26, pp. 305–13.

Fischer, G. H., "Ueber Ziele und Einsatz psychologischer Anthropologie," *National-sozialistischer Volksdienst*, 1942, 9, pp. 1–8.

Fischer, G. H., "Zur Entwicklungslinie der neueren deutschen Psychologie," *Z. päd. Ps.*, 1942a 43, pp. 1–7.

Flik, G., "Wille und Sport auf Grund psychologischer Gutachten von Spitzen-könnern," *Soldatentum*, 1940, 7, pp. 18–21.

Filk, G., "Historisches und Organisatorisches über Spezialistenuntersuchungen," *WPsM*, 1942, 4 (H. 11), pp. 1–5.

Frank, W., *Die deutschen Geisteswissenschaften im Kriege*, Hamburg 1940.

Frankenberger, K., "Gedanken über Führerauslese in Betrieben," *Arbeit und Betrieb*, 1942, 13, pp. 83–9.

Gehlen, A., *Der Mensch, seine Natur und seine Stellung in der Welt*, Berlin 1940.

Gessler, O., "Der Aufbau der neuen Wehrmacht," in *Zehn Jahre deutscher Geschichte, 1918–1928*, Berlin 1928, pp. 87–102.

Giese, F., "Die psychologische Laborantin als Beruf," *Z. päd. Ps.*, 1919, 20, pp. 418–22.

Giese, F., *Das Studium der Psychologie und Psychotechnik*, Dessau 1922 (= Dünn-haupts Studien- und Berufsführer, Bd. 2).

Giese, F., "Laienpsychotechnik," *Psychotechnische Rundschau*, 1922a, 1, pp. 49–52.

Giese, F., *Handbuch psychotechnischer Eignungsprüfungen*, Halle 1925.

Giese, F., *Methoden der Wirtschaftspsychologie*, Berlin/Vienna 1927.

Giese, F., *Psychologie als Lehrfach und Forschungsgebiet auf der Technischen Hochschule*, Halle 1933.

Giese, F., "Stammespsychologie und Persönlichkeitsbegutachtung," in O. Klemm (1935, pp. 199–202).

Gl., F., "Das arbeitspsychologische Institut der DAF," *Betriebsführer und Ver-trauensrat*, 1942, 9, pp. 117f.

Göring, M. H., "Weg und Ziel der Psychotherapie," *Münchener medizinische Wochenschrift*, 1938, 85, pp. 1472f.

Göring, M. H. (Ed.), "1. Sonderheft des Deutschen Instituts für Psychologische Forschung und Psychotherapie," *Zentralblatt für Psychotherapie und ihre Grenzgebiete*, 1940.

Göring, M. H. (Ed.), "Erziehungshilfe," *Zentralblatt für Psychotherapie und ihre Grenzgebiete*, 2nd special issue 1940a.

Goldschmidt, R., "Bericht über den V. Kongress der Gesellschaft für experimentelle Psychologie," *Arch. ges. Psych.*, 1912, 24, pp. 71–97.

Graf, O., "Experimentelle Psychologie und Psychotechnik," *Fortschritte der Neurologie und Psychiatrie*, 1936, 8, pp. 437–54.

Griessbach, –, "Psychologische Prüfung der Offiziersanwärter," *Militär-Wochenblatt*, 1932, pp. 116, pp. 1279f.

"Gründungsbericht der Deutschen Gesellschaft für Tierpsychologie," *Z. ang. Ps.*, 1936, 51, pp. 255f.

Grunwaldt, H. H., "General von Voss. Zehn Jahre deutscher Wehrmachtpsychologie," *Arch. ges. Psych.*, 1939, 103, pp. 273–5.

Grunwaldt, H. H., "Bericht über die Entscheidungen des Generalkommandos bei Sorgenkinder-Begutachtungen, sowie über die Führung der Sorgenkinder bei der Sonderabteilung," *WPsM*, 1941, 3 (H. 6), pp. 50f.

Günther, H. F. K., *Rassenkunde des deutschen Volkes*, Munich 1926.

Günther, H. R. G., "Zwei Soldatenfamilien," *WPsM*, 1939, 1 (H. 4), pp. 11–13.

Günther, H. R. G., *Begabung und Leistung in deutschen Soldatengeschlechtern*, Berlin 1940 (= Wehrpsychologische Arbeiten No. 9).

Gütt, A., "Das Gesetz über die berufsmässige Ausübung der Heilkunde ohne Bestallung (Heilpraktikergesetz) vom 17.12.1939," *Der öffentliche Gesundheitsdienst*, 1938–9, 4, A, pp. 929–36.

Hass, L., *Auswahl und Einsatz der Ostarbeiter. Psychologische Betrachtungen, Leistung und Leistungssteigerung*. Ed. Wehrkreisbeauftragter VII b des Reichsministers für Rüstung und Kriegsproduktion, Saarbrücken 1944.

Härtle, H., "Nationalsozialistische Philosophie?" *Nationalsozialistische Monatshefte*, 1941, 12, pp. 723–41.

Handbuch über den preussichen Staat (1939: *Preussisches Staatshandbuch*), Berlin 137, 1931–141, 1939.

Hartmann, E. von, *Die moderne Psychologie*, Leipzig 1901.

Hartshorne, E. Y., *The German Universities and National Socialism*, Cambridge, Mass. 1937.

Heinz, E., "Der Psychologe," *Frankfurter Zeitung* vom 13. Aug. 1942.

Heiss, R., *Die Lehre vom Charakter*, Berlin 1936.

Heiss, R., *Die Deutung der Handschrift*, Hamburg 1943.

Hellpach, W., "Wirkliche Sozialpsychologie," *Ind. Pst.*, 1935, 12, pp. 33–41.

Helwig, H., "Zur Lage der Psychologie in der völkischen Schule," *Kurhessischer Erzieher*, 1936, 80, pp. 491–3.

Henning, H., *Psychologie der Gegenwart*, Berlin 1925.

Hesse, K., "Praktische Psychologie in der Wehrmacht," *Militär-Wochenblatt*, 1930, 115, pp. 401–4.

Hetzer, H., "Aufgaben der Erziehungsberatung der NSV-Jugendhilfe," *Nationalsozialistischer Volksdienst*, 1940, 7, pp. 236–42.

Hetzer, H., "Die Erziehungsberatung als Mittel der NSV-Jugendhilfe," *Deutsche Jugendhilfe*, 1940a, 32 (Issue 3/4), pp. 33–9.

Hetzer, H., "Der Einsatz des Psychologen in der Erziehungsberatung der NSV-Jugendhilfe," *Die Ärztin*, 1942, 18, pp. 171–5.

Hippius, R., *Die Umsiedlergruppe aus Estland. Ihre soziale, geistige und seelische Struktur*, Posen 1940.

Hippius, R., "Vom russischen Volkscharakter," *Volkswissenschaftliche Feldpostbriefe*, 1944, H. 5, pp. 1–8.

Hippius, R., and I. G. Feldmann, "Siedlungsbereitschaft für den Osten," *Deutsche Wissenschaftliche Zeitschrift im Wartheland*, 1942, 3, H. 5/6.

Hippius, R., I. G. Feldmann, K. Jelinek, and K. Leider, *Volkstum, Gesinnung und Charakter. Bericht über psychologische Untersuchungen an Posener deutsch-polnischen Mischlingen und Polen, Sommer 1942*, Stuttgart/Prague 1943.

Hische, W., *Die öffentliche Berufsberatung*, Bernau 1931.

Hische, W., *Deutscher Arbeitsdienst als Erziehungsgemeinschaft*, Leipzig/Berlin 1935.

Hitschmann, E., "Zehn Jahre Wiener Psychoanalytisches Ambulatorium (1922–1932). Zur Geschichte des Ambulatoriums," *Internationale Zeitschrift für Psychoanalyse*, 1932, 18, pp. 265–71.

Hoffmann, A., "Die psychologische Schulung im Rahmen der Bildungsaufgaben einer pädagogischen Akademie," *Z. päd. Ps.*, 1931, 32, pp. 143–52.

Hoffmann, A., "Der Erziehungspsychologe," *Z. päd. Ps.*, 1941, 42, pp. 197ff.

Hoffmann, H. F., *Die Schichttheorie, eine Anschauung von Natur und Leben*, Stuttgart 1935.

Hofstätter, P. R., "Die Krise der Psychologie. Betrachtungen über den Standort einer Wissenschaft im Volksganzen," *Deutschlands Erneuerung*, 1941, 25, pp. 561–78.

Homann, N., *Der Kampf um die Berufsberatung*, Bernau 1932 (= Der Arbeitsmarkt, Vol. 2).

Huber, H., *Erziehung und Wissenschaft im Kriege*, Schriften für Politik und Auslandskunde Nr. 58. Berlin 1940.

Huth, A., "Seelische Unterschiede in der bayrischen Jugend," *Ind. Pst.*, 1936, 13, p. 224.

Huth, A., "Grundlagen der Persönlichkeitsbegutachtung," *Z. päd. Ps.*, 1936a, 37, pp. 289–94.

Huth, A., *Seelenkunde und Arbeitseinsatz*, Munich 1937.

Jaensch, E., *Die Lage und die Aufgabe der Psychologie. Ihre Sendung in der deutschen Bewegung und an der Kulturwende*, Leipzig 1933.

Jaensch, E., *Der Kampf der deutschen Psychologie*, Leipzig 1934.

Jaensch, E. R., *Der Gegentypus. Psychologisch-anthropologische Grundlagen deutscher Kulturphilosophie, ausgehend von dem, was wir überwinden wollen*, Leipzig 1938 (= *Z. ang. Ps.*, Beiheft 75).

Jaensch, E., "Wege und Ziele der Psychologie in Deutschland," *Z. päd. Ps.*, 1938a, 39, 161–81.

Jaensch, E., "Grundsätze für Auslese, Intelligenzprüfung und ihre praktische Verwirklichung," *Z. ang. Ps.*, 1938b, 55, pp. 1–14.

Jaensch, E., "Wozu Psychologie?" *Der deutsche Erzieher*, 1938c, pp. 213–21 (also in O. Klemm 1939, pp. 7–30).

Jaensch, E., "Wege und Ziele der Psychologie in Deutschland," *Ind. Pst.*, 1938d, 15, pp. 10–19.

Jaensch, E., *Der Hühnerhof als Forschungs- und Aufklärungsmittel in menschlichen Rassenfragen*, Berlin 1939 (also *Zeitschrift für Tierpsychologie*, 1939, 2, pp. 223–58).

"Jahresbericht 1940 des Deutschen Instituts für Psychologische Forschung und Psychotherapie," *Zentralblatt für Psychotherapie und ihre Grenzgebiete*, 1942, 14, pp. 1–62.

"Jahresbericht 1941 des Deutschen Instituts für Psychologische Forschung und Psychotherapie," *Zentralblatt für Psychotherapie und ihre Grenzgebiete*, 1942, 14, pp. 62–77.

Jastrow, J., "Promotionen und Prüfungen," in *Das akademische Deutschland*, Vol. 3. Berlin 1930, pp. 219–44.

Junius, M., *Menschenformen. Volkstümliche Typen. Schilderungen ihres Lebensverhaltens*, Berlin 1943 (= Die Lehre von der praktischen Menschenkenntnis, Part 3, Vol. 1a).

Kafka, G. (Ed.), *Bericht über den 12. Kongress der Deutschen Gesellschaft für Psychologie in Hamburg 1931*, Jena 1932.

Kalender der reichsdeutschen Universitäten und Hochschulen, Leipzig: 114, 1933/34–120, 1941/42 (1937/38 and later *Deutsches Hochschulverzeichnis*).

Kallfelz, W., "Zum Gesetz über die berufsmässige Ausübung der Heilkunde ohne Bestallung vom 17.2.1939," *Deutsches Recht*, Ausg. A, 1939, pp. 692–6.

Kasper, G., H. Huber, K. Kaebsch, and F. Senger (Eds.), *Die deutsche Hochschulverwaltung*, Berlin 1942–3.

Keilhacker, M., "Sprechweise und Persönlichkeit," *Z. ang. Ps.*, 1940, 59, pp. 215–41.
Kelchner, M., "Die Neuordnung des Studiums der Psychologie", *Die Ärztin*, 1942, 18, pp. 168–70.
Keller, H., "Zur Entwicklung und zum gegenwärtigen Stand der Tierpsychologie," *Zeitschrift für Züchtung*, Series B, Tierzüchtung and Züchtungsbiologie, 1936, 36, pp. 1–12.
Kesselring, M., "Die Psychologie im Dienste der neuen Lehrerbildung," *Deutsches Bildungswesen*, 1936, pp. 254–70.
Kienzle, R., "Charakter- und Jugendkunde an der Hochschule für Lehrerbildung," *Der deutsche Erzieher*, 1936, 4, pp. 848f.
Kiessling, A., "Über die Gestaltung des Psychologieunterrichts in der neuen Lehrerbildung," *Z. päd. Ps.*, 1931, 32, pp. 362–6.
Klages, L., *Ausdrucksbewegung und Gestaltungskraft. Grundlegung der Wissenschaft vom Ausdruck*, Leipzig 1923.
Klages, L., *Die Grundlagen der Charakterkunde*, Leipzig 1926.
Klages, L., *Handschrift und Charakter*, Leipzig 1932.
Klemm, O. (Ed.), *Bericht über den 13. Kongress der Deutschen Gesellchaft für Psychologie in Leipzig 1933*, Jena 1934.
Klemm, O. (Ed.), *Bericht über den 14. Kongress der Deutschen Gesellchaft für Psychologie in Tübingen 1934*, Jena 1935.
Klemm, O. (Ed.), *Bericht über den 15. Kongress der Deutschen Gesellchaft für Psychologie in Jena 1936*, Jena 1937.
Klemm, O. (Ed.), *Bericht über den 16. Kongress der Deutschen Gesellchaft für Psychologie in Bayreuth 1938*, Leipzig 1939.
Knoff, P., "Ausbildung der Berufsberater," *Ind. Pst.*, 1926, 3, pp. 60f.
Koenig-Faschsenfeld, O. Freiin von, "Organisation und Ausbau der Erziehungshilfe," M. H. Göring, (1940a, pp. 8–11).
Koenig-Fachsenfeld, O. Freiin von, "Erziehungshilfe," *Die Ärztin*, 1941, 17, pp. 350–2.
Koenig-Fachsenfeld, O. Freiin von, "Erziehungshilfe," *Zentralblatt für Psychotherapie*, 1942, 14, pp. 25–34.
Kraepelin, E., "Arbeitspsychologische Untersuchungen," *Zeitschrift für die gesamte Neurologie und Psychiatrie*, 1921, 70, pp. 230–40.
Krannhals, P., *Das organische Weltbild*, Munich 1928.
Kreipe, K., *Psychologische Gesichtspunkte für Auswahl und Ausbildung von Entfernungsmessern*, Berlin 1936 (= Wehrpsychologische Arbeiten, No. 2).
Kreipe, K., "Über den Aufbau psychologischer Gutachten," in M. Simoneit, E. Zilian, E. Wohlfahrt, and K. Kreipe, *Leitgedanken zur psychologischen Erforschung der Persönlichkeit*, Berlin 1937 (= Die Lehre von der praktischen Menschenkenntnis, Part 1, Vol. 1a), pp. 56–67.
Kreipe, K., "Psychologische Eignungsuntersuchungen in der Wehrmacht," *Volksspiegel*, 1937a, 4, pp. 66–73.
Kreipe, K., "Die Forderung der Truppennähe für das Personal-Eignungsprüfwesen der Luftwaffe," *Wiss. Ber. Erfahr. aus d. Pers. Eig. Prüfwesen d. Luftwaffe*, 1941, H.1, pp. 3–10.
Kröber, W., "Über die Hauptaussprache (Exploration)," *WPsM*, 1942, 4 (H. 1), pp. 26–38.
Kroh, O., *Erziehung im Heere. Ein Beitrag zur Nationalerziehung der Erwachsenen*, Langensalza 1926.

302   *Bibliography*

Kroh, O., "Die Aufgabe der pädagogischen Psychologie und ihre Stellung in der Gegenwart," *Z. päd. Ps.*, 1933, 34, pp. 305–27.

Kroh, O., "Zur Frage der Anwendung in der Psychologie," *Z. päd. Ps.*, 1938, 39, pp. 4–13.

Kroh, O., "Ein bedeutsamer Fortschritt in der deutschen Psychologie. Werden und Absicht der neuen Prüfungsordnung," *Z. Ps.*, 1941, 151, pp. 1–32.

Kroh, O., "Wozu Diplompsychologen?" *Arbeit und Betrieb*, 1942, 13, pp. 4–8.

Kroh, O., "Missverständnisse um die Psychologie," *Deutschlands Erneuerung*, 1943, 27, pp. 21–37.

Kroh, O., "Zum Ausbau der Prüfungsordnung für Diplompsychologen," *Z. Ps.*, 1943a, 155, pp. 1–15.

Kroh, O., "Vierzig Jahre Deutsche Gesellschaft für Psychologie," *Z. Ps.*, 1944, 156, pp. 183–9.

Kroh, O., "Vom Wesen und Aufgabenbereich der pädagogischen Psychologie," *Internationale Zeitschrift für Erziehung*, 1944a, 13, pp. 161–74.

Krueger, F., "Die Aufgaben der Psychologie an den Deutschen Hochschulen," in G. Kafka (1932, pp. 25–73).

Krueger, F., "Die Lage der Seelenwissenschaft in der deutschen Gegenwart," in O. Klemm (1934, pp. 9–36).

"(Ueber die) künftige Pflege der Pädagogik an den deutschen Universitäten," *Z. päd. Ps.*, 1918, 19, pp. 209–34.

Kürschners Deutscher Gelehrtenkalender, 5, 1935ff.

"Kundgebung der Deutschen Gesellschaft für Psychologie: Ueber die Pflege der Psychologie an den deutschen Hochschulen," in H. Volkelt (1930, pp. vii–x).

Läpple, E., "Die Arbeitskurve als charakterologischs Prüfverfahren," *Z. ang. Ps.*, 1940, 60, pp. 1–63.

Lerner, E., "A Reply to Wyatt and Teuber," *Psychological Review*, 1945, 52, pp. 52–4.

Lersch, P., *Gesicht und Seele. Grundlinien einer mimischen Diagnostik*, Munich 1971 (1932).

Lersch, P., *Lebensphilosophie der Gegenwart*, Berlin 1932a.

Lersch, P., "Grundriss einer Charakterologie des Selbst," *Z. ang. Ps.*, 1934, 46, pp. 129–69.

Lersch, P., *Der Aufbau des Charakters*, Leipzig 1938.

Lersch, P., "Praktische Einsatzgebiete der Psychologie," *Deutschlands Erneuerung*, 1943, 27, pp. 54–67.

Liebenberg, R., *Berufsberatung. Methode und Technik*, Leipzig 1925.

Lipmann, O., "Mehr Psychotechnik in der Psychotechnik," *Z. ang. Ps.*, 1930, 37, pp. 188–91.

Lipmann, O., "Grundlagen und Ziele der Psychotechnik und der Praktischen Psychologie," *Z. ang. Ps.*, 1933, 44, pp. 64–79.

Litt, T., *Die Stellung der Geisteswissenschaften im nationalsozialistischen Staate*, Leipzig 1933 (reprinted from *Die Erziehung*, 1933, 8, H. 12).

Löbner, W., "Berufslenkung und Berufserziehung im gegenwärtigen Deutschland," *Donaueuropa. Zeitschrift für die Probleme des europäischen Südostens*, 1942, 2, pp. 507–21.

Lottig, H., "Grundsätzliches über die Grenzen des Experiments bei der Charakterbeurteilung unter besonderer Berücksichtigung der Fliegertauglichkeitsprüfung," *Zeitschrift für die gesamte Neurologie und Psychiatrie*, 1938, 161, pp. 468–82.

Machacek, H., "Zur praktischen Durchführung der Schülerauslese durch Lehrer und Schularzt," *Die deutsche Hauptschule*, 1941, H. 2, pp. 34–9.

Marbe, K., "Die Bedeutung der Psychologie für die übrigen Wissenschaften und die Praxis," in F. Schumann (Ed.), *Ber. V. Kongr. f. exp. Psychol. in Berlin 1912*, Leipzig 1912, pp. 110–13.

Marbe, K., "Die Stellung und Behandlung der Psychologie an den Universitäten," *Preuss. Jahrb.*, 1921, 185, pp. 202–10.

Marbe, K., *Praktische Psychologie der Unfälle und Betriebsschäden*, Munich/Berlin 1926.

Marrenbach, O., *Fundamente des Sieges. Die Gesamtarbeit der Deutschen Arbeitsfront von 1933 bis 1940*, Berlin 1941.

Marx, –, "Die Auswahl der Führer im Heere," *Militär-Wochenblatt*, 1939, 123, pp. 3305–10.

Masuhr, H., *Psychologische Gesichtspunkte fur die Beurteilung von Offizieranwärtern*, Berlin 1937 (= Wehrpsychologische Arbeiten, No. 4).

Masuhr, H., "Ergebnisse einer Bewährungskontrolle für die bei den Prüfstellen des Heeres durchgeführten Prüfungen von SS-Führeranwärtern (SS FA 1935/36)," *WPsM*, 1939, 1 (H. 3) pp. 44f.

Masuhr, H., "Ein interessanter Fall aus der Bewährungskontrolle," *WPsM*, 1939a, 1 (H. 2), pp. 50–3.

Masuhr, H., "Ein weiterer interessanter Fall aus der Bewährungskontrolle," *WPsM*, 1939b, 1 (H. 3), pp. 36–9.

Masuhr, H., "Ein dritter interessanter Fall aus der Bewährungskontrolle," *WPsM*, 1939c, 1 (H. 4), pp. 46f.

Masuhr, H., "Ein weiterer interessanter Fall aus der Bewährungskontrolle," *WPsM*, 1939d, 1 (H. 6), pp. 27f.

Mathieu, J., "Die Poppelreuterschen handwerklichen Primitivkurse (Robinsonkurse) des Deutschen Instituts für Nationalsozialistische Technische Arbeitsforschung und -schulung," in O. Klemm (1935, pp. 258–62).

Mathieu, J., *Möglichkeiten einer betrieblichen Eignungsuntersuchung*, Berlin, no date (Schriftenreihe des Amtes für Berufserziehung und Betriebsführung der DAF; ~ 1938–9).

Meier, M., "Das Studium der Psychologie," *Studentisches Jahrbuch*, Technische Hochschule Darmstadt, 1943, pp. 138–40.

Menschenformen. Volkstümliche Typen, ed. by Inspektion des Personalprüfwesens des Heeres. Berlin 1941 (= Die Lehre von der praktischen Menschenkenntnis, Part 3, Vol. 1).

Messer, A., "Die Bedeutung der Psychologie, für Pädagogik, Medizin, Jurisprudenz und Nationalökonomie," *Jahrbuch der Philosophie*, 1914, 2, pp. 183–218.

Messer, A., *Einführung in die Psychologie und die psychologischen Richtungen der Gegenwart*, Leipzig 1927.

Metz, P., "Zur Organisation der Fliegerpsychologie," *WPsM*, Beilage 1 Luftwaffe, 1939, pp. 1–3.

Metz, P., "Zur Organisation der Psychologie in der Luftwaffe," *WPsM*, Beilage 2 Luftwaffe, 1939a, pp. 1f.

Metz, P., "Der Personal-Eignungsprüfer der Luftwaffe," in *Der Psychologe in der Wehrmacht*, 1942, pp. 15–23.

Metzger, W., "Ganzheit und Gestalt. Ein Blick in die Werkstatt der Psychologie," *Erzieher im Braunhemd*, 1938, 6, pp. 90–3.

Metzger, W., "Lebendiges Denken, nach Schopenhauer und v. Clausewitz," *Erzieher im Braunhemd*, 1938a, 6, pp. 193–6.

Metzger, W., "Der Auftrag der Psychologie in der Auseinandersetzung mit dem Geist des Westens," *Volk im Werden*, 1942, 10, pp. 133–44.

Michligk, P., *Innerbetriebliche Werbung*, Berlin 1942.

Michligk, P., *Die Praxis des betrieblichen Vorschlagswesens*, Berlin 1942a.

Mierke, K., "Organisatorische Richtlinien für die Auswahlprüfungen von Bewerbern der Mannschaftsfachlaufbahnen der Kriegsmarine," *WPsM*, 1940, 2, Supplement No 6.

Mierke, K., "Methodik der psychologischen Auswahl für die Funkerlaufbahn der Kriegsmarine," *WPsM*, 1941, 3, Supplement No. 8.

Mierke, K., "Der Marinepsychologe," in *Der Psychologe in der Wehrmacht*, 1942, pp. 25–35.

Mierke, K., "Psychologische Diagnostik," in N. Ach (1944, pp. 1–79).

Minerva. Jahrbuch der gelehrten Welt, Berlin/Leipzig 31, 1934.

Moede, W., "Kraftfahrer-Eignungsprüfungen beim deutschen Heer 1915–1918," *Ind. Pst.*, 1926, 3, pp. 23–8.

Moede, W., *Lehrbuch der Psychotechnik*, Berlin 1930.

Moede, W., *Zur Methodik der Menschenbehandlung*, Berlin 1930a.

Moede, W., *Arbeitstechnik. Die Arbeitskraft, Schutz – Erhaltung – Steigerung*, Stuttgart 1935.

Moede, W., "Missverständnisse ohne Ende in der angewandten Psychologie der Gegenwart," *Ind. Pst.*, 1936, 13, pp. 289–99.

Moede, W., "Leistungs- und Ausdrucksprinzip bei der Eignungsbegutachtung," in O. Klemm (1939, pp. 158–65).

Moede, W., "Lebenspraktische Psychologie," *Stuttgarter Neues Tageblatt*, 4./5. Feburar 1939a.

Moede, W., "Die notwendige Neuordnung des psychologischen Studiums sowie dessen Richtlinien" (1939b), in O. Klemm (1939, pp. 275–8).

Moede, W., "Anwendung und Erfolg der Psychotechnik in Verkehrswesen, Heer und Industrie," *Grossdeutscher Verkehr*, 1942, pp. 171–80.

Moede, W., "Richtungen der Psychologie in ihrer Bedeutung für die Eignungsbegutachtung in Betrieb und Wirtschaft," *Der Sanitätsdienst bei der Deutschen Reichsbahn*. Supplement to *Die Reichsbahn*, 1942a, 6, pp. 171–6.

Moede, W., "Wirtschaftspsychologie," in N. Ach (1944, pp. 80–121).

Müller, F., "Zur Organisation der Berufsberatung in Deutschland," *Praktische Psychologie*, 1920/21, 2, pp. 115–22.

Naegelsbach, H., "Extreme Sonderfälle," *WPsM*, 1941, 3 (H. 5), pp. 84–7.

Nass, G., "Die Persönlichkeit des Kampfwagenführers," in *Abhandlungen zur Wehrpsychologie*, 2. Folge. Leipzig 1938, *Z. ang. Ps.*, Beiheft 79, pp. 131–48.

Nipper, H. A. (Ed.), *Studienpläne sowie Studien- und Prüfungsordnungen für die Ausbildung von Diplom- und Doktor-Ingenieuren an deutschen Technischen Hochschulen und Bergakademien*, Berlin 1941.

Oelrich, –, "Menschenauslese und Wissenschaft," *WPsM*, 1942, 4 (H. 12), pp. 3–11.

Organisationsbuch der NSDAP, Munich 1938.

Paschukanis, J., "Zur Charakteristik der faschistischen Diktatur," *Unter dem Banner des Marxismus*, 1928, 2, pp. 282–315.

Pauli, R., "Der Arbeitsversuch als ganzheitlicher Prüfungsversuch, insbesondere als charakterologischer Test," *Ind. Pst.*, 1939, 16, pp. 119–23.

Pauli, R., "Der Arbeitsversuch als charakterologisches Prüfverfahren," *Z. ang. Ps.*, 1943, 65, pp. 1–40.

Pauli, R., "Die Arbeitsform der Introvertierten und Extravertierten," *Z. päd. Ps.*, 1944, 45, pp. 33–40.

Pechhold, E., "Stand der Psychotechnik, Unfallverhütung und Berufserziehung in deutschen Eisenhüttenwerken," *Archiv für das Eisenhüttenwesen*, 1938, 12, pp. 41–7.

Petermann, B., *Wesensfragen seelischen Seins*, Leipzig 1938.

Pfahler G., "Die psychologische Ausbildung der Studierenden an der Pädagogischen Akademie in Altona," *Z. päd. Ps.*, 1931, 32, pp. 152–6.

Pfahler, G., *Warum Erziehung trotz Vererbung?* Leipzig 1935.

Pfahler, G., *Rassekerne des deutschen Volkes*, 3 Vols., Munich 1942.

Piorkowski, C., "Die Entwicklung der Psycho-Technik in Deutschland während des Krieges," *Deutsche Politik*, 1918, 3, pp. 498–506.

Poppelreuter, W., "Praktische Psychologie als ärztlicher Beruf," *Münchener medizinische Wochenschrift*, 1921, 68, pp. 1262f.

Poppelreuter, W., *Die Aufgaben des Landesarbeits- und Berufsamtes bei der Organisation praktisch-psychologischer Einrichtungen.* Schriften des Landesarbeits- und Berufsamtes der Rheinprovinz, 1921a, H. 3.

Poppelreuter, W., *Allgemeine methodische Richtlinien der praktisch psychologischen Begutachtung*, Leipzig 1923.

Poppelreuter, W., *Hitler der politische Psychologe*, Langensalza 1934.

"Prüfungsordnung für Studierende der Psychologie, vom 16. Juni 1941," *DWEV*, 1941, 7, pp. 255–9.

(*Der*) *Psychologe in der Wehrmacht.* Akad. Auskunftsamt Berlin in Verbindung mit dem Amt für Berufserziehung und Betriebsführung in der DAF. Berlin 1942.

Renthe-Fink, L. von, "Bericht über die Beispielsammlung von Befehlsreihen-Protokollen. (Atlas E)," *WPsM*, 1941, 3 (H. 4), pp. 49f.

Renthe-Fink, L. von, "Methodische Grundprobleme der diagnostischen Handlungs-untersuchung," *WPsM*, 1942, 4 (H. 1), pp. 45–51.

Rieffert, J. B., "Psychotechnik im Heere," in K. Bühler (Ed.), *Bericht über den 7. Kongress für experimentelle Psychologie in Marburg 1921*, Jena 1922, pp. 79–96.

Rieffert, J. B., "Sprechtypen," in G. Kafka (1932, pp. 409–13).

Rohracher, H., *Kleine Einführung in die Charakterkunde*, Leipzig 1934.

Rothacker, E., *Die Schichten der Persönlichkeit*, Leipzig 1938.

Rupp, H., "6. Tagung des Verbandes der Deutschen Praktischen Psychologen," *Psychotechnische Zeitschrift*, 1929, 4, pp. 21–3.

Rupp, H., "Die sittliche Verpflichtung der Psychotechnik," *Psychotechnische Zeitschrift*, 1930, 5, pp. 103–8.

Rupp, H., "Aufgaben der Psychotechnik in Deutschland," *Psychotechnische Zeitschrift*, 1933, 8, p. 1.

Rupp, H., "General von Voss in den Ruhestand getreten," *Zeitschrift für Arbeitspsychologie*, 1939, 12, pp. 87f.

Sachs, H., *Zur Organisation der Eignungspsychologie*. Schriften zur Psychologie der Berufseignung und des Wirtschaftslebens, Heft 14, Leipzig 1920.

Sander, F., "Die Idee der Ganzheit in der deutschen Psychologie," *Der Thüringer Erzieher*, 1933, 1, pp. 10–12.

Sander, F., "Zum neuen Jahrgang," *Zeitschrift für Jugendkunde*, 1934, 4, pp. 1–3.

Sander, F., "Deutsche Psychologie und nationalsozialistische Weltanschauung," *Nationalsozialistisches Bildungswesen*, 1937, 2, pp. 641–9.

Sander, F., "Wandlungen der deutschen Psychologie," *Deutschlands Erneuerung*, 1943, 27, pp. 14–21.

Saupe, E. (Ed.), *Einführung in die neuere Psychologie*, Osterwieck am Harz 1928.

Schack, –, "Psychologie und Offizier," *Militär-Wochenblatt*, 1935, 120, pp. 628–32.

Schänzle, J., *Der mimische Ausdruck des Denkens*, Berlin 1936 (= Die Lehre von der praktischen Menschenkenntnis, Part 1, Vol. 3).

Schimrigk, –, "Die psychologische Beurteilung Dienstpflichtiger bei Musterung und Aushebung," *Soldatentum*, 1939, 6, pp. 24–7.

Schliebe, G., "Wandlungen der Psychologie," *Nationalsozialistisches Bildungswesen*, 1937, 2, pp. 195–205.

Schmidt, –, "Psychologie und Offizier," *Militär-Wochenblatt*, 1936, 120, pp. 1229–31.

Schmidt, –, "Die Berufung zum Offizier," *Militär-Wochenblatt*, 1938, 123, pp. 321–5.

Scholl, R., "Kinderheime als psychologische Forschungsstätten," *Z. ang. Ps.*, 1939, 57, pp. 86–103.

Scholz-Roesner, –, "Eignungsprüfungen," *Militär-Wochenblatt*, 1933, 117, pp. 1348–50.

Schoor, W., "Ein neues Diplom-Examen," *Der Maschinenmarkt*, 1941, 146 (H. 58), pp. 21f.

Schorn, M., "Die praktische Durchführung eines Ausleseverfahrens für den Ausländereinsatz," *Ind. Pst.*, 1942, 19, pp. 207–16.

Schott, E., "Verstärkter Einsatz der NSV-Erziehungsberatung," *Nationalsozialistischer Volksdienst*, 1940, 7, pp. 25–9.

Schulhof, A., "Die Mitwirkung des Arztes bei der Ausarbeitung und Durchführung psychotechnischer Eignungsprüfungen," *Praktische Psychologie*, 1922–3, 4, pp. 222–4.

Schulte, R. W., *Die Berufseignung des Damenfrisörs*, Leipzig 1921 (= Schriften zur Psychologie der Berufseignung und des Wirtschaftslebens, Heft 17).

Schultz, I. H., "Die Aufhebung der Kurierfreiheit," *Deutsches Ärzteblatt*, 1939, 69, pp. 151–7.

Schulz, E., "Ergebnisse aus der Berufsnachwuchslenkungstätigkeit der Arbeitsämter im Grossdeutschen Reich im Kriege," *RABL*, 1942, 22 (v), pp. 290–5.

Schulz, W., "Erbgut, Erziehung und berufliche Leistung. Ein Beitrag zur Psychologie des Arbeitseinsatzproblems," *Das Werk*, 1936, 16, pp. 483–92 and 531–8.

Schulz, W., "Menschenauslese vor allem in der Eisenhüttenindustrie," *Stahl und Eisen*, 1937, 57, pp. 1133–42.

Schumann, F. (Ed.), *Bericht über den I. Kongress für experimentelle Psychologie in Giessen 1904*, Leipzig 1904.

Simoneit, M., *Wehrpsychologie*, Berlin 1933.

Simoneit, M., "Uber Menschenauslese-Methoden," *Soldatentum*, 1934, 1, pp. 55–60.

Simoneit, M., *Die Bedeutung der Lehre von der praktischen Menschenkenntnis*, Berlin 1934a (= Die Lehre von der praktischen Menschenkenntnis, Part 1, Vol. 1)

Simoneit, M., "Zur Willensuntersuchung bei wehrmachtpsychologischen Eignungs-untersuchungen," in M. Simoneit et al. (1937, pp. 5–13).

Simoneit, M., *Leitgedanken über die psychologische Untersuchung des Offizier-Nachwuchses in der Wehrmacht*, Berlin 1938 (= Wehrpsychologische Arbeiten, No. 6).

Simoneit, M., "Das diagnostische Problem in der praktischen Psychologie," *Z. Ps.*, 1938a, 143, pp. 1–3.

Simoneit, M., "Der Psychiater auf der psychologischen Prüfstelle," *Der deutsche Militärarzt*, 1939, 4, pp. 201–5.

Simoneit, M., "Mutübungen. Wehrpsychologische Gedanken über die seelische Kriegsbereitschaft," *Deutsche Infanterie*, 1939a, H. 5, pp. 18–20.

Simoneit, M., "Allgemeines (Einleitende Bemerkungen zum Heft)," *WPsM*, 1939b, 1 (H. 2), pp. 3f.

Simoneit, M., "Vom Werden der deutschen Wehrmachtpsychologie. Ein geschicht-licher Rückblick," *WPsM*, 1940, Supplement 2.

Simoneit, M., "Gegenwartssorgen der Psychologie," *Z. Ps.*, 1940a, 148, pp. 112–26.

Simoneit, M., "Die ersten Assessorenprüfungen der Wehrmachtpsychologie," *Soldatentum*, 1940b, 7, pp. 67f.

Simoneit, M., "Wissenschaftliche Ergebnisse der Besichtigungsreise 1940," *WPsM*, 1940c, 2 (H. 11), pp. 1–3.

Simoneit, M., "Ueber typische Fälle," in *Menschenformen* 1941, pp. 9–13.

Simoneit, M., "Lehrgang 'Trieb- und Willenserforschung' (1941)," *WPsM*, 1942, 4 (H. 2), pp. 45f.

Simoneit, M., "Der Personalgutachter des Heeres" (1942a), in *Der Psychologe in der Wehrmacht*, 1942, pp. 5–14.

Simoneit, M., K. Kreipe, E. Zilian, and P. Metz, *Wehrpsychologische Willensunter-suchungen*, Langensalza 1937.

Sonnenschein, A., *Unfallverhütung und Invalidenfürsorge im Eisenwerk Witkowitz*, Prague 1931.

Spranger, E., *Lebensformen*, Halle 1930.

Steinwarz, H., *Das betriebliche Vorschlagswesen als nationalsozialistisches Führungs-instrument*, Berlin 1943.

"(Die) Stellung der Psychologie an den deutschen Universitäten," *Z. päd. Ps.*, 1931, 32, pp. 157–60.

Stern, W., *Differentielle Psychologie*, Leipzig 1921 [1911].

Stern, W., "Das Psychologische Laboratorium der Hamburgischen Universität," *Z. päd. Ps.*, 1922, 23, pp. 161–96.

Stern, W., "Die Stellung der Psychologie an den Deutschen Universitäten," *Die Deutsche Schule*, 1931, 35, pp. 74–83.

Stets, W., "Vom Stand der öffentlichen Berufsberatung im September 1922," *RABL*, 1923, pp. 311–14.

Stets, W., "Die psychologischen Eignungsuntersuchungen im Dienste der Nach-wuchslenkung," in *Arbeitseinsatz und Arbeitslosenhilfe*, 1939. Reprinted in *Berufs-beratung, gestern–heute–morgen*, Bielefeld 1959, pp. 33–6.

Stets, W., "Berufsberatung-Berufsnachwuchslenkung," *Die Erziehung*, 1941, 17, pp. 1–10.

Strehle, H., *Analyse des Gebarens*, Berlin 1935 (= Die Lehre von der praktischen Menschenkenntnis, Part 1, Vol. 2).

(Das) Studium der Psychologie. Handbuch für das Hochschulstudium in Deutschland, Heft 9. Ed. by Deutscher Akademischer Austauschdienst, Berlin 1935.

Syrup, F., "Jedem seinen Arbeitsplatz," *Kalender der deutschen Arbeit*, Berlin 1938, pp. 66–8.

"(Die '1.) Tagung der Gruppe für angewandte Psychologie (Gesellschaft für experimentelle Psychologie)' in Berlin (10–14 Oktober 1922)," *Z. ang. Ps.*, 1923, 21, pp. 390–405.

(Die) Technische Hochschule Darmstadt 1836 bis 1936, Darmstadt 1936.

Tenax, —, "Die psychologische Aufgabe des Heeres," *Militär-Wochenblatt*, 1934, 119, pp. 43–7.

Thiele, R., *Person und Charakter*, Leipzig 1940.

Thomae, H., "Kasuistischer Beitrag zur Frage der Motivierung von Brandstiftungen Jugendlicher," *Monatsschrift für Kriminalbiologie und Strafrechtsreform*, 1944, 35, pp. 21–5.

Tiling, E., "Der Psychiater auf der psychologischen Prüfstelle," *Der deutsche Militärarzt*, 1938, 3, pp. 509–14.

Tiling, E., "Der Psychiater auf der psychologischen Prüfstelle," *Der deutsche Militärarzt*, 1939, 4, pp. 205f.

Tramm, A., "Angriffe gegen psychologische und psychotechnische Untersuchungsverfahren," *Ind. Pst.*, 1932, 9, pp. 92–4.

Ulich-Beil, E., "A National System in the Reich," *Occupations*, 1935, 13, pp. 582–91.

Utitz, E., "Charakterologie," in E. Saupe (1928, pp. 403–26).

"Verband der deutschen praktischen Psychologen. Mitteilung Nr. 1," *Ind. Pst.*, 1927, 4, pp. 158f.

Vetter, A., "Testverfahren als Hilfsmittel der psychologischen Diagnostik," *Zentralblatt für Psychotherapie*, 1940, 14, pp. 41–6.

Volkelt, H. (Ed.), *Bericht über den XI. Kongress für experimentelle Psychologie in Wien 1929*, Jena 1930.

Volkelt, H., "Einsatz der Psychologie in deutscher Erziehung und Bildung," in O. Klemm (1939, pp. 30–5).

Wagner, J., "Angewandte Psychologie," in E. Saupe (1928, pp. 203–19).

Walzer, —, "Offiziernachwuchsfragen," *Militär-Wochenblatt*, 1939, 124, pp. 410f.

Wartegg, E., *Gestaltung und Charakter*, Leipzig 1939 (= *Z. ang Ps.*, Supplement 84).

Weber, M., "Wissenschaft als Beruf," in Weber, *Gesammelte Aufsätze zur Wissenschaftslehre*, Tübingen 1922, pp. 524–55.

Weber, W., *Die praktische Psychologie im Wirtschaftsleben*. Leipzig 1927.

Weber, W., "Die rechtliche Stellung des im Wirtschaftsleben praktisch tätigen Psychologen," *Ind. Pst.*, 1929, 6, pp. 346–55.

Weiss, A., "Der Unterführer im Betrieb – Neue Wege zur Leistungssteigerung," *Wirtschaftskurve*, 1944, pp. 18–28.

Wenke, H., "Psychologie als Beruf," *Das Reich*, No. 40, 5. 10. 1941.

Wenke, H., "Psychologie im Gefüge der Wissenschaften," *Deutschlands Erneuerung*, 1943, 27, pp. 37–54.

Wenzl, A., *Graphologie als Wissenschaft*, Leipzig 1937.

Werder, P. von, "Psychologie als deutsche Seelenkunde," *Nationalsozialistische Monatshefte*, 1943, 14, pp. 237–51.

Wer ist's? Berlin 1905–35.

Wissenschaft und Vierjahresplan, Berlin n.d. [1937].

Wohlfahrt, E., "Ein Vergleich der Prüfaufsätze des Zwillingspaares Hans und Rolf H.," *WPsM*, 1939, 1 (H. 2), pp. 10–26.
Wohlfahrt, E., "Die verschiedene Beliebtheit der neuen Bildkarten," *WPsM*, 1939a, 1 (H. 7/8), pp. 28–31.
Wolff, B. von, *Zur Heilpraktikerfrage*. Med. Diss., Göttingen 1941.
Wyatt, F., and H. L. Teuber, "German Psychology under the Nazi-System – 1933– 1940," *Psychological Review*, 1944, 51, pp. 229–47.
Zehnjahres-Statistik des Hochschulbesuches und der Abschlussprüfungen. II. Band: Abschlussprüfungen. Ed. by Reichsminister für Wissenschaft, Erziehung und Volksbildung, Berlin 1943.
Ziehen, T., *Die Grundlagen der Charakterologie*, Langensalza 1930.
Zielasko, G., "Zur Bewährungskontrolle von O. B.-Gutachten," *WPsM*, 1939, 1 (H. 4), pp. 48–51.
Zilian, E., "Ueber die erb- und rassenbedingte seelische Wesenslage, im Hinblick auf die Frage ihrer Erkennbarkeit," *Soldatentum*, 1938, 5, pp. 273–7.
Zilian, E., "Charakter und leibliches Erbgut nach einer Zwillingsuntersuchung im Bereich der Wehrmachtpsychologie," *Soldatentum*, 1938a, 5, pp. 3–15.
Zilian, E., "Rasse und seelenkundliche Persönlichkeitsauslese in der Wehrmacht," *Rasse* 1938b, 5, pp. 321–33.
Zilian, E., "Anlagekundliche Auswertung eines Falles von EZ unter O. B.," *WPsM*, 1939, 1 (H. 2), pp. 32–9.
Zilian, E., "Angewandte Rassenseelenlehre in Ausleseuntersuchungen der Wehrmacht," *Rasse*, 1939a, 6, pp. 1–13.
Zilian, E., "Bericht über die erste Sitzung der Arbeitsgemeinschaft 'Wehrpsychologie' in der 'Deutschen Gesellschaft für Wehrpolitik und Wehrwissenschaften' (Berlin) im Winterhalbjahr 1938/39 am 18. Oktober 1938," *Soldatentum*, 1939b, 6, pp. 43f.
Zilian, E., "Zum Rassendiagnostischen Atlas," *WPsM*, 1939c, 1 (H. 7/8), pp. 38–43.
Zilian, E., "Der Rassendiagnostische Atlas der Wehrmachtpsychologie," *Soldatentum*, 1939d, 6, pp. 275–8.
Zilian, E., "Zur Berücksichtigung rassenkundlicher Feststellungen in den Persönlichkeitsbefunden," *WPsM*, 1939e, 1 (H. 4), pp. 32–6.
Zilian, E., "Wehrpsychologische Arbeitsgemeinschaft über 'Rasse und Soldatentum' in der Deutschen Gesellschaft für Wehrpolitik und Wehrwissenschaften: Rassen des Nahen und Fernen Ostens," *WPsM*, 1939f, 1 (H. 3), pp. 20–5.
Zilian, E., "Ein Fall besonderer Verschiedenheit erbungleicher Zwillinge. 1. Rassenkundlicher Teil," *WPsM*, 1939g, 1 (H. 6), pp. 12–19.
Zilian, E., "Art- und persönlichkeitsgemässe Auslese unter dem Gesichtspunkt der Rasse," *Soldatentum*, 1939h, 6, pp. 45–9.
Zilian, E., "Mitteilungen über den Ansatz einer rassenpsychologischen Forschungsreihe im Anschluss an den rassendiagnostischen Atlas," *WPsM*, 1939i, 1 (H.10), pp. 53–61.

B. Literature after 1945

Abendroth, W. (Ed.), *Faschismus and Kapitalismus*, Frankfurt/M. 1976.
Absolon, R., *Die Wehrmacht im Dritten Reich*. Vols. II–IV, Boppard 1971ff.

Absolon, R., "Das Offizierkorps des Deutschen Heeres 1935–1945," in H. H. Hofmann (1980, pp. 247–68).

Adam, U. D., *Hochschule und Nationalsozialismus. Die Universität Tübingen im Dritten Reich*, Tübingen 1977.

Adler, M., and H. P. Rosemeier, "Analyse der in der deutschen Psychologie benützten Methoden," *Archiv für Psychologie*, 1970, 122, pp. 327–38.

Adorno, T. W., "Was bedeutet: Aufarbeitung der Vergangenheit," in Adorno, *Eingriffe*, Frankfurt/M. 1968, pp. 125–46.

Adorno, T. W., H. Albert et al., *Der Positivismusstreit in der deutschen Soziologie*, Neuwied/Berlin 1970.

Agnoli, J., B. Blanke, and N. Kadritzke, "Einleitung der Herausgeber," in Sohn-Rethel (1973, pp. 7–24).

Aleff, E. (Ed.), *Das Dritte Reich*, Hannover 1976.

Allesch, J. von, "German Psychologists and National Socialism," *Journal of Abnormal and Social Psychology*, 1950, 45, p. 402.

Almond, G. A., "Comparative Political Systems," *Journal of Politics*, 1956, 18, pp. 391–409.

Aly, G., Ebbinghaus, A. et. al., *Aussonderung und Tod. Die klinische Hinrichtung der Unbrauchbaren*, Berlin 1985 (= Beiträge zur nationalsozialistischen Gesundheits- und Sozialpolitik, Vol. 1).

Aly, G., Masuhr, K. F. et. al., *Reform und Gewissen. "Euthanasie" im Dienst des Fortschritts*, Berlin 1985 (= Beiträge zur nationalsozialistischen Gesundheits- und Sozialpolitik, Vol. 2).

Ansbacher, H. L., "Bleibendes und Vergängliches aus der deutschen Wehrmacht-psychologie." *Mitteilung des Berufsverbandes deutscher Psychologen*, 1949, 3 (H. 11), pp. 3–9.

Ansbacher, H. L., "Testing, Management and Reactions of Foreign Workers in Germany during World War II," *American Psychologist*, 1950, 5, pp. 38–49.

Arendt, H., *The Origins of Totalitarianism*, New York 1951.

Arendt, H., *Eichmann in Jerusalem: A Report on the Banality of Evil*, New York 1963 (rev. and enlarged ed. Harmondsworth 1977).

Arnold, W., *Der Psychologe*, Nürnberg 1948.

Arnold, W., *Der psychologische Dienst in der Bundesanstalt*, 1959 (reprinted from: *Handbuch der Arbeitsvermittlung und Berufsberatung*, pp. 296–332).

Arnold, W., "Professor Dr. Bernhard Herwig 70 Jahre," *Psychologie und Praxis*, 1964, 8, pp. 1–3.

Arnold, W., *Angewandte Psychologie*, Stuttgart/Berlin/Cologne/Mainz 1970.

Ash, M. G., "The Struggle Against the Nazis," *American Psychologist*, 1979, 34, pp. 363f.

Ash, M. G., "Wilhelm Wundt and Oswald Külpe on the Institutional Status of Psychology: An Academic Controversy in Historical Context," in W. G. Bringmann and R. D. Tweney (Eds.), *Wundt Studies/Wundt Studien*, Toronto 1980, pp. 396–421.

Ash, M. G., "Wie entstand die Psychologie? Eine Wissenschaft zwischen Philosophie und Experiment," *Psychologie heute*, 1980a, 7 (H. 7), pp. 50–4.

Ash, M. G., "Fragments of the Whole: Documents of the History of Gestalt Psychology in the United States, the Federal Republic of Germany and the German Democratic Republic," in J. Brožek and L. J. Pongratz (1980b, pp. 187–200).

Ash, M. G., "Academic Politics in the History of Science: Experimental Psychology in Germany, 1879–1941," *Central European History*, 1981, 14, pp. 255–86.

Ash, M. G., *The Emergence of Gestalt Theory: Experimental Psychology in Germany 1890–1920*. Unpubl. Ph.D. Diss. Harvard University, Cambridge, Mass. 1982.

Ash, M. G., "The Self-Presentation of a Discipline: History of Psychology in the United States between Pedagogy and Scholarship," in L. Graham et al. (1983, pp. 143–89).

Ash, M. G., "Disziplinentwicklung und Wissenschaftstransfer – Deutschsprachige Psychologen in der Emigration," in *Berichte zur Wissenschaftsgeschichte*, 1984, 7, 207–26.

Ash, M. G., and U. Geuter, "NSDAP-Mitgliedschaft und Universitätskarriere in der Psychologie," in C. F. Graumann (Ed.), *Psychologie und Nationalsozialismus*, Berlin/Heidelberg/New York 1985, 263–78.

Baader, G., "Zur Ideologie des Sozialdarwinismus," in G. Baader and U. Schultz, (1980, pp. 39–51).

Baader, G., and U. Schultz (Eds.), *Medizin und Nationalsozialismus*, Berlin 1980.

Bald, D., *Vom Kaiserheer zur Bundeswehr. Sozialstruktur des Militärs: Politik der Rekrutierung von Offizieren und Unteroffizieren*, Frankfurt/ M./Bern 1981.

Bauer, R. A., *Der neue Mensch in der sowjetischen Psychologie*, Bad Nauheim 1955.

Baumgarten-Tramer, F., "Hellmuth Bogen," *Berufsberatung und Berufsbildung*, 1947, 5/6, pp. 142f.

Baumgarten, F., "German psychologists and recent events," *Journal of Abnormal and Social Psychology*, 1948, 43, pp. 452–65.

Baumgarten-Tramer, F., "Die Berufsberatung als Vorläuferin der Persönlichkeitsforschung," in E. Stern (Ed.), *Die Tests in der klinischen Psychologie*, Vol. 1, 2, Zurich 1955, pp. 757–804.

Baumgarten-Tramer, F., "Psychologie," in S. Kaznelson (Ed.), *Juden im deutschen Kulturbereich*, Berlin 1959, pp. 282–92.

Bayertz, K., "Darwinismus als Ideologie," in *Darwin und die Evolutionstheorie* (= Dialektik, 5), Cologne 1982, pp. 105–20.

Bechstedt, M., " 'Gestalthafte Atomlehre' – Zur 'Deutschen Chemie' im NS-Staat," in H. Mehrtens and S. Richter (1980, pp. 142–65).

Ben David, J., and R. Collins, "Social Factors in the Origins of a New Science: The Case of Psychology," V. S. Sexton and H. Misiak (Eds.), *Historical Perspectives in Psychology: Readings*, Belmont, Calif. 1971, pp. 98–122. (1966, *American Sociological Review*).

Bengeser, G., "Doktorpromotion in Deutschland. Begriff, Geschichte, gegenwärtige Gestalt," in *Hochschulrecht. Promotionsordnungen*, Bonn 1964, pp. 1–128.

Bergius, R., "Oswald Kroh. Sein Weg und sein pädagogisches Werk," *Berliner Lehrerzeitung*, 1955, 9, pp. 526–9.

Bergius, R., "Zum 75. Geburtstag von Adhemar Gelb," *Psychologische Beiträge*, 1962, 7, pp. 360–9.

Bergler, G., *Die Entwicklung der Verbrauchsforschung in Deutschland und die Gesellschaft für Konsumforschung bis zum Jahre 1945*, Kallmünz 1960.

Bergmann, W., W. Dittmar, H. Müggenburg et al., *Soziologie im Faschismus, 1933–1945*, Cologne 1981.

Berman, L. H., "Oral History as a Source Material for the History of Behavioral Sciences," *JHBS*, 1967, 3, pp. 58f.

Berufsverband Österreichischer Psychologen, *Memorandum zum Psychologengesetz,* Vienna 1980.

Besson, W., "Zur Geschichte des nationalsozialistischen Führungsoffiziers (NSFO). Dokumentation," *Vierteljahrshefte für Zeitgeschichte,* 1961, 9, pp. 76–116.

Bettelheim, C., *Die deutsche Wirtschaft unter dem Nationalsozialismus,* Munich 1974.

Beyerchen, A. D., *Scientists under Hitler. Politics and the Physics Community in the Third Reich,* New Haven/London 1977.

Billig, M., *Fascists – A Social Psychological View of the National Front,* London 1978.

Black, M., *Models and Metaphors. Studies in Language and Philosophy,* Ithaca 1968.

Blätter für Berufskunde. Ed. by Bundesanstalt für Arbeit, Bielefeld.

Bleuel, H. -P., *Deutschlands Bekenner. Professoren zwischen Kaiserreich und Diktatur,* Bern/Munich/Vienna 1968.

Bock, G., *Zwangssterilisation im Nationalsozialismus. Studien zur Rassenpolitik und Frauenpolitik,* Opladen 1986.

Böhme, G., W. v. d. Daele, R. Hohlfeld et al., *Die gesellschaftliche Orientierung des wissenschaftlichen Fortschritts,* Frankfurt/M. 1978.

Böhme, G., W. v. d. Daele, and W. Krohn, "Die Finalisierung der Wissenschaft," in W. Diederich (Ed.), *Theorien der Wissenschaftsgeschichte,* Frankfurt/M. 1974, pp. 276–311.

Bollmus, R., *Das Amt Rosenberg und seine Gegner,* Stuttgart 1970.

Bollmus, R., *Handelshochschule und Nationalsozialismus,* Meisenheim 1973.

Bollmus, R., "Zum Projekt einer nationalsozialistischen Alternativuniversität: Alfred Rosenbergs 'Hohe Schule,'" in M. Heinemann (1980, Vol. 2, pp. 125–52).

Boring, E. G., *A History of Experimental Psychology,* Englewood Cliffs 1950.

Bornemann, E., "Probleme und Ergebnisse der Arbeitspsychologie," *Studium Generale,* 1961, 14, pp. 342–54.

Bornemann, E., "Oswald Kroh zum Gedächtnis. Rückblick auf 50 Jahre Deutscher Psychologie," *Ps Rd,* 1967, 18, pp. 223–8.

Bornemann, E., "Einführung," in H. L. Ansbacher and R. R. Ansbacher, *Alfred Adlers Individualpsychologie,* Munich/Basel 1975, pp. 17–39.

Boyd, R., "Metaphor and Theory change. What is 'Metaphor' a Metaphor for?" in A. Ortony (Ed.), *Metaphor and Thought,* Cambridge 1979, pp. 356–408.

Bracher, K. D., "Die Gleichschaltung der deutschen Universität," in *National-sozialismus und die deutsche Universität,* Berlin 1966, pp. 126–42.

Bracher, K. D., "Totalitarismus," in E. Fraenkel and K. D. Bracher (Eds.), *Staat und Politik,* Frankfurt/M. 1968, pp. 328–30.

Bracher, K. D., *Stufen der Machtergreifung,* Frankfurt/Berlin/Vienna 1974.

Bracher, K. D., "Tradition und Revolution im Nationalsozialismus," in M. Funke (1976, pp. 17–29).

Bracken, H. von, "Deutsche Persönlichkeitstheorie im xx. Jahrhundert," in Von Bracken and H. P. David (Eds.), *Perspektiven der Persönlichkeitstheorie,* Bern/Stuttgart 1959, pp. 67–80.

Brecht, K., Friedrich V. et al. (Eds.), *"Hier geht das Leben auf eine sehr merkwürdige Weise weiter..." Zur Geschichte der Psychoanalyse in Deutschland,* Hamburg 1985.

Bringmann, W. G., "Design Questions in Archival Research," *JHBS,* 1975, 11, pp. 23–6.

Bringmann, W. G., and G. A. Ungerer, "An Archival Journey in Search of Wilhelm Wund," in J. Brožek and L. J. Pongratz (1980, pp. 201–40).

Brodthage, H., and S. O. Hoffmann, "Die Rezeption der Psychoanalyse in der Psychologie," in J. Cremerius (Ed.), *Die Rezeption der Psychoanalyse in der Soziologie, Psychologie und Theologie im deutschsprachigen Raum bis 1940*, Frankfurt/M. 1981, pp. 135–253.

Broszat, M., *Der Staat Hitlers*, Munich 1975.

Brožek, J., "Irons in the Fire: Introduction to a Symposium on Archival Research," *JHBS*, 1975, 11, pp. 15–19.

Brožek, J., and L. J. Pongratz (Eds.), *Historiography of Modern Psychology*, Toronto 1980.

Bruder, K. J., "Entwurf der Kritik der bürgerlichen Psychologie," in Bruder (Ed.), *Kritik der bürgerlichen Psychologie*, Frankfurt/M. 1973, pp. 92–217.

Brückner, P., *Zur Sozialpsychologie des Kapitalismus*, Frankfurt/Cologne 1972.

Brückner, P., *Versuch, uns und anderen die Bundesrepublik zu erklären*, Berlin 1978.

Bühler, Ch., "(Selbstdarstellung)," in L. J. Pongratz et al. (1972, pp. 9–42).

Bunk, G. P., *Erziehung und Industriearbeit*, Weinheim/Basel 1972.

Burchardt, L., "Professionalisierung oder Berufskonstruktion? Das Beispiel des Chemikers im wilhelminischen Deutschland," *Geschichte und Gesellschaft*, 1980, 6, pp. 327–48.

Burrichter, C., "Aufgabe und Funktionen einer historischen Wissenschaftsforschung," in Burrichter (Ed.), *Grundlegung der historischen Wissenschaftsforschung*, Basel 1979, pp. 7–21.

Busch, A., *Die Geschichte des Privatdozenten*, Stuttgart 1959.

Buser, R., *Ausdruckspsychologie*, Munich 1973.

Buss, A. R. (Ed.), *Psychology in Social Context*, New York 1979.

Cadwallader, T. C., "Unique Values of Archival Research," *JHBS*, 1975, 11, pp. 27–33.

Camfield, T. M., *Psychologists at War: The History of American Psychology and the First World War*, Ph.D. Diss., Univ. of Texas, 1969.

Camfield, T. M., "The Professionalization of American Psychology, 1870–1917," *JHBS*, 1973, 9, pp. 66–75.

Carlsen, R., *Zum Prozess der Faschisierung und zu den Auswirkungen der faschistischen Diktatur auf die Universität Rostock 1932–1935*, Phil. Diss., Rostock 1965.

Caspar, G. A., *Die sozialdemokratische Partei und das deutsche Wehrproblem in den Jahren der Weimarer Republik*, Berlin/Frankfurt/M. 1959 (= *Wehrwissenschaftliche Rundschau*, Supplement 11).

Chestnut, R. W., "Psychotechnik: Industrial Psychology in the Weimar Republic 1918–1924," *Proceedings of the Annual Convention of the American Psychological Association*, 1972, 7 (2), pp. 781f.

Chroust, P., "Gleichschaltung der Psyche. Zur Faschisierung der deutschen Psychologie am Beispiel Gerhard Pfahlers," *Psychologie- und Gesellschaftskritik*, 1979, 3 (H. 4), pp. 29–40.

Clark, T. N., "The Stages of Scientific Institutionalization," *International Social Science Journal*, 1972, 24, pp. 658–71.

Clever, U., "Die Geschichte der Standesorganisationen und ihre oppositionellen Alternativen," in G. Baader and U. Schultz (1980, pp. 75–82).

Cocks, G. C., *Psyche and Swastika. "Neue deutsche Seelenheilkunde" 1933–1945*, Ph. D. Diss., Univ. of Cal., Los Angeles, 1975.

Cocks, G. C., *Psychotherapy in the Third Reich. The Göring Institute*, New York/Oxford 1985.

Corino, K. (Ed.), *Intellektuelle im Bann des Nationalsozialismus*, Hamburg 1980.

Daele, W. van den, and P. Weingart, *The Utilization of the Social Sciences in the Federal Republic of Germany*. Wissenschaftsforschung, Report 2. Bielefeld (n.d.).

Daele, W. van den, and P. Weingart, "Resistenz und Rezeptivität – zu den Entstehungsbedingungen neuer Disziplimen durch wissenschaftspolitische Steuerung," *Zeitschrift für Soziologie*, 1975, 4, pp. 146–64.

Däumling, A., "25 Jahre Diplompsychologe. Kritische Betrachtungen zur Entwicklung eines Berufsstandes," *PsRd*, 1967, 18, pp. 251–62.

Dahrendorf, R., *Gesellschaft und Demokratie in Deutschland*, Munich 1972.

Danziger, K., "The Social Origins of Modern Psychology," in A. R. Buss (1979, pp. 27–45).

Davis, D. R., *German Applied Psychology*. Bios Final Report No. 970, Item 24.28. Publ. Bd. No. L 63616, London 1946.

Davis, D. R., "Post-mortem on German Applied Psychology," *Occupational Psychology*, 1947, 21, pp. 105–10.

Dehue, T., "Niederländische Psychologie unter deutscher Besetzung 1940–1945," *Psychologische Rundschau*, 1988, 39, p. 39.

Demandt, A., *Metaphern für Geschichte*, Munich 1978.

(*Die*) *deutsche Universität im dritten Reich*, Munich 1966.

Deutsche Wehrmachtpsychologie 1914–1945, Munich 1985.

Doepner, F., "Zur Auswahl der Offizieranwärter im 100 000-Mann-Heer," *Wehrkunde*, 1973, 22, pp. 200–4, 259–63.

Dorsch, F., *Geschichte und Probleme der angewandten Psychologie*, Bern/Stuttgart 1963.

Dorsch, F., "Der Beruf des Psychologen in der Geschichte," in H. Benesch and F. Dorsch (Eds.), *Berufsaufgaben und Praxis des Psychologen*, Munich/Basel 1971, pp. 32–9.

Dräger, K., "Bemerkungen zu den Zeitumständen und zum Schicksal der Psychoanalyse und der Psychotherapie in Deutschland zwischen 1933 und 1949," *Psyche*, 1971, 25, pp. 255–68.

Dunlap, J. W., *Training of Free Gunners in the German Air Force*. Washington, D. C.: U. S. Department of Commerce, Publ. Bd. No. 4364, 1946.

Dunlap, J. W., and J. B. Rieffert, *Tests for Selection of Personnel in German Industry*. U.S. Naval Technical Mission in Europe, Report No. 300–45. Publ. Bd. No. 22933. Sept. 1945.

Ebert, H., "Die Technische Hochschule Berlin und der Nationalsozialismus: Politische 'Gleichschaltung' und rassistische 'Säuberungen,'" in R. Rürup (1979, Vol. 1, pp. 455–68).

Ebert, H., and K. Hausen, "Georg Schlesinger und die Rationalisierungsbewegung in Deutschland," in R. Rürup (1979, Vol. 2, pp. 315–34).

Eckardt, G., "Die Gründung der Psychologischen Anstalt in Jena (1923)," *Wiss. Zeitschrift der Friedr.-Schiller-Univers. Jena, Ges. u. Sprachwiss. Reihe*, 1973, 22, pp. 517–59.

Elliott, P., *The Sociology of the Professions*, London 1972.

Epstein, B., "Cold War in Psychology," *Benjamin Rush Bulletin*, 1950, 1, pp. 43–9.

Erdmann, K. D., *Wissenschaft im Dritten Reich*, Kiel 1967.

Eyferth, K., "Bericht über eine Umfrage über die Ausbildung im zweiten Studienabschnitt des Diplomstudienganges Psychologie," in E. Stephan (1980, pp. 8–21).

Feder, D. D., H. Gulliksen, and H. L. Ansbacher, *German Naval Psychology*. U.S. Naval Publication NavPers 18080. Washington, D.C., Bureau of Naval Personnel, Princeton, N.J., College Entrance Examination Board, April 1948.

Ferber, C. von, *Die Entwicklung des Lehrkörpers der deutschen Universitäten und Hochschulen 1864–1956*, Göttingen 1956.

Fichter, M. M., and H.-U. Wittchen, *"Nicht-ärztliche" Psychotherapie im In- und Ausland*, Weinheim/Basel 1980.

Fischer, W., *Die Wirtschaftspolitik Deutschlands 1918–1945*, Hannover 1961.

Fitts, P. M., "German Applied Psychology during World War II," *American Psychologist*, 1946, 1, pp. 151–61.

Flitner, A. (Ed.), *Deutsches Geistesleben und Nationalsozialismus*, Tübingen 1965.

Focke, H., and U. Reimer, *Alltag unterm Hakenkreuz*, Reinbek 1979.

Forman, P., "Weimar Culture, Causality, and Quantum Theory, 1918–1927: Adaptation by German Physicists and Mathematicians to a Hostile Intellectual Environment," *Historical Studies in the Physical Sciences*, 1971, 3, pp. 1–115.

Freidson, E., *Dominanz der Experten. Zur sozialen Struktur medizinischer Versorgung*, Munich/Berlin/Vienna 1975.

Fried, J. H. E., *The Exploitation of Foreign Labour by Germany*, Montreal 1945.

Friedrich, C. J. (Ed.), *Totalitarianism*, Cambridge/Mass. 1954.

Friedrich, C. J., *Totalitäre Diktatur*, Stuttgart 1957.

Fritsche, C., *Wilhelm Wirth*. Unpubl. paper, Department of Psychology, Karl-Marx-University, Leipzig 1976.

Fritsche, C., *Untersuchung zur Krise der bürgerlichen Psychologie anhand ihrer zeitgenössischen Reflexion von 1897–1945*. Unpubl. Dissertation, Department of Psychology, Karl-Marx-University Leipzig 1981.

Funke, M. (Ed.), *Hitler, Deutschland und die Mächte. Materialien zur Aussenpolitik des Dritten Reichs*, Kronberg/Düsseldorf 1976.

Gay, P., *Weimar Culture*, London 1969.

Gay, P., *Die Republik der Aussenseiter. Geist und Kultur in der Weimarer Zeit: 1918–1933*, Frankfurt/M. 1970.

Gerathewohl, S. J., *Psychological Examination for Selection and Training of Fliers*. German Aviation Medicine in World War II, Vol. II. Department of the Air Force, U.S. Government Printing Office. Washington, D. C. 1950, pp. 1027–52.

"(Die) Germanisation polnischer Kinder im Lichte von Dokumenten" (pol.). *Bulletin der Hauptkommission zur Erforschung der Naziverbrechen in Polen* (Biuletyn Głównej Komisji Badania Zbrodni Niemieckich w Polsce), 1949, 5, pp. 9–122.

Geuter, U., "Der Leipziger Kongress der Deutschen Gesellschaft für Psychologie 1933," *Psychologie- und Gesellschaftskritik*, 1979, 3 (H. 4), pp. 6–25.

Geuter, U., "Institutionelle und professionelle Schranken der Nachkriegsauseinandersetzungen über die Psychologie im Nationalsozialismus," *Psychologie- und Gesellschaftskritik*, 1980, 4 (H. 1/2), pp. 5–39.

Geuter, U., "Die Zerstörung wissenschaftlicher Vernunft. Felix Krueger and die Leipziger Schule der Ganzheitspsychologie," *Psychologie heute*, 1980a, 7 (H. 4), pp. 35–43.

Geuter, U., "Psychologiegeschichte," "Psychotechnik," "Rasse/Rassenpsychologie," in G. Rexilius and S. Grubitzsch (1981, pp. 824–38, 869–73, 882–8).

Geuter, U., "The Uses of History for the Shaping of Field: Observations on German Psychology," in L. Graham et al., (1983, pp. 191–228).

Geuter, U., "Der Nationalsozialismus und die Entwicklung der deutschen Psychologie," in G. Lüer (1983a, pp. 99–106).

Geuter, U., "'Gleichschaltung' von oben? Universitätspolitische Strategien und Verhaltensweisen in der Psychologie während des Nationalsozialismus," *Psychologische Rundschau*, 1984, 35, pp. 198–213.

Geuter, U., "The Eleventh and Twelfth International Congresses of Psychology – A Note on Politics and Science Between 1936 and 1948," in H. Carpintero (Ed.), *Psychology in Its Historical Context. Essays in Honour of Professor Joseph Brožek*, Valencia 1984a, pp. 127–40.

Geuter, U., "Das Ganze und die Gemeinschaft. Wissenschaftliches und politisches Denken in der Ganzheitspsychologie Felix Kruegers," in C. F. Graumann (Ed.), *Psychologie im Nationalsozialismus*, Berlin/Heidelberg/New York 1985, pp. 55–87.

Geuter, U., "Nationalsozialistische Psychologie und Ideologie," in M. G. Ash and U. Geuter (Eds.), *Geschichte der deutschen Psychologie im 20. Jahrhundert. Ein Überblick*, Opladen 1985a, pp. 172–200.

Geuter, U. (Ed.), *Daten zur Geschichte der deutschen Psychologie*. Vol. 1, Göttingen 1986.

Geuter, U., "Psychologie in der Zeit des Nationalsozialismus," in H. E. Lück et al., *Sozialgeschichte der Psychologie*, Opladen 1987, pp. 61–140.

Geuter, U., "Das Institut für Arbeitspsychologie und Arbeitspädagogik der Deutschen Arbeitsfront. Eine Forschungsnotiz," *1999. Zeitschrift für Sozialgeschichte des 20. und 21. Jahrhunderts*, 1987a, 2, pp. 87–95.

Geuter, U., "Mitgliederverluste in der deutschen Gesellschaft für Psychologie 1928 bis 1932 – Ausdruck des Protestes von Experimentalpsychologen oder der Verselbständigung der Disziplin?" *Psychologische Rundschau*, 1990, 41, pp. 144–54.

Geuter, U., and B. Kroner, "Militärpsychologie," in G. Rexilius and S. Grubitzsch (1987, pp. 672–89).

Geuter, U., and P. Mattes, "Historiography of Psychology in West Germany – Approaches from Social History: A Review of the Past Ten Years," *Storia e critica della psicologia*, 1984, 5, pp. 111–26.

Gilbert, G. M., *Nürnberger Tagebuch*, Frankfurt/M. 1977.

Giles, G. J., "University Government in Nazi Germany: Hamburg," *Minerva*, 1978, 16, pp. 196–221.

Goode, W. J., "Encroachment, Charlatanism and the Emerging Professions: Psychology, Sociology and Medicine," *American Sociological Review*, 1960, 25, pp. 902–14.

Goode, W. J., "Professionen und die Gesellschaft. Die Struktur ihrer Beziehungen," in T. Luckmann and W. M. Sprondel (1972, pp. 157–67).

Graessner, S., "Neue soziale Kontrolltechniken durch Arbeits- und Leistungsmedizin," in G. Baader and U. Schultz (1980, pp. 145–51).

Graham, L., W. Lepenies, and P. Weingart (Eds.), *Functions and Uses of Disciplinary Histories*, Dordrecht/Boston/Lancaster 1983.

Grassel, H., "Zur Entwicklung der Psychologie an der Universität Rostock," *Wiss. Zeitschrift der Univers. Rostock, Ges. u. Sprachwiss. Reihe*, 1971, 20, pp. 155–63.

Graumann, C. F., "Theorie und Geschichte," in G. Lüer (1983, pp. 64–75).

Grimm, G., "Die deutschen Universitäten von 1939 bis 1945," *Politische Studien*, 1969, 20, pp. 222–30.

Grünwald, H., *Die sozialen Ursprünge psychologischer Diagnostik*, Darmstadt 1980.

Grünwald, H., "Ueber die Vernachlässigung der Psychologiegeschichtsschreibung und wie und zu welchem besseren Ende Historiographie der Psychologie betrieben werden sollte," in W. Michaelis (1981, pp. 139–45).

Grunberger, R., *A social history of the Third Reich*, Harmondsworth 1974.

Güse, H.-G., and N. Schmacke, *Psychiatrie zwischen bürgerlicher Revolution und Faschismus*, Vol. 2. Kronberg 1976.

Güse, H.-G., and N. Schmacke, "Psychiatrie und Faschismus," in G. Baader and U. Schultz (1980, pp. 86–94).

Gundlach, H., "Willy Hellpach; Attributionen," in C. F. Graumann (1985, pp. 165–95).

Gundlach, H., "Willy Hellpachs Sozial- und Völkerpsychologie unter dem Aspekt der Auseinandersetzung mit der Rassenideologie," in Klingemann (1987, pp. 242–76).

Habermas, J., *Technik und Wissenschaft als Ideologie*, Frankfurt/M. 1968.

Habermas, J., "Gegen einen positivistisch halbierten Rationalismus," in T. W. Adorno et al. (1970, pp. 235–66).

Hallgarten, G. W. F., and J. Radkau, *Deutsche Industrie und Politik von Bismarck bis in die Gegenwart*, Reinbek 1981.

Hantel, W., "Die Eliminierung der Behinderten und die Ausrichtung der Psychologie im Hitler-Faschismus – ein historisches Paradigma," *Psychologie- und Gesellschaftskritik*, 1979, 3 (H. 4), pp. 42–59.

Hanvik, J. E., *Luftfahrt-Forschungsanstalt München, Medizinisches Institut Garmisch-Partenkirchen*. Washington D.C., U.S. Department of Commerce, 1946.

Hardin, B., *The Professionalization of Sociology*, Frankfurt/New York 1977.

Hartmann, H., "Arbeit Beruf, Profession," in T. Luckmann and W. M. Sprondel (1972, pp. 36–52).

Hartung, D., R. Nuthmann, and W. D. Winterhager, *Politologen im Beruf – zur Aufnahme und Durchsetzung neuer Qualifikationen im Beschäftigungssystem*, Stuttgart 1970.

Hartwig, A., "Die Entwicklung der öffentlichen Berufsberatung in Deutschland," *Arbeitsblatt für die Britische Zone*, 1947, 1, pp. 203–6.

Hartwig, A., *Die Entwicklung der öffentlichen Berufsberatung in Deutschland*, Düsseldorf 1948.

Hartwig, A., "Die Entwicklung der Berufsberatung in Deutschland," in *Handbuch der Arbeitsvermittlung und Berufsberatung*. Part 2. Stuttgart 1959, pp. 29–59.

Haug, W. F., *Der hilflose Antifaschismus*, Cologne 1977.

Hearnshaw, L. S., *A Short History of British Psychology*, 1840–1940, London 1964.

Heckhausen H., "Zur Lage der Psychologie," in G. Lüer (1983, pp. 3–27).

Heiber, H., *Walter Frank und sein Reichsinstitut für Geschichte des neuen Deutschland*, Stuttgart 1966.

Heinemann, M. (Ed.), *Erziehung und Schulung im Dritten Reich*, 2 Vols., Stuttgart 1980.

Henle, M., "One Man Against the Nazis – Wolfgang Köhler," *American Psychologist*, 1978, 33, pp. 939–44.

Hennig, E., "Zum Verhältnis von Industrie und Faschismus in Deutschland," R. Kühnl (1974, pp. 140–63).

Herrmann, T., "Zur Geschichte der Berufseignungsdiagnostik," *Arch. ges. Psych.*, 1966, 118, pp. 253–78.

Hesse, H. A., *Berufe im Wandel*, Stuttgart 1972.

Hiebsch, H., "Die Bedeutung des Menschenbildes für die Theoriebildung in der Psychologie," *Probleme und Ergebnisse der Psychologie*, 1961, 1, pp. 5–29.

Hillebrand, M.-J., "Begriffsbestimmung und geschichtliche Entwicklung der Pädagogischen Psychologie," *Handbuch der Psychologie*, Vol. X. Göttingen 1959, pp. 44–59.

Hillel, M., and C. Henry, *Au nom de la race*, no place, 1975.

Hinrichs, P., *Um die Seele des Arbeiters*, Cologne 1981.

Hirsch, W., "The Autonomy of Science in Totalitarian Societies: The Case of Nazi Germany," in K. D. Knorr, H. Strasser, and H. G. Zilian (Eds.), *Determinants and Controls of Scientific Development*, Dordrecht/Boston 1975, pp. 343–66.

Hofer, W. (Ed.), *Wissenschaft im totalen Staat*, München 1964.

Hofmann, H. H. (Ed.), *Das deutsche Offizierkorps, 1860–1960*, Boppard 1980.

Hofstätter, P. R., *Psychologie*, Frankfurt/M. 1967.

Hofstätter, P. R., and D. Wendt, *Quantitative Methoden der Psychologie*, Munich 1966.

Holzkamp, K., "Zur Geschichte und Systematik der Ausdruckstheorien," *Handbuch der Psychologie*, Vol. 5. Göttingen 1964, pp. 39–113.

Holzkamp, K., *Sinnliche Erkenntnis*, Frankfurt/M. 1973.

Homze, E., *Foreign Labor in Nazi Germany*, Princeton N.J., 1967.

Horkheimer, M., *Zur Kritik der instrumentellen Vernunft*, Frankfurt/M. 1967.

Hrabar, R., Z. Tokarz, and J. E. Wilczur, *Kinder im Krieg – Krieg gegen Kinder. Die Geschichte der polnischen Kinder 1939–1945*, Reinbek 1981.

Huber, C. (Ed.), *Kurt Huber zum Gedächtnis*, Regensburg 1947.

Hürten, H., "Das Offizierkorps des Reichsheeres," in H. H. Hofmann 1980, pp. 231–45.

Hüttenberger, P., "Nationalsozialistische Polykratie," *Geschichte und Gesellschaft*, 1976, 2, pp. 417–42.

Hüttenberger, P., *Bibliographie zum Nationalsozialismus*, Göttingen 1980.

Hughes, E. C., "Psychology: Science and/or Profession," *American Psychologist*, 1952, 7, pp. 441–3.

Hughes, T. P., "Scientists in the Mainstream: Alan Beyerchen's 'Scientists under Hitler,'" *Central European History*, 1979, 12, pp. 392–8.

Jackson, J. A. (Ed.), *Professions and Professionalization*, Cambridge 1970.

Jaeger, S., and I. Staeuble, *Die gesellschaftliche Genese der Psychologie*, Frankfurt/M./New York 1978.

Jaeger, S., and I. Staeuble, "Die Psychotechnik und ihre gesellschaftlichen Entwicklungsbedingungen," in *Die Psychologie des 20. Jahrhunderts*, Vol. XIII. Zurich 1981, pp. 53–95.

Jenak, R., *Der Missbrauch der Wissenschaft in der Zeit des Faschismus. Dargestellt am Beispiel der Technischen Hochschule Dresden 1933–1945*, Phil. Diss., Berlin 1964.

Johnson, T. J., *Professions and Power*, London 1972.

Kalikow, T. J., "Die ethologische Theorie von Konrad Lorenz: Erklärung und Ideologie, 1938 bis 1943," in H. Mehrtens and S. Richter (1980, pp. 189–214).

Kalisch, J., and G. Voigt, "Reichsuniversität Posen," in *Juni 1941. Beiträge zur Geschichte des hitlerfaschistischen Überfalls auf die Sowjetunion*, Berlin 1961, pp. 188–206.

Kaminski, G., and H. Märtin, "Oswald Kroh (1887–1955) und sein Lebenswerk," *Psychologische Beiträge*, 1956, 2, pp. 226–38.

Kardorff, E. v., and E. Koenen (Eds.), *Psyche in schlechter Gesellschaft*, Munich/Vienna/Baltimore 1981.

Kater, M. H., *Das "Ahnenerbe" der SS 1935–1945. Ein Beitrag zur Kulturpolitik des Dritten Reiches*, Stuttgart 1974.

Kater, M. H., "Sozialer Wandel in der NSDAP im Zuge der nationalsozialistischen Machtergreifung," in W. Schieder (Ed.), *Faschimus als soziale Bewegung*, Göttingen 1983, pp. 25–67.

Katz, D., "Fünf Jahrzehnte im Dienst der psychologischen Forschung," *Psychologische Beiträge*, 1953, 1, pp. 470–91.

Kelly, R. C., *National Socialism and German University Teachers: The NSDAP's Efforts to Create a National Socialist Professoriate and Scholarship*. Ph.D. Diss., Univ. of Washington 1973.

Kent, G. O., "Research Opportunities in West and East German Archives for the Weimar Period and the Third Reich," *Central European History*, 1979, 12, pp. 38–67.

Keupp, H., and M. Zaumseil, *Die gesellschaftliche Organisierung psychischen Leidens*, Frankfurt/M. 1978.

Kienreich, W., "Die Wiener Gesellschaft für Rassenpflege im Lichte ihrer Nachrichten," *Psychologie- und Gesellschaftskritik*, 1979, 3 (H. 4), pp. 61–73.

Kipp, M., "Zentrale Steuerung und planmässige Durchführung der Berufserziehung in der Luftwaffenrüstungsindustrie des Dritten Reiches," in M. Heinemann (1980, Vol. 1, pp. 310–33).

Kipp, M., and G. Miller, "Anpassung, Ausrichtung und Lenkung: Zur Theorie und Praxis der Berufserziehung im Dritten Reich," *Argument*, Special issue 21, Berlin 1978, pp. 248–66.

Kirchhoff, R., "Zur Geschichte des Ausdrucksbegriffs," *Handbuch der Psychologie*, Vol. 5. Göttingen 1964, pp. 9–38.

Kliem, K., J. Kammler, and R. Griepenburg, "Einleitung. Zur Theorie des Faschismus," in W. Abendroth (1976, pp. 5–18).

Klingemann, C., "Heimatsoziologie oder Ordnungsinstrument? Fachgeschichtliche Aspekte der Soziologie in Deutschland zwischen 1933 und 1945," *Kölner Zeitschrift für Soziologie und Sozialpsychologie*, Special issue 23, 1981, pp. 273–307.

Klingemann, C., "Soziologie im NS-Staat. Vom Unbehagen an der Soziologiegeschichtsschreibung," *Soziale Welt*, 1985, 36, pp. 366–88.

Klingemann, C., "Soziologie an Hochschulen im NS-Staat. Teil 1," *Zeitschrift für Hochschuldidaktik*, 1985a, 9, pp. 403–27; Teil II, 1986, 10, pp. 127–55.

Klingemann, C., "Vergangenheitsbewältigung oder Geschichtsschreibung? Unerwünschte Traditionsbestände deutscher Soziologie zwischen 1933 und 1945," in S. Papcke (Ed.), *Ordnung und Theorie. Beiträge zur Geschichte der Soziologie in Deutschland*, Darmstadt 1986, pp. 223–79.

Klingemann, C. (Ed.), *Rassenmythos und Sozialwissenschaften in Deutschland*, Opladen 1987.

Kluth, H., "Das Studium der Soziologie an den Hochschulen der Bundesrepublik Deutschland," *Kölner Zeitschrift für Soziologie und Sozialpsychologie*, 1966, 18, pp. 671–80.

Koch, H. R., "Diplompädagogen im Beruf," *Neue Praxis*. 1977, 7, pp. 9–51.

Koch, H. R., "Die Qualifikation, die niemand kannte," *UNI-Berufswahlmagazin*, 1980, 4 (H. 5), pp. 13–18.

Kreipe, K., *Evaluation and Procedure in the Characterological Selection of Fliers*. *German Aviation Medicine in World War II*, Vol. II. Department of the Air Force, U.S. Government Printing Office. Washington, D.C. 1950, pp. 1053–8.

Kroeber-Keneth, L., *Fetzen aus meinen Tagebüchern: Lehr- und Lesestücke ohne Moral*, Frankfurt/M. 1976.

Krohn, W., "Zur soziologischen Interpretation der neuzeitlichen Wissenschaft," in E. Zilsel, *Die sozialen Ursprünge der neuzeitlichen Wissenschaft*, ed. by Wolfgang Krohn, Frankfurt/M. 1976, pp. 7–43.

Kuczynski, J., *Die Geschichte der Lage der Arbeiter in Deutschland von 1800 bis in die Gegenwart*, Berlin, Vol. 1 1946, Vol. II 1948.

Kudlien, F., *Aerzte im Nationalsozialismus*, Cologne 1985.

Kügele, U., *Der deutsche Arzt im Zweiten Weltkrieg – im Spiegel des Deutschen Ärzteblattes*, Med. Diss., Düsseldorf 1974.

Kühne, K., *Das Psychologiestudium in der Schweiz*, Bern 1977.

Kühnl, R., *Formen bürgerlicher Herrschaft – Liberalismus und Faschismus*, Reinbek 1971.

Kühnl, R. (Ed.), *Texte zur Faschismusdiskussion* I, Reinbek 1974.

Kühnl, R., *Faschismustheorien*, Reinbek 1979.

Kuhn, T. S., *The Structure of Scientific Revolutions*, Chicago 1970.

Kuhn, T. S., *Die Entstehung des Neuen*, Frankfurt/M. 1978.

Kuklick, H., "Restructuring the Past: Toward an Appreciation of the Social Context of Social Science," *Sociological Quarterly*, 1980, 21, pp. 5–21.

Kuklick, H., "Boundary Maintenance in American Sociology: Limitations to Academic 'Professionalization,'" *JHBS*, 1980a, 16, pp. 201–19.

Lepenies, W., "Wissenschaftsgeschichte und Disziplingeschichte," *Geschichte und Gesellschaft*, 1978, 4, pp. 437–51.

Lepenies, W., *Das Ende der Naturgeschichte. Wandel kultureller Selbstverständlichkeiten in den Wissenschaften des 18. und 19. Jahrhunderts*, Frankfurt/M. 1978a.

Lepsius, R., "Die Entwicklung der Soziologie nach dem Zweiten Weltkrieg 1945 bis 1967," *Kölner Zeitschrift für Soziologie und Sozialpsychologie*, Special issue 21, 1979, pp. 25–70.

Lersch, P., *Der Mensch in der Gegenwart*, Munich 1947.

Lersch, P., "Nachruf auf Oswald Kroh," *PsRd*, 1956, 7, pp. 55–7.

Levine, H. S., *Hitler's Free City. A History of the Nazi Party in Danzig 1925–1939*, Chicago/London 1973.

Lindner, H., "'Deutsche' und 'gegentypische' Mathematik. Zur Begründung einer 'arteigenen' Mathematik im 'Dritten Reich' durch Ludwig Bieberbach," in H. Mehrtens and S. Richter (1980, pp. 88–115).

Lockot, R., *Erinnern und Durcharbeiten. Zur Geschichte der Psychoanalyse und Psychotherapie im Nationalsozialismus*, Frankfurt 1985.

Lohmann, H.-M., and L. Rosenkötter, "Psychoanalyse in Hitlerdeutschland. Wie war es wirklich?" *Psyche*, 1982, 36, 961–88.

Lohmann, H.-M. (Ed.), *Psychoanalyse und Nationalsozialismus. Beiträge zur Bearbeitung eines unbewältigten Traumas*, Frankfurt 1984.

Losemann, V., *Nationalsozialismus und Antike*, Hamburg 1977.

Luckmann, T., and W. M. Sprondel (Eds.), *Berufssoziologie*, Köln 1972.

Łuczak, C. (Ed.), *Diskriminierung der Polen in Wielpolska zur Zeit der Hitler-Okkupation. Dokumentenauswahl*, Wydawnictwo Poznańskie 1966.

Lüer, G. (Ed.), *Bericht über den 33. Kongress der Deutschen Gesellschaft für Psychologie in Mainz 1982*, Göttingen/Toronto/Zurich 1983.

Ludwig, K.-H., *Technik und Ingenieure im Dritten Reich*, Königstein/Düsseldorf 1979.

Lukács, G., *Die Zerstörung der Vernunft*, Vol. 2, Darmstadt/Neuwied 1974.

Lundgreen, P. (Ed.), *Wissenschaft im Dritten Reich*, Frankfurt 1985.

MacLeod, R., "Changing Perspectives in the Social History of Science," in I. Spiegel-Rösing and D. de Solla Price (Eds.), *Science, Technology, and Society*, London/Beverly Hills 1977, pp. 149–95.

Maetze, G., "Psychoanalyse in Deutschland. Mit einem Nachwort von D. Eicke," in *Die Psychologie des 20. Jahrhunderts*, Vol. II. Zurich 1976, pp. 1145–79.

Maikowski, R., P. Mattes, and G. Rott, *Psychologie und ihre Praxis. Materialien zur Geschichte und Funktion einer Einzelwissenschaft in der Bundesrepublik*, Frankfurt/M. 1976.

Mandler, J. M., and G. Mandler, "The Diaspora of Experimental Psychology: The Gestaltists and Others," in D. Fleming and B. Bailyn (Eds.), *The Intellectual Migration. Europe and America 1930–1960*, Cambridge, Mass. 1969, pp. 371–419.

Märtin, H., "Oswald Kroh zum Gedenken," in *Bericht über den 20. Kongress der Deutschen Gesellschaft für Psychologie 1955*, Göttingen 1956, pp. 9–12.

Mason, T., *Sozialpolitik im Dritten Reich*, Opladen 1978.

Mason, T., E. Czichon, D. Eichholz, and K. Gossweiler, *Faschismusdiskussion. Das Argument*, Studienheft 6, 1978.

Mattes, P., "Profession bei Fuss – Wehrmachtspsychologie nach 1945," *Psychologie- und Gesellschaftskritik*, 1980, 4 (H. 1/2), pp. 40–6.

Matthes, J., "Soziologie ohne Soziologen?" *Zeitschrift für Soziologie*, 1973, 2, pp. 47–58.

Matzerath, H., *Nationalsozialismus und kommunale Selbstverwaltung*, Stuttgart/Berlin/Cologne/Mainz 1970.

McPherson, M. W., "Some Values and Limitations of Oral Histories," *JHBS*, 1975, 11, pp. 34–6.

Mehrtens, H., "Die Naturwissenschaften im Nationalsozialismus," in R. Rürup (1979, Vol. 2, pp. 427–43).

Mehrtens, H., "Das Dritte Reich in der Naturwissenschaftsgeschichte: Literaturbericht und Problemskizze," in H. Mehrtens and S. Richter (1980, pp. 15–87).

Mehrtens, H., and S. Richter (Eds.), *Naturwissenschaft, Technik und NS-Ideologie. Beiträge zur Wissenschaftsgeschichte des Dritten Reichs*, Frankfurt/M. 1980.

Meier-Welcker, H., "Der Weg zum Offizier im Reichsheer der Weimarer Republik," *Militärgeschichtliche Mitteilungen*, 1976, 19, pp. 147–80.

Meiser, H., *Die deutsche Berufsberatung*, Berlin/Cologne/Mainz 1978.

Merz, F., "Amerikanische und deutsche Psychologie," *Psychologie und Praxis*, 1960, 4, pp. 78–91.

322 Bibliography

Messerschmidt, M., *Die Wehrmacht im NS-Staat. Zeit der Indoktrination*, Hamburg 1969.

Messerschmidt, M., "Politische Erziehung der Wehrmacht. Scheitern einer Strategie," in M. Heinemann (1980, Vol. 2, pp. 261–84).

Messerschmidt, M., and U.v. Gersdorff, *Offiziere im Bild von Dokumenten aus drei Jahrhunderten. Beiträge zur Militär- und Kriegsgeschichte*, Vol. 6. Stuttgart 1964.

Metzger, W., "The Historical Background for National Trends in Psychology: German Psychology," *JHBS*, 1965, 1, pp. 109–15.

Metzger, W., "Gestalttheorie im Exil," in *Die Psychologie des 20. Jahrhunderts*, Vol. I, Zurich 1976, pp. 659–83.

Metzger, W., "Gestaltpsychologie – ein Ärgernis für die Nazis," *Psychologie heute*, 1979, 6 (H. 3), pp. 84f.

Meyer-Thurow, G., "Zum unprofessionellen Umgang mit Professionalisierungsprozessen," *Geschichte und Gesellschaft*, 1980, 6, pp. 586–97.

Michaelis, W., "Bestimmungen über die Ausbildung in der Psychologie an wissenschaftlichen Hochschulen," in E. Stephan (1980, pp. 33–62).

Michaelis, W. (Ed.), *Bericht über den 32. Kongress der Deutschen Gesellschaft für Psychologie in Zürich 1980*, Göttingen/Toronto/Zurich 1981.

Miehe, G., *Zur Rolle der Universität Rostock in der Zeit des Faschismus in den Jahren 1935 bis 1945*. Phil. Diss., Rostock 1968.

Misiak, H., and V. S. Sexton, *History of Psychology. An Overview*, New York/London 1966.

Moede, W., "Die Eignung zum Führen von Kraftfahrzeugen und ihre Begutachtung," *Zeitschrift für Verkehrssicherheit*, 1955, 2, pp. 3–20.

Mommsen, H., "Ausnahmezustand als Herrschaftstechnik des NS-Regimes," in M. Funke (1976, pp. 30–45).

Müller, M. K., "Berufsfeldforschung: Zur Situation des Diplom-Pädagogen," *Neue Praxis*, Special issue 1979, Literaturrundschau, pp. 96–105.

Müller-Hill, B., *Tödliche Wissenschaft. Die Aussonderung von Juden, Zigeunern und Geisteskranken 1933–1945*, Reinbek 1984.

Müller-Hillebrand, B., *Das Heer 1933–1945*, Vol. 2, Frankfurt/M. 1956.

Myers, C. R., "The Collection, Preservation and Use of Oral History Materials," *Canadian Psychological Review*, 1975, 16, pp. 130–3.

Napoli, D. S., *The Architects of Adjustment: The Practice and Professionalization of American Psychology, 1920–1945*. Ph.D. Diss., Univ. of California, 1975; revised in *Architects of Adjustment*, Port Washington, N.Y. 1981.

Nationalsozialismus und die deutsche Universität, Berlin 1966.

Neubert, A., "Geschichte, Aufgabe und Organisation der Berufsberatung," *Handbuch der Berufspsychologie*, Göttingen 1977, pp. 401–25.

Neumann, F., *Behemoth*, London/New York 1942 (German edition Cologne 1977).

Nolte, E., "Zur Typologie des Verhaltens der Hochschullehrer im Dritten Reich," *Aus Politik und Zeitgeschichte*. Supplement to *Parlament*, 1965, No. 46.

Nowotny, H., "Deprofessionalisierung," *Psychosozial*, 1979, 2 (H. 2), pp. 12–18.

Nussbaum, A., and H. Feger, "Analyse des deutschsprachigen psychologischen Zeitschriftensystems," *PsRd*, 1978, 19, pp. 91–112.

O'Donnell, J. M., "The Crisis of Experimentation in the 1920s: E. G. Boring and his Uses of History," *American Psychologist*, 1979, 34, pp. 289–95.

Oldfield, R. C., "Great Britain," *Gawein*, 1956, 4, pp. 93–7.

Opitz, R., "Über Faschismustheorien und ihre Konsequenzen," in R. Kühnl (1974, pp. 219–40).

Ottersbach, H.-G., *Der Professionalisierungsprozess in der Psychologie. Berufliche Strategien der Psychotherapieverbände*, Weinheim/Basel 1980.

Ottweiler, O., *Die Volksschule im Nationalsozialismus*, Weinheim/Basel 1979.

Parsons, T., "Professions," *International Encyclopedia of the Social Sciences*, Vol. 12, 1968, pp. 536–47.

Pauly, P., "The Political Structure of the Brain: Cerebral Localization in Bismarckian Germany," *International Journal of Neuroscience*, 1983, 21, pp. 145–50.

Pechhold, E., "Psychotechnisches aus Deutschland," in F. Baumgarten (Ed.), *Progrès de la Psychotechnique 1939–1945*, Bern (n.d.), pp. 310f.

Pechhold, E., *50 Jahre REFA*, Darmstadt 1974.

Peterson, D. R., "Need for the Doctor of Psychology Degree in Professional Psychology," *American Psychologist*, 1976, 31, pp. 792–8.

Peterson, D. R., "Is Psychology a Profession?" *American Psychologist*, 1976a, 31, pp. 572–81.

Petry, C., *Studenten aufs Schafott. Die Weisse Rose und ihr Scheitern*, Munich 1968.

Pfahlmann, H., *Fremdarbeiter und Kriegsgefangene in der deutschen Kriegswirtschaft 1939–1945*, Darmstadt 1968.

Pieper, J., *Noch wusste es niemand. Autobiographische Aufzeichnungen*, Munich 1976.

Pinn, I., "Die rassistischen Konsequenzen einer völkischen Anthropologie. Zur Anthropologie Erich Jaenschs," in C. Klingemann (1987, pp. 212–41).

Poliakov, L., and J. Wulf, *Das Dritte Reich und seine Denker. Dokumente*, Berlin 1959.

Pongratz, L. J., "Germany," in V. S. Sexton and H. Misiak (1976, pp. 154–81).

Pongratz, L. J., "German Historiography of Psychology, 1808–1970," in J. Brožek and L. J. Pongratz (1980, pp. 74–89).

Pongratz, L. J., "Die historische Entwicklung des Berufsfeldes Klinische Psychologie," in V. Birtsch und D. Tscheulin (Eds.), *Ausbildung in Klinischer Psychologie und Psychotherapie*, Weinheim/Basel 1980a, pp. 23–36.

Progratz, L. J., W. Traxel, and E. G. Wehner (Eds.), *Psychologie in Selbstdarstellungen*, Bern, Vol. I 1972, Vol. II 1979.

Prahl, H.-W., *Sozialgeschichte des Hochschulwesens*, Munich 1978.

Prinz, W., "Ganzheits- und Gestaltpsychologie und Nationalsozialismus," in P. Lundgreen (Ed.), *Wissenschaft im III. Reich*, Frankfurt/M. 1985 pp. 55–81.

Psychoanalyse unter Hitler. Dokumentation einer Kontroverse. Published by the editors of the journal *Psyche*, Frankfurt 1984.

Psychologen-Taschenbuch, Göttingen 1955ff.

"Psychology in Europe," *Gawein*, 1956, 4, H. 3/4.

Rammstedt, O., "Theorie und Empirie des Volksfeindes. Zur Entwicklung einer 'deutschen Soziologie,'" in P. Lundgreen (1985, pp. 253–313).

Rammstedt, O., *Deutsche Soziologie 1933–1945. Die Normalität einer Anpassung*, Frankfurt 1986.

Rauch, M., "Wehrpsychologie," in R. Zoll, E. Lippert, and T. Rössler (Eds.), *Bundeswehr und Gesellschaft*, Opladen 1977, pp. 332–6.

Reimann, H., and K. Kiefer, *Soziologie als Beruf*, Heidelberg 1962.

Reisman, J. M., *A History of Clinical Psychology*, New York 1976.

Revers, W. J., "Philosophisch orientierte Theorien der Person und Persönlichkeit," *Handbuch der Psychologie*, Vol. 4. Göttingen 1960, pp. 391–436.

324   *Bibliography*

Rexilius, G., and S. Grubitzsch (Eds.), *Handbuch psychologischer Grundbegriffe*, Reinbek 1981 (revised ed. 1987).

Richter, S., "Die 'Deutsche Physik,'" in H. Mehrtens and S. Richter (1980, pp. 116–41).

Riedesser, P., Verderber, A., *Aufrüstung der Seelen. Militärpsychologie und Militärpsychiatrie in Deutschland und Amerika*, Freiburg 1985.

Rieffert, J. B., "Summary Report to Management on the Aptitude Tests Used at Rheinmetall-Borsig-AG," in J. W. Dunlap and J. B. Rieffert (1945, pp. 9–17).

Ringer, F. K., *The Decline of the German Mandarins*, Cambridge, Mass. 1969.

Roth, E., "Zur Lage der Psychologie," in W. Michaelis (1981, pp. 3–14).

Rothacker, E., *Heitere Erinnerungen*, Frankfurt/M./Bonn 1963.

Rudert, J., "(Selbstdarstellung)," in L. J. Pongratz et al. (1972, pp. 288–308).

Rürup, R. (Ed.), *Wissenschaft und Gesellschaft. Beiträge zur Geschichte der Technischen Universität Berlin 1879–1979*, 2 vols., Berlin/Heidelberg/New York 1979.

Rüschemeyer, D., "Professionalisierung. Theoretische Probleme für die vergleichende Geschichtsforschung," *Geschichte und Gesellschaft*, 1980, 6, pp. 311–25.

Salewski, M., "Das Offizierkorps der Reichs- und Kriegsmarine," in H. H. Hofmann (1980, pp. 211–29).

Samelson, F., "Putting Psychology on the Map: Ideology and Intelligence Testing," in A. R. Buss (1979, pp. 103–68).

Schäfer, G., "Franz Neumanns Behemoth und die heutige Faschismusdiskussion," Afterword to German edition, F. Neumann, Cologne 1977, pp. 663–776.

Schelsky, H., "Zur Entstehungsgeschichte der bundesdeutschen Soziologie. Ein Brief an Rainer Lepsius," *Kölner Zeitschrift für Soziologie und Sozialpsychologie*, 1980, 32, pp. 417–56.

Scherzer, O., "Physik im totalen Staat," in Flitner (1965, pp. 47–58).

Schieder, W. (Ed.), *Faschismus als soziale Bewegung*, Hamburg 1976.

Schlottmann, U., "Soziologen im Beruf," *Kölner Zeitschrift für Soziologie und Sozialpsychologie*, 1968, 20, pp. 572–97.

Schmid, R., *Intelligenz- und Leistungsmessung. Geschichte und Funktion psychologischer Tests*, Frankfurt/M./New York 1977.

Schmitz, E., and P. Weingart, *Knowledge, Qualifications and Credentials: Changing Patterns of Occupations – An Analysis of Six Cases of Credentialling in Germany*. Wissenschaftsforschung, Report 5. Bielefeld (n.d).

Schoenbaum, D., *Hitler's Social Revolution. Class and Status in Nazi Germany, 1933–39*, London 1967.

Schoenbaum, D., *Die braune Revolution. Eine Sozialgeschichte des Dritten Reichs*, Cologne/Berlin 1968.

Schultz, D. P., *A History of Modern Psychology*, New York/London 1969.

Schumak, R., "Der erste Lehrstuhl für Pädagogik an der Universität München. Ein Beitrag zur Institutionalisierung und zur Geschichte der Pädagogik als Universitätsdisziplin (1893–1945)," in L. Boehm and J. Spörl (Eds.), *Die Ludwig-Maximilians-Universität in ihren Fakultäten*, Vol. II, Berlin 1980, pp. 303–44.

Schunter-Kleemann, S., "Zwischen bürgerlicher und sowjetischer Ideologie. Psychologie in der DDR (1945–1960)," *Psychologie heute*, 1980, 7 (H. 6), pp. 74–81.

Schunter-Kleemann, S., "Die Nachkriegsauseinandersetzung in der DDR über die Psychologie im deutschen Faschismus," *Psychologie- und Gesellschaftskritik* 1980a, 4 (H. 1/2), pp. 47–67.

Schwartz, C., "Professionalisierung im Bereich Klinische Psychologie," in H. Keupp and M. Zaumseil (Eds.), *Die gesellschaftliche Organisierung psychischen Leidens*, Frankfurt/M. 1978, pp. 298–325.

Seebohm, H.-B., *Otto Selz. Ein Beitrag zur Geschichte der Psychologie*. Phil. Diss., Heidelberg 1970.

Seeger, F., *Relevanz und Entwicklung der Psychologie*, Darmstadt 1977 (= Psychologie und Gesellschaft Vol. I).

Seier, H., "Der Rektor als Führer. Zur Hochschulpolitik des Reichserziehungsministeriums 1934–1945," *Vierteljahreshefte für Zeitgeschichte*, 1964, 12, pp. 105–46.

Seubert, R., *Berufserziehung und Nationalsozialismus*, Weinheim/Basel 1977.

Sexton, V. S., and H. Misiak (Eds.), *Psychology around the World*, Monterey, Calif. 1976.

Siefer, G., "Das Studium der Soziologie in der Bundesrepublik Deutschland," *Soziologie*, 1972–73, 1, pp. 29–53.

Simon, L. R., *German Research in World War II*, New York/London 1947.

Simoneit, M., "Einige Tatsachen zur ehemaligen deutschen Wehrpsychologie, die für Heer, Kriegsmarine und Luftwaffe tätig war," *Wehrwissenschaftliche Rundschau*, 1954, 4, pp. 138–41.

Simoneit, M., "Deutsche Wehrmachtpsychologie von 1927–1942," *WPsM*, 1972, 6 (H. 2), pp. 71–110.

"(Zur) Situation in der westdeutschen Psychologie," *Forum*, 1960, 14 (H. 23). Wissenschaftliche Beilage, pp. 1–4.

Sohn-Rethel, A., *Ökonomie und Klassenstruktur des deutschen Faschismus*. Edited with an introduction by J. Agnoli, B. Blanke and N. Kadritzke. Frankfurt/M. 1973.

Solla Price, D. J. de, *Little Science, Big Science. Von der Studierstube zur Grossforschung*, Frankfurt/M. 1974.

Sontheimer, K., *Antidemokratisches Denken in der Weimarer Republik*, Munich 1978.

Sosnowski, K., *The Tragedy of Children under Nazi Rule*, Poznań 1962.

Speer, A., *Erinnerungen*, Berlin 1972.

Spiegel-Rösing, I., "Disziplinäre Strategien der Statussicherung," *Homo*, 1974, 25, pp. 11–37.

Sporer, S. L., "A Brief History of the Psychology of Testimony," *Current Psychological Reviews*, 1982, 2, 323–40.

Spur, G., and H. Grage, "75 Jahre Institut für Werkzeugmaschinen und Fertigungstechnik der Technischen Universität Berlin," in R. Rürup (1979, Vol. 2, pp. 107–31).

Staeuble, I., "Politischer Ursprung und politische Funktionen der pragmatischen Sozialpsychologie," in H. Nolte and I. Staeuble, *Zur Kritik der Sozialpsychologie*, Munich 1972, pp. 7–57.

Stephan, E. (Ed.), *Ausbildung und Weiterbildung in Psychologie*, Weinheim/Basel 1980.

Stets, W., "Von den Ursprüngen der Berufsberatung," *Blätter für Berufsberatung*. Beilage zu *Berufskundliche Mitteilungen* No. 13, 18. 9. 1963, pp. 1–11.

Stiebitz, F., "50 Jahre Psychologie im Dienste der Polizei," *Die Polizei*, 1974, 65, pp. 298–301.

Stocking, G. W. Jr., "On the Limits of 'Presentism' and 'Historicism' in the Historiography of the Behavioral Sciences," *JHBS*, 1965, 1, pp. 211–18.

Stollberg, G., "Der vierköpfige Behemoth. Franz Neumann und die moderne

326   *Bibliography*

Auffassung vom pluralistischen Herrschaftssystem des Faschismus," in *Gesellschaft. Beiträge zur Marxschen Theorie*, 1976, 6, pp. 92–117.

Stolper, G., K. Häuser, and K. Borchardt, *Deutsche Wirtschaft seit 1870*, Tübingen 1966.

Thalheimer, A., "Uber den Faschismus," in W. Abendroth (1976, pp. 19–38).

Thielen, M., *Sowjetische Psychologie und Marxismus. Geschichte und Kritik*, Frankfurt/M. 1984.

Thomae, H., *Psychologie in der modernen Gesellschaft*, Hamburg 1977.

Timasheff, N. S., "Business and the Professions in Liberal, Fascist, and Communist Society" (1940); extract as "Professions and Business Management," in H. M. Vollmer and D. L. Mills (Eds.), *Professionalization*, Englewood Cliffs, N.J. 1966, pp. 59–62.

Traxel, W., "Mitgliederstand und Mitgliederbewegung in der Gesellschaft für experimentelle Psychologie und der Deutschen Gesellschaft für Psychologie von 1904 bis 1939," in G. Lüer (1983, pp. 97–9).

Treuheit, L. J., "Dissertationen als Parameter im System der Psychologie in Deutschland: 1885–1969," *PsRd*, 1973, 24, pp. 171–205.

Undeutsch, U., *Die Entwicklung der gerichtspsychologischen Gutachtertätigkeit*, Göttingen 1954.

Viteles, M. S., and L. D. Anderson, *Training and Selection of Supervisory Personnel in the I. G. Farbenwerke, Ludwigshafen*. FIAT Final Report No. 930. Washington, D.C., U.S. Department of Commerce, 1947.

Völker, K.-H., *Die deutsche Luftwaffe 1933–1939*, Stuttgart 1967.

Volkamer, M., "Psychologie," in P. Rohs, M. Volkamer et al., *Geschichte der Christian Albrechts Universität Kiel, 1665–1965*, Vol. V, Part I: Geschichte der Philosophischen Fakultät, pp. 105–19.

Voss, H. von, "(Zur Kontroverse Simoneit-Moede)," *Mitteilungen des Berufsverbandes deutscher Psychologen*, 1949, 3 (H. 11), pp. 10f.

Warsewa, G., and M. Neumann, "Zur Bedeutung der 'Rassenfrage' in der NS-Industrieforschung," in C. Klingemann (1987, pp. 345–69).

Wehner, E. G., *Gustav Kafka. Ein Beitrag zur Geschichte der Psychologie*, Phil. Diss. Würzburg 1964.

Weimer, W. B., "The History of Psychology and Its Retrieval from Historiography: I. The Problematic Nature of History," *Science Studies*, 1974, 4, pp. 235–58.

Weingart, P., *Wissensproduktion und soziale Struktur*, Frankfurt/M. 1976.

Weingart, P., "Eugenik – Eine angewandte Wissenschaft. Utopien der Menschenzüchtung zwischen Wissenschaftsentwicklung und Politik," in P. Lundgreen (1985, pp. 314–49).

Weinreich, M., *Hitler's Professors*, New York 1946.

Weisenborn, G., *Der lautlose Aufstand*, Frankfurt/M. 1974.

Weiss, W. W., "Ergebnisse der Planlosigkeit – Eine Analyse von Ursachen und Auswirkungen des Bemühens, einen Studiengang für Diplom-Pädagogen einzurichten," *Neue Praxis*, 1976, 6, pp. 256–67.

Wellek, A., "Oswald Kroh," *Jahrbuch für Psychologie und Psychotherapie*, 1955, 3, pp. 325f.

Wellek, A., "West- und ostdeutsche Reform des Diplom-Psychologen," *PsRd*, 1956, 7, pp. 66–70.

Wellek, A., "Deutsche Psychologie und Nationalsozialismus," *Psychol. und Praxis*, 1960, 4, 177–82.

Wellek, A., "Der Einfluss der deutschen Emigration auf die Entwicklung der amerikanischen Psychologie," *PsRd*, 1964, 15, pp. 239–62.

Wer ist's? Berlin 1955.

Westmeyer, H., "Zur Paradigmadiskussion in der Psychologie," in W. Michaelis (1981, pp. 115–26).

Whitley R., "Cognitive and Social Institutionalization of Scientific Specialties and Research Areas," in Whitley (Ed.), *Social Processes of Scientific Development*, London/Boston 1974, pp. 69–95.

Wies, B., *Zur Entstehung von Berufsfeldern praktisch-orientierter Psychologen besonders im industriellen Bereich. Ein Beitrage zur Geschichte der angewandten Psychologie in Deutschland zwischen den Weltkriegen*, unpub. diploma diss., Psych. Inst., Universität Heidelberg 1979.

Wilensky, H. L., "Jeder Beruf eine Profession?" in T. Luckmann and W. M. Sprondel (1972, pp. 198–215) (Engl.: 1964).

Witt, T. E. J. de, *The Nazi Party and Social Welfare, 1919–1939*, Ph.D. Diss., Univ. of Virginia, 1972.

Wittchen, H.-U., and M. M. Fichter, *Psychotherapie in der Bundesrepublik*, Weinheim/Basel 1980.

Witte, W., "Willy Hellpach. Zu seinem 80. Geburtstag am 26. Februar 1957," *Psychologische Beiträge*, 1957, 3, pp. 3–20.

Wohlwill, J. F., "German Psychological Journals under National Socialism: A History of Contrasting Paths," *Journal of the History of the Behavioral Sciences*, 1987, 23, pp. 169–85.

Wolsing, T., *Untersuchungen zur Berufsausbildung im Dritten Reich*, Kastellaun 1977.

Wright, M. J., "CPA: The First Ten Years," *Canadian Psychologist*, 1974, 15, pp. 112–31.

Wróblewska, T., "Die Rolle und Aufgaben einer nationalsozialistischen Universität in den sogenannten östlichen Reichsgebieten am Beispiel der Reichsuniversität Posen 1941–1945," *Informationen zur erziehungs- und bildungspolitischen Forschung*, 1980, 14, pp. 225–52.

Wuttke-Groneberg, W., "Von Heidelberg nach Dachau," in G. Baader and U. Schultz (1980, pp. 113–38).

Young, R. M., "Scholarship and the History of the Behavioral Sciences," *History of Science*, 1966, 5, pp. 1–51.

Zapp, G., *Psychoanalyse und Nationalsozialismus*, Med. Diss., Kiel 1980.

Zillmer, H., *Psychologen im Beruf*, Weinheim/Basel 1980.

Zipfel, F., "Krieg und Zusammenbruch," in E. Aleff (1976, pp. 177–240).

Zneimer, R., "The Nazis and the Professors: Social Origin, Professional Mobility, and Political Involvement of the Frankfurt University Faculty, 1933–1939," *Journal of Social History*, 1978, 12, pp. 147–58.

Index

174, 234, 238–9, 243, 261, 294
Gerhards, Karl, 46
German Association for Practical
  Psychology, 184–5; *see also* Association
  of Practical Psychologists
German Institute for Psychological
  Research and Psychotherapy, 144, 149,
  194ff., 245–6, 251
German Institute for Technical Training,
  88, 131, 136, 151
German Labor Front 145, 148
  Institute for Work Psychology and Work
    Pedagogy, 37, 93, 136, 150ff., 254, 264,
    273, 288
  Office for Vocational Training and
    Works Management, 92, 151, 173, 179,
    195–6
  Work Science Institute, 152
German Philosophical Society, 72
German Research Society, 170
German Society for Animal Psychology, 255
German Society for Psychology, 40–1, 51,
  55–6, 180, 191, 193, 195–6, 199, 205,
  207, 210–11, 243, 252, 255–6, 274,
  286–7; *see also* Society for Experi-
  mental Psychology
Gessler, Otto, 137–8
Gestalt psychology, xvii, 9, 55, 89, 168,
  216n
Gestapo, 246
Giese, Fritz, 46, 54, 61, 80, 87–8, 91, 119,
  123, 126–7, 144, 165, 179, 183, 189–90,
  191
Giessen University, 62, 65, 75ff., 187n, 221–2
Goldmeier, Erich, 294
Goldschmidt, Richard Hellmuth, 54, 137,
  215
Goldstein, Kurt, 144
Göppert, Friedhilde, 209n
Göring, Hermann, 144, 233n, 234
Göring, Matthias Heinrich, 144, 146, 194,
  228, 245
Göttingen University, 48, 65, 187n, 211
Gottschaldt, Kurt, 65, 70, 81, 160, 250
Graf, Otto, 123
graphology, 95ff., 115, 222
Graz University, 66, 211–12
Greifswald University, 62, 65, 75ff., 188, 211
Gruhle, Hans, 200
Günther, Hans F. K., 121–2, 178
Günther, Hans R. G., 72, 74n, 217

habilitation, 188–9
Haeckel, Ernst, 112
Haenisch, Konrad, 47
Halle, Provincial Institute for Practical
  Psychology, 144
Halle University, 55, 61, 64–5, 75ff., 170,

173, 187, 219–20, 221–2, 249
Hamburg University, 48, 55–6, 74, 77ff.,
  82, 187n, 211, 213
Handrick, Johannes, 135, 183–4
Hannover Technical College, 46, 80
Hantel, Erika, 149, 151
Harmjanz, Heinrich, 59, 73, 195, 205–6, 230–1
Härtle, Heinrich, 170
Hartmann, Eduard von, 99
Hartmann, Nicolai, 102, 110
Hartnacke, Wilhelm, 167
Health Practitioners Law, 145
Heidegger, Martin, 217
Heidelberg University, 75ff., 187, 211, 214
Heidenreich and Harbeck firm, 151
Heinkel Works, 151
Heisenberg, Werner, 282
Heiss, Robert, 75ff., 112, 160, 215ff., 273
Hellpach, Willy, xvii, 46, 78, 214, 225, 287
Henning, Hans, 46, 54, 61, 81, 84n, 89, 199
Herwig, Bernhard, 42, 46, 54, 80, 160, 179,
  223, 250, 252
Hess, Rudolf, 58, 62, 175
Hetzer, Hildegard, 143, 244ff., 294
Heyer, Gustav R., 146, 195
Hilgenfeldt, Erich, 249, 258
Himmler, Heinrich, 235, 246–7, 282
Hippius, Maria, 294
Hippius, Rudolf, 65–6, 74n, 160, 254–5
Hische, Wilhelm, 46, 54, 80, 129, 133, 136,
  179, 191n, 223
Hitler, Adolf, 58, 119, 153, 163, 175, 182,
  215, 234ff., 240, 245, 280, 282
Hitler Youth, 91, 135, 170n, 245
Hochheimer, Wolfgang, 147, 234, 243, 261,
  294
Hochstetter, Erich, 250
Hoesch firm, 92, 150
Hoffmann, Arthur, 143, 146, 195
Hoffmann, H. F., 109
Hofstätter, Peter R., 217, 250–1
holistic psychology (*Ganzheitspsychologie*),
  xvii, 89, 112, 168, 251
Honecker, Martin, 217
Hönigswald, Richard, 52
Horkheimer, Max, 270n, 285
Hornbostel, Erich von, 54
Huber, Hans, 207
Huber, Kurt, xvii, 268
Husserl, Edmund, 112
Huth, Albert, 91, 134

I. G. Farben, 148ff.
Innsbruck University, 66, 211
Inspectorate for Aptitude Testing, 37, 69,
  114, 120, 178, 196, 206, 224
Inspectorate of the Infantry, 287
Inspectorate of Military Aeronautics, 48

Supreme Command of the Wehrmacht,
    161n, 175, 233n

teacher training, academic, 48
teacher training, psychology in, xviii, 39,
    143
Teuber, Hans Lukas, 7, 12, 261
theology and psychology, 50, 68, 72, 217–18
Thomae, Hans, 250, 294
Thurnwald, Richard, 221
Todt, Fritz, 92, 282
Tramm, K. A., 87, 131, 181
Tübingen University, 65–6, 75ff., 169–70,
    173, 187n, 213, 221, 249
Tumlirz, Otto, 66, 160, 168, 195n
types, theory of (typology), 104, 111,
    121ff., 179, 273

Undeutsch, Udo, 236, 274
Utitz, Emil, 61, 98

Vahlen, Theodor, 59
Valentiner, Theodor, 119, 133, 142, 179
Vetter, August, 150–1
Vienna University, 66, 211, 218
Villinger, C. G. W., 230
Vögler, Albert, 131
Volkelt, Hans, 51, 59, 71, 73, 168, 170,
    219n, 220, 250, 255
Voss, Hans von, 69ff., 160–1, 171, 185,
    188, 193, 195, 206, 210–11, 235, 265–6

Wacke, Otto, 59
Wagner, Julius, 84n

Wallach, Hans, 294
Wallichs, Adolf, 46, 80
Walther, Rudolf H., 193
Wartegg, Ehrig, 98
Weber, Josef, 133
Weber, Max, 57n
Wehrmacht Center for Psychology and
    Race Studies, 159, 175–6; *see also*
    Inspectorate for Aptitude Testing
Weiss, Albrecht, 148, 150
Wellek, Albert, 2n, 63, 70, 74, 76–7, 100,
    189, 220, 274, 287
Wenke, Hans, 60, 65, 67–8, 76, 160, 226, 250–1
Werner, Heinz, 9, 54–5, 201
Wertheimer, Max, 9, 42, 51, 53, 76, 213
Wilde, Kurt, 59, 77, 160n, 170, 219–20,
    249–50, 274
will, research on the psychology of, 97,
    102–3, 106ff.
Wirth, Wilhelm, 47, 173, 219n, 221
Wittmann, Johannes, 78
Wohlfahrt, Erich, 142
Wundt, Max, 49
Wundt, Wilhelm, 38, 45, 57, 71, 110
Würzburg University, 62, 75ff., 82, 187n,
    221
Wuth, Otto, 229ff.
Wyatt, Frederick, 7, 12, 261, 294

Zeddies, Adolf, 136
Zeise, Ludwig, 150
Ziehen, Theodor, 64, 99
Zilian, Erich, 140, 178, 243
Zschintzsch, Werner, 70